Mesenchymal Stem Cells and Craniofacial Regeneration

Edited By

Jing Wang
Department of Stomatology
Shanghai Tenth People's Hospital
Tongji University School of Medicine
P.R. China
&
YunFeng Lin
State Key Laboratory of Oral Diseases
West China School of Stomatology
Sichuan University
China

Mesenchymal Stem Cells and Craniofacial Regeneration

Editors: Jing Wang & YunFeng Lin

ISBN (eBook): 978-1-68108-315-5

ISBN (Print): 978-1-68108-316-2

Published by Bentham Science Publishers – Sharjah, UAE.

First published in 2016.

Acknowledgement:

Declared none.

BENTHAM SCIENCE PUBLISHERS LTD.

End User License Agreement (for non-institutional, personal use)

This is an agreement between you and Bentham Science Publishers Ltd. Please read this License Agreement carefully before using the ebook/echapter/ejournal (**"Work"**). Your use of the Work constitutes your agreement to the terms and conditions set forth in this License Agreement. If you do not agree to these terms and conditions then you should not use the Work.

Bentham Science Publishers agrees to grant you a non-exclusive, non-transferable limited license to use the Work subject to and in accordance with the following terms and conditions. This License Agreement is for non-library, personal use only. For a library / institutional / multi user license in respect of the Work, please contact: permission@benthamscience.org.

Usage Rules:

1. All rights reserved: The Work is the subject of copyright and Bentham Science Publishers either owns the Work (and the copyright in it) or is licensed to distribute the Work. You shall not copy, reproduce, modify, remove, delete, augment, add to, publish, transmit, sell, resell, create derivative works from, or in any way exploit the Work or make the Work available for others to do any of the same, in any form or by any means, in whole or in part, in each case without the prior written permission of Bentham Science Publishers, unless stated otherwise in this License Agreement.
2. You may download a copy of the Work on one occasion to one personal computer (including tablet, laptop, desktop, or other such devices). You may make one back-up copy of the Work to avoid losing it. The following DRM (Digital Rights Management) policy may also be applicable to the Work at Bentham Science Publishers' election, acting in its sole discretion:

- 25 'copy' commands can be executed every 7 days in respect of the Work. The text selected for copying cannot extend to more than a single page. Each time a text 'copy' command is executed, irrespective of whether the text selection is made from within one page or from separate pages, it will be considered as a separate / individual 'copy' command.
- 25 pages only from the Work can be printed every 7 days.

3. The unauthorised use or distribution of copyrighted or other proprietary content is illegal and could subject you to liability for substantial money damages. You will be liable for any damage resulting from your misuse of the Work or any violation of this License Agreement, including any infringement by you of copyrights or proprietary rights.

Disclaimer:

Bentham Science Publishers does not guarantee that the information in the Work is error-free, or warrant that it will meet your requirements or that access to the Work will be uninterrupted or error-free. The Work is provided "as is" without warranty of any kind, either express or implied or statutory, including, without limitation, implied warranties of merchantability and fitness for a particular purpose. The entire risk as to the results and performance of the Work is assumed by you. No responsibility is assumed by Bentham Science Publishers, its staff, editors and/or authors for any injury and/or damage to persons or property as a matter of products liability, negligence or otherwise, or from any use or operation of any methods, products instruction,

Bentham Science Publishers Ltd.
Executive Suite Y - 2
PO Box 7917, Saif Zone
Sharjah, U.A.E.
Email: subscriptions@benthamscience.org

BENTHAM
SCIENCE

CONTENTS

PREFACE i

LIST OF CONTRIBUTORS iv

CHAPTER 1 FUNDAMENTAL CONCEPTS AND FEATURES OF MESENCHYMAL STEM CELLS: PROLIFERATION, DIFFERENTIATION, MIGRATION AND IMMUNOMODULATORY CHARACTERISTICS 3

Shuwen Deng, Shu Zhang, Ke Sun, RaorRao Wang, Jing Wang and *Yunfeng Lin*

1. INTRODUCTION 4
- 1.1. Overview 4
- 1.2. History 4
- 1.3. Definition 5

2. THE IDENTIFICATION OF MSCs: BIOMARKERS, MORPHOLOGY 6

3. THE MAIN SOURCE OF MSCs: LOCATION AND ISOLATION 7
- 3.1. Genetic Phenotypes of MSCs 9
- 3.2. Procedures for MSC Isolation 10

4. THE PROLIFERATION OF MSCs: REGULATORY FACTORS INVOLVED IN THEIR MAINTENANCE, GROWTH AND SURVIVAL 10

5. THE FACTORS DETERMINING OR REGULATING MSC MIGRATION 11
- 5.1. Homing of Mesenchymal Stem Cells 11
- 5.2. Growth Factors Influence the Ability of MSCs to Migrate 12
  - *5.2.1. PDGF* 12
  - *5.2.2. IGF* 12
  - *5.2.3. BMP-2, BMP-4, BMP-7, TGF-β1, TGF-β3* 13
  - *5.2.4. EGF, HB-EGF, TGFa* 13
  - *5.2.5. FGF-2* 14
  - *5.2.6. HGF* 14
  - *5.2.7. VEGF-A* 14
- 5.3. Chemotactic Factors Influence the Ability of MSCs to Migrate 15
  - *5.3.1. SDF-1* 15
  - *5.3.2. MCP-1, MIP-1a, IL-8* 15
  - *5.3.3. RANTES, MDC* 16
  - *5.3.4. Fractalkine* 16
  - *5.3.5. SLC, TARC* 17
- 5.4. Other Chemoattractants 17
  - *5.4.1. LPA, S1P* 17
  - *5.4.2. HMGB-1* 18
  - *5.4.3. TNF-a* 18
  - *5.4.4. Ligands for Toll-Like Receptors* 18
- 5.5. Signaling Pathways Influence the Ability of MSCs to Migrate 19
  - *5.5.1. Mitogen-Activated Protein Kinase Pathway (MAPK)* 19
  - *5.5.2. PI-3k/AKT (Phosphatidylinositol-3-Kinase)* 19
  - *5.5.3. Jak/STAT Signaling Pathway* 20

CONFLICT OF INTEREST 20

ACKNOWLEDGEMENTS 20

REFERENCES 21

CHAPTER 2 THREE-DIMENSIONAL PRINTING IN DENTISTRY: AN ADVANCED TECHNOLOGY FOR CRANIOFACIAL REGENERATION 33

Maobin Yang, Qian Zeng, Maria Pereira Vieira and *Bernardo Olsson*

2.1. INTRODUCTION ... 33
2.2. THE HISTORY OF 3DP ... 34
2.3. CATEGORY OF 3DP TECHNOLOGY ... 36
2.3.1. Stereolithography (SLA) ... 36
2.3.2. Fused Deposition Modeling (FDM) ... 36
2.3.3. Selective Laser Sintering (SLS) ... 37
2.3.4. Inkjet Printing ... 37
2.3.5. Laminated Object Manufacturing (LOM) ... 38
2.3.6. Polyjet Printing ... 38
2.4. COMPARISON OF 3DP AND SUBTRACTIVE MANUFACTURING TECHNIQUE ... 39
2.5. CURRENT APPLICATION OF 3DP IN DENTISTRY ... 39
2.5.1. 3DP In Clinical Dentistry ... 40
2.5.1.1. Endodontics ... 40
2.5.1.2. Oral Surgery ... 41
2.5.1.3. Prosthodontics ... 43
2.5.1.4. Periodontics ... 44
2.5.1.5. Orthodontics ... 45
2.5.1.6. Implant ... 47
2.5.2. 3DP in Bioengineering Research in Dentistry ... 48
CONFLICT OF INTEREST ... 50
ACKNOWLEDGEMENTS ... 50
REFERENCES ... 50

CHAPTER 3 INDUCED PLURIPOTENT STEM CELLS: PROLIFERATION, MIGRATION, MICRORNA, SIGNALING MOLECULES ... 60
Yuanzhi Xu, Feiyu Wang, Jing Wang, Yun Lu, Yun Shen, Songtao Wu, Zhaozhao Chen, Runyi Mao and *Raorao Wang*
INTRODUCTION ... 61
2. IPS CELLS: METHODS OF ACQUISITION, PROMOTING GROWTH AND INCREASING INDUCED EFFICIENCY ... 62
2.1. Selection of Proper Recipient Cells ... 62
2.2. Alternative Vector Systems ... 63
2.3. Induction Genes ... 64
2.4. Acquisition of iPS Cells ... 66
2.5. Culture of Recipient and Induced Cells ... 68
2.6. Increased Induced Efficiency ... 69
3. IPS CELLS: METHODS FOR IDENTITY AND SAFETY FOR REGENERATIVE MEDICINE ... 70
3.1. Cellular Biological Properties ... 70
3.2. Pluripotency ... 73
3.3. The Restrictions and Safety Issues for Regenerative Medicine ... 75
4. IPS CELLS: METHODS OF INDUCTION TOWARD DIFFERENT CELL TYPES, FACTORS INFLUENCING THE POTENTIAL OF DIFFERENTIATION, AND MECHANISMS DETERMINING THE FATE OF IPS CELL SAFTER INDUCTION ... 77
5. APPLICATION OF INDUCED IPS CELLS IN TISSUE ENGINEERING: CELL SOURCE, UNDERSTANDING OF CELL ULARPLURIPOTENCY AND DIFFERENTIATION ... 82
CONFLICT OF INTEREST ... 84
ACKNOWLEDGEMENTS ... 84
REFERENCES ... 84

CHAPTER 4 CRANIOFACIAL DEFECTS AND THEIR REGENERATION: DESTRUCTION AND REGENERATION OF THE PERIODONTIUM, CRANIOFACIAL TUMORS, TRAUMA, AND CONGENITAL DEFECTS ... 90
Jingang Xiao, Lin Liu, Lei Wang, Kui Huang, Yong Li and *Zhihao Yao*

4.1. INTRODUCTION ... 91
4.2. CRANIOFACIAL DEFECTS: PERIODONTIUM, CRANIOFACIAL TUMORS, TRAUMA, AND CONGENITAL DEFECTS ... 94
4.2.1. Periodontium Defects ... 95
4.2.2. Craniofacial Tumors ... 96
4.2.2.1. Soft Tissue Defects (Lingual Defect, Penetrating Defect of Cheek, Palatal Defect, Tongue and Lip Defect, Parotid Gland and Duct Defect, Facial Nerve Defect) ... 96
4.2.2.2. Maxillofacial Combined Resection (Soft Tissue and Jaw Bone Defects) ... 97
4.2.3. Craniofacial Trauma ... 100
4.2.3.1. Soft Tissue Injuries ... 101
4.2.3.2. Fractures of the Jaws (Mandible and Maxilla) ... 101
4.2.3.3. Other Maxillofacial Fractures ... 103
4.2.4. Congenital Disease ... 105
4.2.4.1. Classification of Cleft Lip and Palate ... 105
4.2.4.2. Craniofacial Dysplasia ... 107
4.2.5. Discussion ... 110
4.2.6. Conclusions ... 111
4.3. CURRENT TECHNIQUES AND LIMITATIONS OF THERAPY FOR CRANIOFACIAL DEFECTS: COMPOSITE MATERIALS AND SELF-TRANSPLANTATION ... 111
4.3.1. Autologous Tissue Transplants ... 111
4.3.1.1. Characteristics of Autogenous Bone Grafts ... 112
4.3.1.2. Characteristics of Free Flaps ... 119
4.3.1.3. Flap Classification (Local Flaps, Pedicled Flaps, Vascularized Flaps, and Microvascular Free Tissue Transfer) ... 121
4.3.2. Use of Composite Materials for Craniofacial Defects ... 125
4.3.2.1. Features of HA ... 125
4.3.2.2. Types and Characteristics of Different Implants ... 127
4.4. *IN VITRO* STUDIES OF CRANIOFACIAL REGENERATION ... 129
4.4.1. Cell Source ... 130
4.4.1.1. ESCs ... 131
4.4.1.2. iPSCs ... 132
4.4.1.3. Adult Stem Cells ... 134
4.4.2. Tissue Regeneration ... 145
4.4.2.1. Bone Regeneration ... 145
4.4.2.2. Cartilage Tissue Regeneration ... 149
4.4.2.3. Oral Mucosa Regeneration ... 153
4.4.2.4. Fat Tissue Regeneration ... 153
4.4.3. Compatibility of the Regeneration Material ... 155
4.4.3.1. Evolution of Current Concepts of Biocompatibility ... 155
4.4.3.2. Growth Factors and Biologicals ... 156
4.5. *IN VIVO* STUDIES OF CRANIOFACIAL REGENERATION ... 157
4.5.1. Animal Models ... 158
4.5.1.1. Bone Defect Models ... 158
4.5.1.2. TMJ Defect Model ... 164
4.5.1.3. Cleft Lip and Palate Model ... 166
4.5.1.4. Facial Nerve Injury and Regeneration Model ... 166
4.5.2. Cell Source ... 167
4.5.2.1. BMSCs ... 168
4.5.2.2. ASCs ... 170
4.5.2.3. UCMSCs ... 172
4.5.2.4. DPSCs ... 173

4.5.3. Scaffolds 174
4.5.3.1. Classification of Scaffold Materials 175
4.5.3.2. Three-Dimensional Structure 178
4.5.4. Growth Factors 180
4.5.4.1. BMP-2 180
4.5.4.2. bFGF 181
4.5.4.3. TGF-β1 183
4.5.4.4. VEGF 183
4.5.4.5. PRP 184
CONFLICT OF INTEREST 186
ACKNOWLEDGEMENTS 186
REFERENCES 186

CHAPTER 5 CELL SHEETS ENGINEERING AND TRANSPLANTATION IN MSCs REGENERATION 231
Jing Wang, Renhuan Huang, Shengcai Qi, Rui Zhang, Liyan Lu, Fan Jian, Dan Wang, Yuanzhi Xu, Hua Wang, Haiyan Zhu and *Xu Zhang*
1. INTRODUCTION 232
2. FACTORS INFLUENCE THE TRANSPLANTING ABILITY OF MSCs IN CELL SHEETS 236
2.1. Cell Source and Donor-Specific Factors 237
2.2. Substrate Characteristics and Soluble Factors 238
2.3. 3D-Culture and Pre-Vascularization 241
2.4. Immune Responses to Transplanted Cell Sheet 245
2.5. Combination With Scaffold Materials 246
3. METHODS FOR FABRICATION OF CELL SHEETS 247
3.1. Fabricate Cell Sheet Using Temperature-Responsive System 247
3.1.1. Electron Beam Lithographic (EBL) System 248
3.1.2. Plasma Treated System 250
3.1.3. Ultrathin Responsive System 251
3.1.4. Cell Adhesion Protein Coated Thermoresponsive Surfaces System 251
3.1.5. Spin-Coated System 252
3.1.6. Initiated Chemical Vapor Deposition (iCVD) System 252
3.2. Electroactive Responsive System 253
3.2.1. Electrochemical Self-Assembled Monolayer Gold Surface 253
3.2.2. Polyelectrolyte Surface 253
3.2.3. Sulfenyl-Gold Surface 255
3.2.4. Electrochemical Modified Micro-Patterning Surface 255
3.3. Light Responsive Cell Sheets Technology 256
3.3.1. Oxide Treated Surface Method 256
3.3.2. Light Sensitive Molecule/Material Coatings Method 257
3.3.3. Light Sensitive Molecular Combined With Protein Coatings Method 258
3.3.4. Nanoengineered Gold Coating Surface 259
3.3.5. PMB-PL Surface 259
3.4. PH Responsive Material Coating Method 260
3.4.1. Indium Tin Oxide Surface 260
3.4.2. Chitosan Surfaces 260
3.5. Magnetite Nanoparticles and Magnetic Force Method 262
3.6. Others 263
3.6.1. Varies in Culture Medium 263
3.6.2. Automatic Fabrication of Cell Sheets Using Manipulator Technique 264
3.7. The Future and Development of The Cell Sheet Technologies 265

4. ANIMAL MODELS AND CLINICAL TRIALS OF CELLS SHEETS APPLICATION IN REGENERATION MEDICINE 266
4.1. Introduction 266
4.2. Animal Models 266
4.2.1. Corneal Epithelial 267
4.2.2. Retinal Pigment Epithelium (RPE) 268
4.2.3. Middle Ear Mucosal 269
4.2.4. Periodontal Tissue 270
4.2.5. Tooth 271
4.2.6. Calvarial Defect 272
4.2.7. Jaw Defect 273
4.2.8. Esophageal Mucosa 273
4.2.9. Tracheal Replacement 274
4.2.10. Thyroid Gland 275
4.2.11. Articular Cartilage 276
4.2.12. Lung Surface (Sealing of Lung Air Leak) 277
4.2.13. Myocardium 278
4.2.14. Hepatic Tissue 280
4.2.15. Pancreatic Islets 281
4.2.16. Renal Tissue 282
4.2.17. Urothelial Tissue 283
4.3. Clinical Trails 285
4.3.1. Corneal Reconstruction 285
4.3.2. Oral Mucosa 287
4.3.3. Esophageal Mucosa 287
4.3.4. Myocardium 288
CONFLICT OF INTEREST 289
ACKNOWLEDGEMENTS 289
REFERENCES 289

CHAPTER 6 NEW METHODS OF SCAFFOLD FABRICATION: CELL-INSTRUCTIVE SCAFFOLDS, STRUCTURAL SCAFFOLDS, SCAFFOLDS FABRICATED FROM SMART MATERIALS ABLE TO RESPOND SENSITIVELY TO ENVIRONMENTAL CUES 312
Jing Wang, Yuehua Liu, Jing Chen, Yun Shen, Yiming Zhang, Jinglin Zhou, Shu Zhang, Chenyuan Xu and *Raorao Wang*
1. INTRODUCTION 312
2. CELL-INSTRUCTIVE SCAFFOLDS 313
2.1. Synthesis and Fabrication of Cell-instructive Scaffolds 313
2.1.1. Spinning 314
2.1.2. Self-assembly 315
2.1.3. Thermally Induced Phase Separation (TIPS) 316
2.1.4. Printing 317
2.1.5. Ultraviolet (UV) Photolithography 317
2.2. Related Factors of Cell-instructive Scaffolds for Cell Behavior 318
2.2.1. Scaffold Mechanics 318
2.2.2. Scaffold Composition 318
2.2.3. Scaffold Topography 319
2.3. Methods for the Loading of Bioactive Molecules 319
2.3.1. Adsorption 320
2.3.2. Controlled-release Particles 321
2.3.3. Stimulus-based Delivery 322
3. STRUCTURAL SCAFFOLDS IN TISSUE ENGINEERING 322
3.1. Features of Different Tissue Matrices 323
3.1.1. Bone Matrix 323
3.1.2. Cartilage Matrix 324

3.1.3. Skin Matrix ... 325
3.1.4. Scaffolds in Tooth Tissue Engineering ... 326
3.2. Materials and Features of Biomimicry Scaffolds ... 326
3.2.1. Essential Features of the Scaffolds ... 326
3.2.2. Biomaterials used in Scaffold Manufacture ... 329
3.3. Strategy for Fabricating and Assembling Scaffolds which Mimic Natural ECM ... 331
3.3.1. Materials Available for Fabrication ... 331
3.3.2. Synthesis of Polymer Scaffolds ... 336
CONFLICT OF INTEREST ... 339
ACKNOWLEDGEMENTS ... 339
REFERENCES ... 339

CHAPTER 7 TOOTH REGENERATION: DENTIN REGENERATION, PERIODONTAL LIGAMENT REGENERATION, DENTAL PULP REGENERATION ... 349
Guo Li, Na Fu, Xueqin Wei, Tao Zhang, Xiaoru Shao, Qian Huang, Ke Sun, Shiyu Lin, Sirong Shi and *Yunfeng Lin*
1. INTRODUCTION ... 349
2. DENTIN REGENERATION ... 352
2.1. The Stem Cells Used in Dentin Regeneration ... 352
2.1.1. Dental Pulp Stem Cells ... 352
2.1.2. Stem Cells from Human Exfoliated Deciduous Teeth (SHEDs) ... 353
2.1.3. Stem Cells from the Dental Apical Papilla (SCAPs) ... 353
2.1.4. Non-dental Stem Cells ... 354
2.2. The Scaffolds Used in Dentin Regeneration ... 354
2.3. The Cytokines/growth Factors Used in Dentin Regeneration ... 356
2.4. The Regulation Mechanism of Dentin Regeneration ... 359
3. PERIODONTAL LIGAMENT REGENERATION ... 360
3.1. The Stem Cells Used in Periodontal Ligament Regeneration ... 360
3.1.1. Periodontal Ligament Stem Cells (PDLSCs) ... 360
3.1.2. Dental Follicle Stem Cells (DFSCs) ... 361
3.2. The Scaffolds Used in Periodontal Ligament Regeneration ... 361
3.3. The Cytokines/Growth Factors and Regulation Mechanism in Periodontal Ligament Regeneration ... 363
4. DENTAL PULP REGENERATION ... 364
4.1. The Stem Cells Used in Dental Pulp Regeneration ... 364
4.1.1. DPSCs ... 364
4.1.2. SCAPs ... 366
4.2. The Scaffolds Used in Dental Pulp Regeneration ... 366
4.3. The Cytokines/Growth Factors and Regulation Mechanism in Dental Pulp Regeneration ... 367
CONFLICT OF INTEREST ... 368
ACKNOWLEDGEMENTS ... 369
REFERENCES ... 369

SUBJECT INDEX ... 380

PREFACE

In humans, tooth loss cannot only lead to physical and mental suffering, but also affect the aesthetics, which compromise an individual's self-esteem and life quality. There are several dental diseases that can eventually result in tooth loss. With the development of tissue engineering, regenerating a whole tooth is now considered a promising strategy to treat tooth loss.

In addition, cranial and maxillofacial trauma is notoriously difficult to repair and presents a significant burden to the healthcare system. Current therapies usually utilize autologous tissue transplantation and artificial composite material reparation. However, limitations exist. Tissue engineering as an alternative to autologous transplantation provides new hope for restoration and reconstruction of missing tissue and avoids immunosuppression. It has been used to successfully regenerate many tissue types, such as bone and cartilage. The aim of regenerative medicine is to restore and replace tissue that has been damaged or lost through trauma.

Tissue engineering is an interdisciplinary technique that creates a biological substitute through application of the principles of engineering to biology and medicine to restore, maintain, and improve tissue function. The emerging discipline of regenerative medicine and tissue engineering is committed to a rational approach to maintain and preserve the microenvironment, which is based on growth factors for tissue induction, stem cells, and scaffolds.

Stem cell-based tissue engineering approaches have the capacity to regenerate damaged and diseased tissues. Induced pluripotent stem cell (iPS) cells are a type of pluripotent stem cell that is generated from somatic cells, contrary to natural course of cell differentiation. iPS cells are derived from differentiated cells or tissue, reprogrammed to an embryonic-like state. Reprogrammed human induced pluripotent stem cells (hiPS) are similar to human embryonic stem (hES) cells in many aspects, such as morphology, epigenetic status of pluripotent genes, surface antigen expression, telomerase activity and proliferation rate.

Mesenchymal Stromal Progenitor/Stem Cells (MSCs) are not a large rare population of non-hematopoietic stromal cells.They are present in the bone marrow and most connective tissues. MSCs can be isolated from a variety of tissues, such as endometrial polyps, umbilical cord, endometrial polyps, adipose tissue, menses blood, bone marrow, adipose tissue, *etc*. These sources are most practical for experimental and possible clinical applications due to the ease of harvest and large obtainable quantity. In recent decades, cell sheet technology (CST) has been exploited as a novel promising scaffold-free approach for cell culturing and delivering in tissue regeneration. Studies have demonstrated that MSCs facilitate cell viability and function in the CST co-culture system to improve transplantation efficiency. There are extensive interests in developing strategies to enhance the formation of cell sheets with MSC for downstream applications.

Growth factors are naturally occurring proteins induced by cell surface receptor-mediated chemokines, mitosis, and angiogenesis. They participate in the growth, development, repair, reconstruction, and regeneration of tissues and organs.

The scaffold is a key factor in tissue regeneration. It provides a desired three-dimensional (3D) space that allows stem cells to adhere, proliferate, migrate, and differentiate. In the field of regenerative medicine and tissue engineering, biomaterials are critical for the encouragement and maintenance of biological activities. Scaffolds can be cell-instructive, structural and fabricated from smart materials. Design considerations include the creation of well-defined materials, cell-material interactions, and controlled-release bioactive agents. Since three-dimensional printing (3DP) technology was first invented in 1980s, it has been used in a range of fields, including architecture, manufacturing, engineering, education and medicine. In dentistry, 3DP provides a rapid and precise technology that helps diagnosis, treatment planning and treatment of various dental diseases. In craniofacial regeneration, 3DP helps to design and fabricate dental prosthesis and implants, and it also improves craniofacial tissue engineering. Not only scaffolds can be fabricated by 3DP, biologics such as stem cells and growth factors can also be delivered. Thus, 3DP technology has become a major tool in craniofacial regeneration.

Overall this book brings together research spanning from diverse backgrounds ranging from materials science and engineering, biology and clinical medicine, to discuss various aspects of craniofacial and dental regeneration. We hope this book will add further insight for basic and applied researchers as well as clinicians involved in tissue engineering and regenerative medicine, thus contributing to further advances in dental medicine.

This book provides an in-depth overview of current knowledge about mesenchymal stem cell and its applicant in craniofacial regeneration, including concepts and features of mesenchymal stem cells, induced pluripotent stem cells,craniofacial regeneration, new methods on scaffold fabrication, tooth regeneration and three-dimensional printing in dentistry. We sincerely hope that the publication of this book will provide an overview for researches and clinician who are concentrated on the field related to this promising subject .

Jing Wang
Department of Stomatology
Shanghai Tenth People's Hospital
Tongji University School of Medicine

P.R. China
&
YunFeng Lin
State Key Laboratory of Oral Diseases
West China School of Stomatology
Sichuan University
China

List of Contributors

J. Chen	Department of Orthodontics, School & Hospital of Stomatology, Tongji University, Shanghai Engineering Research Center of Tooth Restoration and Regeneration, Shanghai, China
Z. Chen	State Key Laboratory of Oral Diseases, West China School of Stomatology, Sichuan University, Chengdu, China
S Deng	Department of Stomatology, Shanghai Tenth People's Hospital, Tongji University School of Medicine, Shanghai , China
N. Fu	State Key Laboratory of Oral Diseases, West China School of Stomatology, Sichuan University, Chengdu, China
K. Huang	Department of Oral and Maxillofacial Surgery, Hospital of Stomatology, Luzhou Medical College, Luzhou, China
R. Huang	State Key Laboratory of Oral Diseases, West China School of Stomatology, Sichuan University, Chengdu, China
Q. Huang	State Key Laboratory of Oral Diseases, West China School of Stomatology, Sichuan University, Chengdu, China
F. Jian	State Key Laboratory of Oral Diseases, West China School of Stomatology, Sichuan University, Chengdu, China
G. Li	State Key Laboratory of Oral Diseases, West China School of Stomatology, Sichuan University, Chengdu, China
Y. Li	Department of Oral and Maxillofacial Surgery, Hospital of Stomatology, Luzhou Medical College, Luzhou, China
S. Lin	State Key Laboratory of Oral Diseases, West China School of Stomatology, Sichuan University, Chengdu, China
Y. Lin	State Key Laboratory of Oral Diseases, West China School of Stomatology, Sichuan University, Chengdu, China
L. Liu	Department of Oral and Maxillofacial Surgery, Hospital of Stomatology, Luzhou Medical College, Luzhou, China
Y. Liu	Department of Orthodontics, Shanghai Stomatological Hospital, Shanghai, China
L. Lu	Department of Stomatology, Shanghai Tenth People's Hospital, Tongji University School of Medicine, Shanghai, China
Y. Lu	Department of Orthodontics, College of Stomatology, Dalian Medical University, Dalian, China
R. Mao	Department of Oral and Maxillofacial Surgery, Ninth People's Hospital, College of Stomatology, Shanghai Jiao Tong University School of Medicine, Shanghai, China

B. Olsson	Federal University of Santa Catarina, Florianópolis, Brazil
S. Qi	Department of Stomatology, Shanghai Tenth People's Hospital, Tongji University School of Medicine, Shanghai, China
X. Shao	State Key Laboratory of Oral Diseases, West China School of Stomatology, Sichuan University, Chengdu, China
Y. Shen	Department of Stomatology, Shanghai Tenth People's Hospital, Tongji University School of Medicine, Shanghai, China
S. Shi	State Key Laboratory of Oral Diseases, West China School of Stomatology, Sichuan University, Chengdu, China
K. Sun	State Key Laboratory of Oral Diseases, West China School of Stomatology, Sichuan University, Chengdu, China
M. Pereira Vieira	Federal University of Mato Grosso do Sul, Campo Grande, Mato Grosso do Sul, Brazil
D. Wang	Department of Stomatology, Shanghai Tenth People's Hospital, Tongji University School of Medicine, Shanghai, China
F. Wang	Department of Oral Surgery, Ninth People's Hospital, Shanghai Jiao Tong University School of Medicine, Shanghai, China
H. Wang	Department of Stomatology, Shanghai Tenth People's Hospital, Tongji University School of Medicine, Shanghai, China
L. Wang	Department of Oral and Maxillofacial Surgery, Hospital of Stomatolo, Luzhou Medical College, Luzhou, China
J. Wang	Department of Stomatology, Shanghai Tenth People's Hospital, Tongji University School of Medicine, Shanghai, China
R. Wang	Department of Stomatology, Shanghai Tenth People's Hospital, Tongji University School of Medicine, Shanghai, China
X. Wei	State Key Laboratory of Oral Diseases, West China School of Stomatology, Sichuan University, Chengdu, China
S. Wu	State Key Laboratory of Oral Diseases, West China School of Stomatology, Sichuan University, Chengdu, China
J. Xiao	Department of Oral and Maxillofacial Surgery, Hospital of Stomatology, Luzhou, China Orofacial Reconstruction and Regeneration Laboratory, Hospital of Stomatology, Luzhou Medical College, Luzhou, China
C. Xu	Department of Stomatology, Shanghai Tenth People's Hospital, Tongji University School of Medicine, Shanghai, China
Y. Xu	Department of Stomatology, Shanghai Tenth People's Hospital, Tongji University School of Medicine, Shanghai, China

M. Yang	Kornberg School of Dentistry, Temple University, Philadelphia, USA
Z. Yao	Department of Oral and Maxillofacial Surgery, Hospital of Stomatology, Luzhou Medical College, Luzhou, China
Q. Zeng	Sun Yat-sen University, Guanghua School of Stomatology, Guangzhou, China
S. Zhang	State Key Laboratory of Oral Diseases, West China School of Stomatology, Sichuan University, Chengdu, China
R. Zhang	Department of Stomatology, Shanghai Tenth People's Hospital, Tongji University School of Medicine, Shanghai, China
T. Zhang	State Key Laboratory of Oral Diseases, West China School of Stomatology, Sichuan University, Chengdu, China
X. Zhang	Department of Stomatology, Shanghai Tenth People's Hospital, Tongji University School of Medicine, Shanghai, China
Y. Zhang	Department of Stomatology, Shanghai Tenth People's Hospital, Tongji University School of Medicine, Shanghai, China
J. Zhou	State Key Laboratory of Oral Diseases, West China School of Stomatology, Sichuan University, Chengdu, China
H. Zhu	Department of Stomatology, Shanghai Tenth People's Hospital, Tongji University School of Medicine, Shanghai, China

Mesenchymal Stem Cells and Craniofacial Regeneration

2

CHAPTER 1

Fundamental Concepts and Features of Mesenchymal Stem Cells: Proliferation, Differentiation, Migration and Immunomodulatory Characteristics

Shuwen Deng[1], **Shu Zhang**[2], **Ke Sun**[2], **RaorRao Wang**[1], **Jing Wang**[1,*] and **Yunfeng Lin**[2]

[1] *Department of Stomatology, Shanghai Tenth People's Hospital, Tongji University School of Medicine, Shanghai, China*

[2] *State Key Laboratory of Oral Diseases, West China School of Stomatology, Sichuan University, Chengdu, China*

Abstract: Mesenchymal Stromal Progenitor/Stem Cells (MSCs) are not a large population of non-hematopoietic stromal cells. This type of cells is usually present in the most connective tissues and bone marrow of the body. They can be isolated from a lot of tissues, such as endometrial polyps, adipose tissue, umbilical cord, menses blood, bone marrow, *etc*. MSCs have been defined following isolation and culture expansion, by their expression of different molecules including CD90, CD73 and CD105 and the negative markers like CD45, CD34 and CD14. MSCs have the proliferation ability in culture in uncommitted state, while retaining their pluripotency, which makes them attractive effector for biological cell-based tissue repair approaches. MSCs could be differentiated to several cell lineages. Furthermore, the exclusive biological properties of MSCs are mediated by paracrine mechanisms and by intensive immunomodulatory activity. The MSCs use for treatment has been fund and evaluated to be very useful in some pre-clinical animal models and in clinical trials. MSCs and MSC-like cells have low rates of homing to target tissues and organs, limiting therapeutic efficacy. Thus, researches of the homing mechanisms and homing factors of MSCs may lead to the

* **Corresponding author Jing Wang:** Department of Stomatology, Shanghai Tenth People's Hospital, Tongji University School of Medicine, No. 301, Middle Yanchang Road, Shanghai 200072, P.R. China; Tel: 86-21-66301726; Fax: 86-21-66301725; E-mail: dentist_tracy@hotmail.com.

Jing Wang & YunFeng Lin (Eds.)

development of therapies with the potential to promote clinical applications of MSCs.

Keywords: Cell migration, Differentiation, Homing, Immune, Mesenchymal stem cell (MSC), Stem cells.

1. INTRODUCTION

1.1. Overview

Mesenchymal stem cells are self-renewing, multipotent somatic cells that play vital roles in tissue homeostasis and repair and, as such, exhibit tremendous potential for tissue engineering and regenerative medicine. As one of the first few stem cell types isolated, mesenchymal stem cells are among the most-wel--characterized and have since been harvested from a variety of tissues. However, clinical translation of these cells is still hampered by issues including cell heterogeneity, maintenance of stem cell phenotype with extended *in vitro* culture and identification of the optimal strategies for controlling mesenchymal stem cell behavior to promote tissue healing and regeneration. To address this, we summarize the history as well as review the essential characteristics and properties of mesenchymal stem cells, including the isolation procedures of these cells, methods for maintaining their proliferation and stemness *ex vivo* as well as growth factors and signaling pathways essential to mesenchymal stem cell migration, chemotaxis and homing. The purpose of this chapter is to provide readers with essential concept of MSC isolation, maintenance and biology for optimal deployment of these cells in tissue engineering applications.

1.2. History

The history of mesenchymal stem cells began with their isolation from bone marrow tissue before additional populations were isolated from other tissues. In 1970, Friedenstein and colleagues showed that bone marrow comprise hematopoietic non-adherent cells along with an uncommon population of plastic-adherent cells, after entire bone marrow cells were placed in plastic culture dishes with medium supplemented with 10% fetal calf serum. These cells have the ability to form colonies derived from single cells. After two days, these adherent cells begin to proliferate and can differentiate into mature mesenchymal cell lineages [1]. The term Colony Forming Unit–fibroblasts (CFU-f) refers to the round colonies of fibroblastic cells generated from the primary clones of adherent cells. Friedenstein also found part of the colony can differentiate into small region bone or cartilage. They studied the proliferative capacities and phenotypic qualities of CFU-f [2] and found that these cells present multipotential property which can differentiate into chondrocytes, adipocytes, osteoblasts and myoblasts. Dependent on the age and health situation of the bone marrow donor, very donor shows a particular frequency of CFU-f. According to Caplan's propose [3], since the cells differentiate into mesenchymal cell lineage, or mesenchymal stromal cells, they are currently categorized as mesenchymal stem cells (MSC) [4, 5]. Whether these cells ought to be viewed as pure stem cells or as multipotent progenitors of mesenchymal lineages has been very widely discussed. Thus, adoption of the expression "multipotent mesenchymal stromal cells" has been proposed in place of "mesenchymal stem cells" [6, 7]. In 1974, Friedenstein and his fellows firstly identified the MSCs [8], despite the fact that Caplan already initially applied the term "mesenchymal stem cell" in 1991 [3]. Subsequent to MSC isolation from bone marrow, MSCs were isolated from a lot of tissues including adipose tissue, cutaneous tissue, fetal hepatic and pulmonary tissue.

1.3. Definition

The definition of MSCs depends singularly on the examination of cell populations *in vitro*. This is because, despite years of investigation, the natural position and role of MSCs within the body origin tissue remains unclear. Confounding such analysis is the absence of specific MSC markers to permit their unambiguous identification [7, 9, 10]. Since MSCs are removed from their natural environment and grown on artificial plastic dishes in a combination of exogenous chemical and physical growth conditions, the MSC phenotypes and abilities in both *in vivo* and *in vitro* settings may not necessarily be the same. For example, MSCs are rearrangement *in vitro* phenotype operating experience, gain new markers while losing the expression of others [11]. Because of increasing interest in the use of MSCs as a part of cell-based therapy [12, 13], the need to recognize MSCs definitively is not only scientific but also clinical and administrative interest [11, 14]. As described above, no single marker can unequivocally identify the MSC and other cell types differentiate. In 2006, the International Society for Cell

Therapy (ISCT) proposed the following criteria for the minimal identification of human MSCs [15]: adherence to plastic in standard culture conditions; CD73+, CD90+, CD105+, CD34-, CD45-, HLA-DR-, CD14- or CD11b-, CD79a- or CD19-cell phenotype as evaluated by FACS analysis; and/or *in vitro* differentiation into osteoblasts, adipocytes and chondroblasts (demonstrated by staining of *in vitro* cell culture). These criteria only used for review of a cell population containing MSCs yet do not permit the purification of MSCs. In addition, these standards are not fully effective across and within species. In mouse models, MSCs vary not only in humans but also between strains in marker expression and behavior in culture [16], the significant contrasts being in the expression of CD34 and CD105 [17]. These criteria are excessively subject to excessive expansion and culture of MSC populations and, therefore, are less likely to be extrapolated to local cells. Hence, MSCs are extended under the typically considered kept MSC differentiation, but do not retain what is presently considered to be the native MSC phenotype [10, 18].

2. THE IDENTIFICATION OF MSCs: BIOMARKERS, MORPHOLOGY

The ISCT has attempted to build the system further for MSC definition with the recent publication of a working proposal intended to make the first moves toward standardizing MSCs based on their immunomodulatory properties [15]. The publication recommends various measures being utilized to characterize and validate MSCs based on their immunomodulatory intensity so as to move past simple analysis of the MSC surface phenotype [19].

MSC isolation by traditional bone marrow aspiratory techniques is generally clear, especially in model organisms, for example, the mouse. For quite a while, the plastic adherence method of MSC isolation was the 'gold standard'. Nonetheless, real differences in the behaviors and phenotypes of the MSCs being isolated may be identified by inconspicuous differences in the strategies. In 2006, the ISCT published a position statement defining the three minimum criteria required to define a cell as a human MSC (hMSC) [15]: 1) cells must be plastic-adherent, 2) cells must be capable of tri-lineage differentiation (osteogenic, chondrogenic and adipogenic) and 3) cells must express a particular pattern of cell-surface antigens. Relative to the last point, hMSCs must express CD105,

CD73 and CD90, while being negative for CD45, CD34, CD14 or CD11b, CD79α or CD19, and human leukocyte antigen-D-related (HLA-DR) [1]. It should be noted that this surface profile not only distinguishes MSCs from other cells, but also, combined with the other components for recognition, represents the best current approach for the identification of hMSCs [15].

As the field progresses, new isolation methods are being described for the more specific separation of MSCs, and although techniques are becoming more complex, they are also more reproducible [20].

3. THE MAIN SOURCE OF MSCs: LOCATION AND ISOLATION

Expectedly, MSCs have been isolated from the bone marrow, despite the fact that they have now been recognized in (and isolated from) muscle, fat and connective tissue among a wide variety of other tissues and organs [21 - 26].

Generally, the vast majority of the research result on MSC properties are based on the bone marrow isolated MSCs isolated. However, MSCs have also been found in and isolated from different adult and fetal organs and tissues, after they were first isolated from the bone marrow, (Fig. **1**). In fetal tissues, they can be isolated from blood, liver and bone marrow [27], the lung [28], amniotic fluid [2] and the pancreas [21]. In adult tissues, MSCs can be isolated from Wharton's jelly of the umbilical cord [23], placenta [24],umbilical cord blood [22], , peripheral blood [25, 26], adipose tissue [16], skeletal muscle [20], dermis [19], dental pulp [10], synovial membrane [31], heart [32], deciduous teeth [29], scalp tissue [30], and both vein [33] and artery walls [34]. In fact, MSCs can be isolated from almost all post-natal organs, including the cerebrum, spleen, liver, lung, kidney, thymus and pancreas [35]. Despite the fact that any adult tissue could be the sources of MSCs, other tissues may have more obvious advantages. Both umbilical cord blood/cord and placenta once considered as medical waste, but nowadays they are one of the idea tissues for MSC isolation. After a moderately simple and low level invasive subcutaneous lipoaspiration procedure, MSCs can be easily isolated from the fat tissue. A study of 59 donors with a median age of 49 years concluded that, about 2% of all nucleated cells are MSCs in lipoaspirate [36]. In bone marrow tissue, the MSC frequency has been evaluated as 1 in 10,000 nucleated cells [37]. The number of MSCs which isolated from marrow decreases fundamentally with increasing age of donor. MSC frequency is assessed as 1 in 10,000 cells at birth, as1 in 100,000 cells in adolescents, as 1 in 250,000 cells at 30 years of age, 1 in 400,000 cells at 50 years of age as, and as 1 in 2 million cells at 80 years of age [38]. From donors under local anesthesia, the average volume of human marrow is about 40 mL, but a more volume of fat tissue can be isolated from the same donors. It has been assessed that 200 mL of lipoaspirate would normally isolated more than 1 million stem cells, or around 40 times that obtained from 40 mL of bone suction in the most favorable conditions [36]. In Wharton's jelly, the MSC frequency from the umbilical cord has been evaluated as 1 in 333 nucleated cells [39]. Only 40% of processed cord blood samples can isolate stem cell to culture [40].

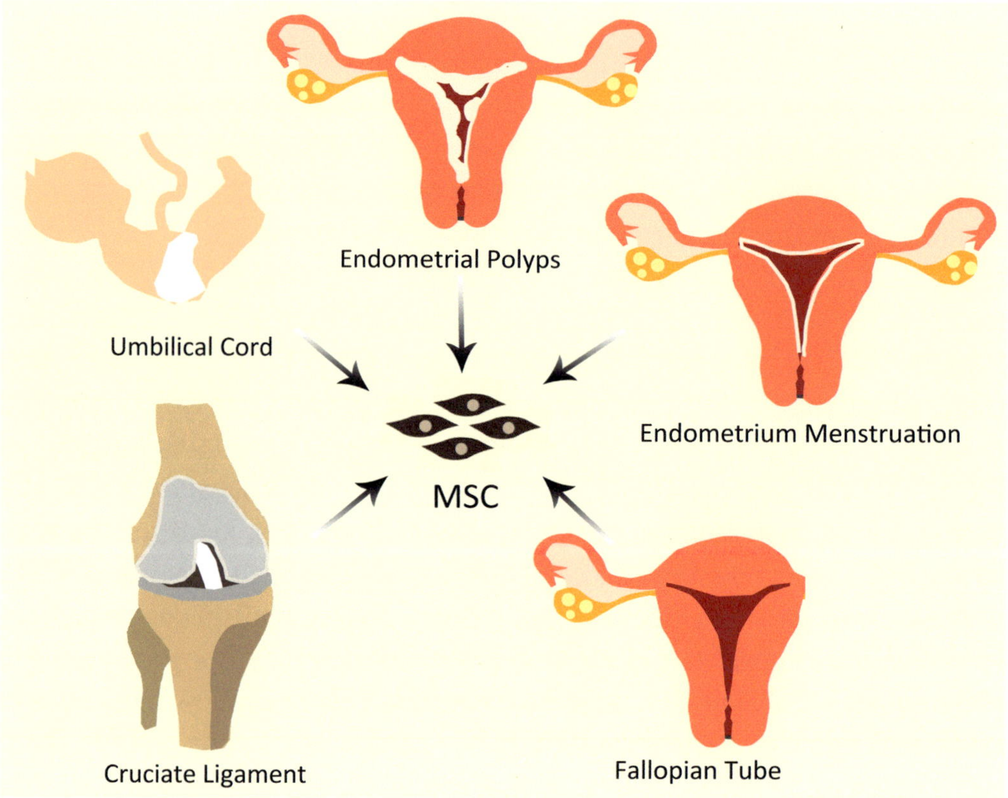

Fig. (1). Various new sources of MSCs. Cells are isolated and cultured and can differentiate into three germ layers for transplantation purposes.

3.1. Genetic Phenotypes of MSCs

In spite of the fact that from different tissues there are no obvious phenotypic differences antigen which expressed in surface of MSCs. The most valuable tool to research characterization of MSCs is a global gene expression profile. Because different laboratories use different processes to isolate and expand variable cells, making it occasionally hard to analyze molecular signatures and properties among MSCs of distinctive origin. In MSCs isolated from umbilical cord blood, adipose tissue and bone marrow, many differentially expressed genes based on their individual development sources. However, 25 genes are overlapped and when human fibroblasts control to all MSC arrangements raised [41]. Full gene expression differences in genetic research group also showed that bone marrow-derived MSC expression of a unique set of genes, and with any embryonic stem cells (ES cells) share any meaningful overlap with hematopoietic stem cells. All 92 genes of a particular molecular signature has been characterized as both umbilical cord blood and bone marrow-derived MSC, and expression of genes have been at least 50 times higher when comparing with those by the division of bone marrow-derived mononuclear cells obtained [42]. Comparing the MSCs from bone marrow and synovial membrane, 46 out of 268 genes were found different in the array of transcriptional profiles [43]. MSCs from the adipose tissue, umbilical cord blood and bone marrow were compared in a recently study. No significant difference was found in cell morphology and surface antigen expression [44].

Morikawa *et al.* [10] studied mMSCs isolated by enzymatic separation from bone marrow and subsequently selected by cell sorting. Platelet-derived growth factor receptor-α (PDGFRα) and stem cell antigen-1 (Sca-1) were positively expressed in isolated cells, while being negative for expression of the hematopoietic markers CD45 and TER119 [10]. The isolated cells expanded without senescence, created a higher number of colonies upon expansion and did not contain any cells capable of forming hematopoietic colonies [10]. Importantly, these cells have the capacity to differentiate into cells of the mesenchymal lineage when examined post-transplantation both *in vitro* and *in vivo* [10]. Investigations of this nature represent a critical step toward the validation of MSC isolation techniques in the mouse.

3.2. Procedures for MSC Isolation

A general procedure was used to separate the various MSC. Dulbecco's modified Eagle's medium (DMEM) was used to dissolved collagenase, digestion time is limited to one hour maximum, at 37 °C. Stem cell separation, as soon as the following euthanasia, culture media was used at a temperature not lower than room temperature, [45]. Adherence to culture plastic was the original MSC isolation protocols. However, such a cell is programmed to generate significant heterogeneity of the cell culture, and relatively time consuming. Therefore, new techniques for MSC isolation and enhancement to avoid these restriction are be considered. Several procedures based on selective affinity for the extracellular matrix stem cells have been developed, for example, MSC adhesion and growth of the choice of collagen type I [46] and fibrin [47]. Antibodies such as CD49a [48], CD105 [49], CD146 [50] and Stro-1 [51] were applied to antibody-based series of elective procedures. Measurement by the colony-forming unit (CFU) [52, 53], CD73 is also easy to use up to 1000-fold enrichment of MSC [54]. MSC has a high proliferative potential can be isolated by selection based on CD133 mobilization from peripheral blood and umbilical cord blood [55]. MSC can also be enriched from the bone marrow by the expression of stage-specific embryonic antigen SSEA-4[56]. MSC from bone marrow production has been concentrated in the expression of the low affinity nerve growth factor receptor cells exhibited (NGFR, CD271) [57]. In addition, enrichment MSC-containing 60 times the population can be obtained by selection of CD10/CD29 double positive cells to achieve [58]. A very promising completely new approach is based on highly specific DNA aptamer technology to select MSC [59].

4. THE PROLIFERATION OF MSCs: REGULATORY FACTORS INVOLVED IN THEIR MAINTENANCE, GROWTH AND SURVIVAL

Somatic cells cease proliferating and become senescent after extensive *in vitro* culture [60]. Many ways have been proposed to participate in the acquisition by the MSC cellular senescence phenotype. Serial passaging process MSC gradually lose their telomeres when cultured *in vitro*, however, MSCs cultured telomerase activity has not been found [61, 62]. MSC cultured under the standard conditions with high oxygen tension also could induced the premature MSC senescence [63].

Many experimental strategies which already reported that can effectively extend the life of MSC *in vitro*. [64]. According a recently research, under the FGF-2 exists circumstances, the life span of MSCs could be extended to more than 70 PD [65]. Dickkopf-1 (DKK-1) is one of Wnt signaling inhibitors. When the medium was added DKK1, cell cycle and cell proliferation of the MSC have undergone significant changes [66]. In contrast, another study showed that using the Wnt pathway activator may stimulate MSC proliferation. However, only a low dosage of Wnt3A stimulates proliferation, in other side a high dosage of Wnt3A may induces osteogenic differentiation [67]. With a retrovirus vector containing the human telomerase (hTERT gene) catalytic subunit gene transduction of MSCs in the MSC proliferative potential may extend over 260 PD. Cells do not form tumors, and maintain the osteogenic differentiation ability [68]. It has been reported MSCs proliferation can be significantly increased from the human papilloma virus or DNA tumor virus SV40 large T antigen derived as the presence of E6 and E7 oncogenes. Even more surprising, carrying the virus oncogene MSCs showed no signs of malignancy [69, 70].

5. THE FACTORS DETERMINING OR REGULATING MSC MIGRATION

5.1. Homing of Mesenchymal Stem Cells

Although MSCs have been found from the non-hematopoietic stem cells in bone marrow, in recent years, scholars have isolated MSCs from fat, bone, muscle, cord blood, umbilical cord placenta, liver, pancreas, lungs and other tissues. They found that MSCs were a kind of stem cell, rather than non-hematopoietic stem cells from bone marrow.

Not only do MSCs have multiple differentiation potential, but also they can be targeted to migrate to the damaged tissue, chronic inflammation or tumor site [59, 71], which makes them the research focus in cell therapy and tissue engineering. The migration of MSCs under physiological or pathological conditions has also been of significant concern, for its possible influence on the factors or mechanisms of MSC migration to control the homing of MSCs. Thus far, several factors and pathways have been shown to be related to the homing of MSCs. In addition, studies have suggested that chemokine-induced migration may depend on the systemic and local inflammatory state and that MSC receptors may vary with tissue source.

5.2. Growth Factors Influence the Ability of MSCs to Migrate

5.2.1. PDGF

Platelet-derived growth factor (PDGF) plays important roles in the successive stages of embryogenesis and in wound healing. PDGF is a polypeptide dimer with four homodimers (PDGF-AA, PDGF-BB, PDGF-CC, PDGF-DD) and one heterodimer (PDGF-AB). PDGF-CC and -DD are not discussed here. PDGF has two receptor dimers, a and b. PDGF-AA binds a/a receptors, PDGF-AB binds a/a and a/b receptors and PDGF-BB binds a/a, a/b and b/b receptors. *In vitro*, PDGF-AA, PDGF-BB and PDGF-AB have been shown to be potent chemoattractants for human MSCs, with PDGF-BB generating the strongest response. PDGF-AB also stimulated chemotaxis by ovine MSCs. Migration was enhanced when MSCs were co-cultured with PDGF-BB-activated fibroblasts in a modified scratch assay. Blocking basic fibroblast growth factor (bFGF) and epithelial neutrophil activating peptide-78 produced by the fibroblasts inhibited the enhanced migration. The co-culture study underscores the fact that migration may result from the integration of several chemotactic signals. In an *in vivo* tracking study in nude rats, PDGF-AA recruited nanoparticle-labeled human bMSCs, which were injected into the proximal femoral medullary canal, to a fibrin scaffold in the distal femur, underscoring the promise of PDGF for regenerative therapies [72, 73].

5.2.2. IGF

Insulin-like growth factors (IGF) are regulators of fetal and post-natal growth and development and of bone growth, remodeling and repair [74]. The IGF-1R receptor binds both IGF-1 and -2.56. *In vitro*, IGF-1 and -2 stimulated migration by human BMSCs in a dose-dependent manner. IGF binding protein 5 (IGFBP-5) also induced migration. Adding IGFBP-5 to IGF-1 further enhanced IGF-1-stimulated migration; adding IGFBP-3 had no effect. Blocking the IGF-1R receptor inhibited migration completely. Overexpression of IGF-1 in rat bMSCs resulted in enhanced recruitment to the infarcted heart and improved cardiac function in a rat model. Overexpression also promoted expression of hepatocyte growth factor (HGF), bFGF, vascular endothelial growth factor (VEGF) and stromal-cell-derived factor (SDF)-1a by MSCs and promoted cell survival in culture. IGF-1 also promotes cell proliferation, making it a strong candidate for tissue regeneration therapies [75, 76].

5.2.3. BMP-2, BMP-4, BMP-7, TGF-β1, TGF-β3

The bone morphogenetic proteins (BMP) and transforming growth factors-β (TGF-β) are members of the TGF-β superfamily of signaling molecules. The BMPs and TGF-βs are regulators of embryogenesis with essential roles in gastrulation, establishment of the dorsal-ventral axis, angiogenesis and organogenesis, particularly skeletal formation. Postnatally, they play a prominent role in endochondral ossification, fracture healing and bone remodeling [77]. Receptors for the BMPs and TGF-βs are classified as type I and type II. Upon ligand binding, the receptors form a heterotetrameric complex composed of type I and type II receptor-pairs. BMPs-2, -4 and -7 are expressed during fracture healing. *In vitro*, BMPs-2, -4 and -7 stimulated chemotaxis by human BMSCs, which suggests that their role in fracture healing may include MSC recruitment. TGF-β1 and -β3 are also expressed during fracture healing. *In vitro*, TGF-β1 stimulated chemotaxis by human adipose-derived MSCs (ASCs), but not by BMSCs. TGF-β3 stimulated chemotaxis by human ASCs, synovium-derived MSCs and BMSCs. *In vivo*, sufficient cartilage was regenerated to cover the surface of a rabbit humeral head using only a scaffold infused with TGF-β3. This cartilage regeneration was achieved by endogenous MSC recruitment [78, 79].

5.2.4. EGF, HB-EGF, TGFa

Epidermal growth factor (EGF; EGFR) has pleiotropic effects that include migration, proliferation, differentiation and dedifferentiation [80]. EGFR is expressed by epithelial and stromal cells and by some glial and smooth muscle cells. EGFR is either the sole or the overwhelmingly predominant receptor for several ligands, including EGF, heparin-binding EGF (HB-EGF) and TGF-α. *In vitro*, EGF stimulated chemotaxis by human, rat, rabbit and ovine BMSCs.

Human ASCs also migrated in response to EGF. HB-EGF stimulated chemotaxis by human and rabbit BMSCs. TGF-α is structurally and functionally similar to EGF, sharing a 40% amino acid sequence identity. TGF-α is up-regulated in response to injury and induces inflammation with ectopic expression. It also induces production of VEGF by human BMSCs. *In vitro*, TGF-α stimulated chemotaxis by human BMSCs. A significant advantage of EGF is the stability of its native and recombinant forms, even under harsh conditions [75].

5.2.5. FGF-2

FGF-2, or b-FGF, is a member of the FGF family that includes 22 structurally related proteins. FGF-2 plays a role in limb development, wound healing and angiogenesis, and it contributes to cell survival, migration, proliferation and differentiation [81]. Alternative splicing for FGF receptors (FGFR) 1-3 produces two isoforms, "b" and "c". FGF-2 preferentially activates the "c" isoforms (FGFR-1c, -2c, -3c) and has activity with FGFR-β1 and FGFR-4. The interaction of FGF with heparin and heparin sulfate plays an important role in FGF bio-activity. Heparin binding protects FGF from proteolysis and thermal denaturation, limits the diffusion of FGF in the interstitial space and results in a store of FGF that is released upon heparin cleavage. *In vitro*, FGF-2 stimulated chemotaxis by murine, rabbit, ovine and human BMSCs and human ASCs [82, 83].

5.2.6. HGF

HGF (c-Met) can promote the mitogenic and motogenic activity of hepatocytes and other cells. HGF plays a key role in organogenesis during embryonic development and in wound healing [84, 85]. *In vitro*, HGF stimulated chemotaxis by BMSCs. However, HGF administration also resulted in marked inhibition of BMSC proliferation.

5.2.7. VEGF-A

The VEGF family has five members: VEGF-A through D and placental growth factor [86] VEGF-A, which binds to the receptors VEGFR-1 and -2. VEGFs are essential for vasculogenesis and angiogenesis in the embryo and angiogenesis in the adult. VEGF-induced angiogenesis plays a key role in wound healing [87].

There is considerable variation in the results of *in vitro* studies that have evaluated VEGF-A as a chemoattractant for BMSCs. VEGF-A-induced migration was mediated by the structurally similar PDGF receptors. Additionally, exposure of murine BMSCs to hypoxia increased expression of VEGFR-1 and enhanced the chemotactic response to VEGF-A *in vitro* after hypoxia treatment. Delivery of VEGF from a collagen gel in endodontically treated human teeth implanted subcutaneously into a murine host resulted in regeneration of dental-pulp-like tissue [88]. These results demonstrate the potential of VEGF for tissue regeneration based on endogenous stem cell recruitment.

5.3. Chemotactic Factors Influence the Ability of MSCs to Migrate

5.3.1. SDF-1

SDF-1 (CXCL12; CXCR4) is expressed during embryonic development at injury sites, and in response to hypoxia [89 - 93]. There are two similar SDF-1 isoforms, SDF-1a and -1b. *In vitro*, SDF-1b stimulated chemotaxis by human BMSCs [94]. SDF-1 is an important chemokine for MSC recruitment. In a murine bone fracture model, SDF-1 promoted bone regeneration by recruiting systemically delivered MSCs to the fracture site [91]. Regeneration was inhibited by an anti-SDF-1 antibody and by a CXCR4 antagonist. In a murine model of myocardial infarction, systemically delivered BMSCs migrated to infarcted tissue [93]. Elevation of SDF-1a expression by adenoviral gene delivery significantly increased MSC recruitment; a CXCR4 antagonist diminished recruitment. SDF-1a expression did not increase MSC recruitment in the absence of infarction. SDF-1 has also been exploited to recruit endogenous BMSCs, and to promote their migration within biomaterials scaffolds [95].

5.3.2. MCP-1, MIP-1a, IL-8

Monocyte chemoattractant protein-1 (MCP-1, CCL2; CCR2), macrophage inflammatory protein-1a (MIP-1a, CCL3; CCR1,5) and interleukin-8 (IL-8, CXCL8; CXCR1, 2) are expressed in ischemic cerebral tissue [96]. Systemically infused BMSCs migrated to ischemic cerebral tissue and reduced neurological functional deficits [97]. *In vitro*, extracts of ischemic brain tissue and recombinant MCP-1, MIP-1a and IL-8 stimulated chemotaxis by human BMSCs. MCP-1 is also transiently up-regulated in the infarcted heart [98]. In a murine model with MCP-1 specifically expressed in the heart, systemically infused murine BMSCs engrafted in the hearts of transgenic mice, but not in those of wild-type controls. CCR2 antibodies inhibited MCP-1-induced migration. MCP-1 or PDGF-AA was released from a heparin-conjugated fibrin scaffold in a distal femoral osteochondral defect. Both chemoattractants recruited BMSCs to the scaffold; however, recruitment by MCP-1 was weak relative to that of PDGF-AA. These studies suggest that MCP-1, MIP-1a and IL-8 may be valuable chemoattractants for regenerative therapies; however, growth factors may induce stronger responses.

5.3.3. RANTES, MDC

RANTES (regulated on the activation of normal T-cell-expressed and -secreted CCL5; CCR1, 3, 4, 5) is a chemoattractant for T-cells, monocytes and eosinophils. Macrophage-derived chemokine (MDC, CCL22; CCR4) is a chemoattractant for monocytes, monocyte-derived dendritic cells and natural-killer cells. *In vitro*, both RANTES and MDC stimulated chemotaxis by human BMSCs. Preincubation of human BMSCs with tumor necrosis factor-α (TNF-α), an inflammatory cytokine, substantially increased the chemotactic response. CCR5 and its ligands are transiently up-regulated in murine skin wounds. Wound healing was delayed in CCR5-/- mice; however, systemic infusion of CCR5+ bone-marrow-derived cells restored normal healing. MDC did not induce *in vitro* migration by human periosteum-derived MSCs [99, 100].

5.3.4. Fractalkine

Fractalkine (CX3CL1; CX3CR1) mediates chemotaxis and the adhesion of inflammatory cells and pain sensation, and is involved in several inflammatory conditions [101]. High levels of fractalkine and its receptor are expressed at wound sites [99, 102]. *In vitro*, fractalkine stimulated chemotaxis by BMSCs; however, the potential of fractalkine for *in vivo* MSC recruitment has not been investigated.

5.3.5. SLC, TARC

Secondary lymphoid tissue chemokine (SLC, CCL21; CCR7) is a potent chemoattractant for lymphocytes and some T-cell lines [103]. Thymus and activation-regulated chemokine (TARC, CCL17; CCR4) is expressed by several cell types, including dendritic cells, endothelial cells, keratinocytes and fibroblasts, and it is constitutively expressed in the thymus [104]. SLC and TARC are up-regulated in several inflammatory skin conditions. In a murine skin wound model, local injections of SLC recruited systemically infused BMSCs to the wound and accelerated healing.

5.4. Other Chemoattractants

5.4.1. LPA, S1P

Lysophospholipids, such as lysophosphatic acid (LPA) and sphingosine 1-phosphate (S1P), are bioactive signaling molecules that affect a variety of cell functions, including survival, migration, proliferation and differentiation [105]. LPA and S1P are released by activated platelets, and they may be as important as cytokines and growth factors in orchestrating tissue repair [106]. Thus far, there are five confirmed LPA receptors (LPA1–5) and S1P receptors (S1P1–5). *In vitro*, LPA stimulated chemotaxis by human and murine BMSCs [107, 108]. Human ASCs also migrated in response to LPA. Migration was inhibited by Ki6425, an antagonist specific for LPA1 and LPA3. In murine and rat cutaneous wound models, topical LPA application resulted in an accelerated healing response. S1P plays a key role in neovascularization for wound healing and for cancer tumor growth [109]. Binding S1P1 and S1P2 had opposing effects on tumor angiogenesis in a murine study, with S1P1 stimulating angiogenesis and S1P2 inhibiting angiogenesis. *In vitro*, S1P stimulated a chemotactic response by murine bMSCs. The migration response for S1P was stronger than that for LPA [108]. In a murine model of liver fibrosis, S1P was elevated in the liver and serum after liver injury, and it stimulated migration of intravenously delivered BMSCs to the liver. Receptors S1P1–3 were abundant on the BMSCs recruited [110].

5.4.2. HMGB-1

High mobility group box-1 (HMGB-1) is a non-histone DNA-binding cytokine that stabilizes nucleosomes and facilitates transcription [111]. HMGB-1 is passively released by injured and necrotic cells, but not by apoptotic cells, and it is secreted by activated macrophages and monocytes [112]. HMGB-1 in the extracellular compartment is recognized as a signal of tissue injury and a trigger to initiate tissue regeneration. However, HMGB-1 also contributes to pathological conditions, such as sepsis, rheumatoid arthritis, lupus and cancer. HMGB-1 binds to the receptor for advanced glycation end-products and Toll-like receptors-2 and -4. *In vitro*, HMGB-1 stimulated transmigration by mesoangioblasts, a vessel-associated stem cell, through an endothelial barrier [113]. Beads containing HMGB-1 implanted into muscle resulted in recruitment of intravenously injected mesoangioblasts. HMGB-1 also stimulated migration by human BMSCs *in vitro* [114]. A better understanding of HMGB-1 signaling will be needed to harness its regenerative potential.

5.4.3. TNF-a

TNF-α is a proinflammatory cytokine produced by macrophages during acute inflammation of injured tissue [115]. TNF-α binds two receptors, TNFR-1 and -2 that are thought to activate distinct signaling pathways. *In vitro*, TNF-α stimulated a strong dose-dependent migration response by BMSCs and ASCs [116 - 118]. Preincubation of the ASCs with TNF-α enhanced the migration response to individual chemokines and growth factors [116]. Incubation of human BMSCs or ASCs with TNF-α stimulated production of VEGF, HGF and IGF-1 by the MSCs [119]. Neutralizing TNFR-2 significantly inhibited migration; neutralizing TNFR-1 had no effect on migration. In a rat ischemia model, preincubation of rat BMSCs with TNF-α before systemic infusion was associated with a greater BMSC accumulation in the ischemic muscle [120]. Additional research will be needed to determine if the migratory effects of TNF-α can be harnessed without introducing damaging inflammatory side-effects.

5.4.4. Ligands for Toll-Like Receptors

Toll-like receptors (TLR) are a large family (TLR1-TLR11) of evolutionarily conserved receptors activated by intracellular components, such as heat shock proteins or RNA, that are released during wounding and by invading pathogens. These signals recruit immune cells to tissue injuries, and it has recently been discovered that they may also recruit MSCs. *In vitro*, human BMSC migration was stimulated by ligands for TLR2-4 and TLR9 [121 - 123]. Preincubation with the ligands enhanced the migration response. However, a longer preincubation of TLR2 was shown to inhibit BMSC migration [123, 124]. Interestingly, TLR signaling was also shown to regulate differentiation down mesodermal pathways for human BMSCs. TLR priming of a heterogenous BMSC population resulted in a nearly homogeneous phenotype.

5.5. Signaling Pathways Influence the Ability of MSCs to Migrate

5.5.1. Mitogen-Activated Protein Kinase Pathway (MAPK)

Mitogen-activated protein kinase (MAPK) is a typical signaling pathway of cell growth regulation, including ERK1/2, JNK, p38MAPK, *etc*. They form a huge, complicated and delicate network, involving cell growth. The physiological activities of the division, differentiation and death at various stages regulate the series of complex life. Yuan and co-workers previously demonstrated that higher shear stress (>2 Pa) hindered human MSC migration, whereas lower shear stress (0.2 Pa) induced cell migration through MAPK pathways [125]. The results also indicated that the SDF-1/CXCR4 axis mediates shear-stress-induced hMSC migration through JNK and p38-MAPK pathways and that the difference in migration speeds between upstream and downstream cells may be due to ERK1/2 activation [125].

Hong *et al.* found that inhibition of p38 MAPK signal transfer by mesenchymal stem cells (MSC) migration mediated inhibition of primary tumor and metastasis, where the cancer-associated fibroblasts differentiate into cells to promote tumor metastasis site [126].

5.5.2. PI-3k/AKT (Phosphatidylinositol-3-Kinase)

The PI-3k/AKT signaling pathway plays an important role in maintaining cell survival and can promote cell growth and inhibit apoptosis. The PI3K/Akt signaling pathway has a central role in regulating tumor growth, and several MSC secreted factors stimulate activation of this pathway. MSCs could be regulated directly, including migration ability, and they may have played a dual role in various stages of tumorigenesis [127]. Some biological material by activating the PI-3k/AKT mediated pathway can increase the functionality of MSC [128]. Further, the transcription factor snail could increase the migration of BMSCs *in vitro*, and the expression of that factor is attributed to activated PI-3k/AKT and MAPK signal pathways [129].

5.5.3. Jak/STAT Signaling Pathway

Jak/STAT signaling pathways can regulate cell proliferation, differentiation, apoptosis, *etc*. Zgheib and co-workers investigated the JAK/STAT signal transducing pathway that may have an impact on CSF (colony-stimulating factors) transcription, which plays an important role in the recruitment of MSCs to hypoxic tumors, where they signal paracrine-regulated angiogenesis [130]. Early in 2005, Silver *et al.* found that JAK/STAT signaling is necessary and sufficient for the specification and migration of a group of cells called border cells in the Drosophila ovary [131]. SDF-1 stimulation activates Jak2/STAT3 as well as MEK/ERK1/2 signaling, which in turn promotes the migration of MSCs toward tumor cells [1, 7, 9, 10, 53, 132].

CONFLICT OF INTEREST

The authors declare that they have no financial or other relationships with other people or organizations that might lead to a conflict of interest.

ACKNOWLEDGEMENTS

We thank Dr. Dan Wang for valuable comments and advice. This work is financially supported by the Natural Science Foundation of Shanghai (15ZR1432700), the National Natural Science Foundation of China (81500808, 81470721, 31170929) and the Sichuan Science and Technology Innovation Team (2014TD0001).

REFERENCES

[1] Friedenstein AJ. Precursor cells of mechanocytes. Int Rev Cytol 1976; 47(4): 327-59. [http://dx.doi.org/10.1016/S0074-7696(08)60092-3] [PMID: 11195]

[2] Caplan AI, Bruder SP. Mesenchymal stem cells: building blocks for molecular medicine in the 21st century. Trends Mol Med 2001; 7(6): 259-64. [http://dx.doi.org/10.1016/S1471-4914(01)02016-0] [PMID: 11378515]

[3] Caplan AI. Mesenchymal stem cells. J Orthop Res 1991; 9(5): 641-50. [http://dx.doi.org/10.1002/jor.1100090504] [PMID: 1870029]

[4] Mahadik BP, Pedron Haba S, Skertich LJ, Harley BA. The use of covalently immobilized stem cell factor to selectively affect hematopoietic stem cell activity within a gelatin hydrogel. Biomaterials 2015; 67: 297-307. [http://dx.doi.org/10.1016/j.biomaterials.2015.07.042] [PMID: 26232879]

[5] Bianco P. "Mesenchymal" stem cells. Annu Rev Cell Dev Biol 2014; 30(6): 677-704. [http://dx.doi.org/10.1146/annurev-cellbio-100913-013132] [PMID: 25150008]

[6] Horwitz EM, Le Blanc K, Dominici M, *et al.* International society for cellular therapy clarification of the nomenclature for MSC: the international society for cellular therapy position statement. Cytotherapy 2005; 7(5): 393-5. [http://dx.doi.org/10.1080/14653240500319234] [PMID: 16236628]

[7] Bianco P, Robey PG, Simmons PJ. Mesenchymal stem cells: revisiting history, concepts, and assays. Cell Stem Cell 2008; 2(4): 313-9. [http://dx.doi.org/10.1016/j.stem.2008.03.002] [PMID: 18397751]

[8] Friedenstein AJ, Deriglasova UF, Kulagina NN, *et al.* Precursors for fibroblasts in different populations of hematopoietic cells as detected by the *in vitro* colony assay method. Exp Hematol 1974; 2(2): 83-92. [PMID: 4455512]

[9] Jones E, McGonagle D. Human bone marrow mesenchymal stem cells *in vivo.* Rheumatology (Oxford) 2008; 47(2): 126-31. [http://dx.doi.org/10.1093/rheumatology/kem206] [PMID: 17986482]

[10] Morikawa S, Mabuchi Y, Kubota Y, *et al.* Prospective identification, isolation, and systemic transplantation of multipotent mesenchymal stem cells in murine bone marrow. J Exp Med 2009; 206(11): 2483-96. [http://dx.doi.org/10.1084/jem.20091046] [PMID: 19841085]

[11] Jones EA, Kinsey SE, English A, *et al.* Isolation and characterization of bone marrow multipotential mesenchymal progenitor cells. Arthritis Rheum 2002; 46(12): 3349-60. [http://dx.doi.org/10.1002/art.10696] [PMID: 12483742]

[12] Barry FP, Murphy JM. Mesenchymal stem cells: clinical applications and biological characterization. Int J Biochem Cell Biol 2004; 36(4): 568-84. [http://dx.doi.org/10.1016/j.biocel.2003.11.001] [PMID: 15010324]

[13] De Bari C, Dell'accio F. Cell therapy: a challenge in modern medicine. Biomed Mater Eng 2008; 18(1) (Suppl.): S11-7.

[PMID: 18334719]

[14] Capelli C, Pedrini O, Valgardsdottir R, Da Roit F, Golay J, Introna M. Clinical grade expansion of MSCs. Immunol Lett 2015; 168(2): 222-7.
[http://dx.doi.org/10.1016/j.imlet.2015.06.006] [PMID: 26092523]

[15] Dominici M, Le Blanc K, Mueller I, *et al.* Minimal criteria for defining multipotent mesenchymal stromal cells. the international society for cellular therapy position statement. Cytotherapy 2006; 8(4): 315-7.
[http://dx.doi.org/10.1080/14653240600855905] [PMID: 16923606]

[16] Peister A, Mellad JA, Larson BL, Hall BM, Gibson LF, Prockop DJ. Adult stem cells from bone marrow (MSCs) isolated from different strains of inbred mice vary in surface epitopes, rates of proliferation, and differentiation potential. Blood 2004; 103(5): 1662-8.
[http://dx.doi.org/10.1182/blood-2003-09-3070] [PMID: 14592819]

[17] Fiorina P, Jurewicz M, Augello A, *et al.* Immunomodulatory function of bone marrow-derived mesenchymal stem cells in experimental autoimmune type 1 diabetes. J Immunol 2009; 183(2): 993-1004.
[http://dx.doi.org/10.4049/jimmunol.0900803] [PMID: 19561093]

[18] Jo CH, Ahn HJ, Kim HJ, Seong SC, Lee MC. Surface characterization and chondrogenic differentiation of mesenchymal stromal cells derived from synovium. Cytotherapy 2007; 9(4): 316-27.
[http://dx.doi.org/10.1080/14653240701291620] [PMID: 17573607]

[19] Krampera M, Galipeau J, Shi Y, Tarte K, Sensebe L. MSC Committee of the International Society for Cellular Therapy (ISCT). Immunological characterization of multipotent mesenchymal stromal cells--The International Society for Cellular Therapy (ISCT) working proposal. Cytotherapy 2013; 15(9): 1054-61.
[http://dx.doi.org/10.1016/j.jcyt.2013.02.010] [PMID: 23602578]

[20] Houlihan DD, Mabuchi Y, Morikawa S, *et al.* Isolation of mouse mesenchymal stem cells on the basis of expression of Sca-1 and PDGFR-α. Nat Protoc 2012; 7(12): 2103-11.
[http://dx.doi.org/10.1038/nprot.2012.125] [PMID: 23154782]

[21] Malgieri A, Kantzari E, Patrizi MP, Gambardella S. Bone marrow and umbilical cord blood human mesenchymal stem cells: state of the art. Int J Clin Exp Med 2010; 3(4): 248-69.
[PMID: 21072260]

[22] Cao H, Qian H, Xu W, *et al.* Mesenchymal stem cells derived from human umbilical cord ameliorate ischemia/reperfusion-induced acute renal failure in rats. Biotechnol Lett 2010; 32(5): 725-32.
[http://dx.doi.org/10.1007/s10529-010-0207-y] [PMID: 20131083]

[23] Summer R, Fitzsimmons K, Dwyer D, Murphy J, Fine A. Isolation of an adult mouse lung mesenchymal progenitor cell population. Am J Respir Cell Mol Biol 2007; 37(2): 152-9.
[http://dx.doi.org/10.1165/rcmb.2006-0386OC] [PMID: 17395889]

[24] Huang Y, Johnston P, Zhang B, *et al.* Kidney-derived stromal cells modulate dendritic and T cell responses. J Am Soc Nephrol 2009; 20(4): 831-41.
[http://dx.doi.org/10.1681/ASN.2008030310] [PMID: 19297559]

[25] Rodriguez AM, Elabd C, Amri EZ, Ailhaud G, Dani C. The human adipose tissue is a source of

multipotent stem cells. Biochimie 2005; 87(1): 125-8.
[http://dx.doi.org/10.1016/j.biochi.2004.11.007] [PMID: 15733747]

[26] Tomar GB, Srivastava RK, Gupta N, *et al.* Human gingiva-derived mesenchymal stem cells are superior to bone marrow-derived mesenchymal stem cells for cell therapy in regenerative medicine. Biochem Biophys Res Commun 2010; 393(3): 377-83.
[http://dx.doi.org/10.1016/j.bbrc.2010.01.126] [PMID: 20138833]

[27] Friedenstein AJ, Chailakhyan RK, Latsinik NV, Panasyuk AF, Keiliss-Borok IV. Stromal cells responsible for transferring the microenvironment of the hemopoietic tissues. Cloning *in vitro* and retransplantation *in vivo*. Transplantation 1974; 17(4): 331-40.
[http://dx.doi.org/10.1097/00007890-197404000-00001] [PMID: 4150881]

[28] Kohyama J, Abe H, Shimazaki T, *et al.* Brain from bone: efficient "meta-differentiation" of marrow stroma-derived mature osteoblasts to neurons with Noggin or a demethylating agent. Differentiation 2001; 68(4-5): 235-44.
[http://dx.doi.org/10.1046/j.1432-0436.2001.680411.x] [PMID: 11776476]

[29] Ren G, Zhang L, Zhao X, *et al.* Mesenchymal stem cell-mediated immunosuppression occurs *via* concerted action of chemokines and nitric oxide. Cell Stem Cell 2008; 2(2): 141-50.
[http://dx.doi.org/10.1016/j.stem.2007.11.014] [PMID: 18371435]

[30] Ren G, Su J, Zhang L, *et al.* Species variation in the mechanisms of mesenchymal stem cell-mediated immunosuppression. Stem Cells 2009; 27(8): 1954-62.
[http://dx.doi.org/10.1002/stem.118] [PMID: 19544427]

[31] Le Blanc K, Tammik C, Rosendahl K, Zetterberg E, Ringdén O. HLA expression and immunologic properties of differentiated and undifferentiated mesenchymal stem cells. Exp Hematol 2003; 31(10): 890-6.
[http://dx.doi.org/10.1016/S0301-472X(03)00110-3] [PMID: 14550804]

[32] Griffin MD, Ryan AE, Alagesan S, Lohan P, Treacy O, Ritter T. Anti-donor immune responses elicited by allogeneic mesenchymal stem cells: what have we learned so far? Immunol Cell Biol 2013; 91(1): 40-51.
[http://dx.doi.org/10.1038/icb.2012.67] [PMID: 23207278]

[33] Camp DM, Loeffler DA, Farrah DM, Borneman JN, LeWitt PA. Cellular immune response to intrastriatally implanted allogeneic bone marrow stromal cells in a rat model of Parkinson's disease. J Neuroinflammation 2009; 6(6): 17.
[http://dx.doi.org/10.1186/1742-2094-6-17] [PMID: 19500379]

[34] Huang XP, Sun Z, Miyagi Y, *et al.* Differentiation of allogeneic mesenchymal stem cells induces immunogenicity and limits their long-term benefits for myocardial repair. Circulation 2010; 122(23): 2419-29.
[http://dx.doi.org/10.1161/CIRCULATIONAHA.110.955971] [PMID: 21098445]

[35] Isakova IA, Dufour J, Lanclos C, Bruhn J, Phinney DG. Cell-dose-dependent increases in circulating levels of immune effector cells in rhesus macaques following intracranial injection of allogeneic MSCs. Exp Hematol 2010; 38(10): 957-967.e1.
[http://dx.doi.org/10.1016/j.exphem.2010.06.011] [PMID: 20600575]

[36] Eliopoulos N, Stagg J, Lejeune L, Pommey S, Galipeau J. Allogeneic marrow stromal cells are

immune rejected by MHC class I- and class II-mismatched recipient mice. Blood 2005; 106(13): 4057-65.
[http://dx.doi.org/10.1182/blood-2005-03-1004] [PMID: 16118325]

[37] Seifert M, Stolk M, Polenz D, Volk HD. Detrimental effects of rat mesenchymal stromal cell pre-treatment in a model of acute kidney rejection. Front Immunol 2012; 3: 202.
[http://dx.doi.org/10.3389/fimmu.2012.00202] [PMID: 22826709]

[38] Consentius C, Reinke P, Volk HD. Immunogenicity of allogeneic mesenchymal stromal cells: what has been seen *in vitro* and *in vivo?* Regen Med 2015; 10(3): 305-15.
[http://dx.doi.org/10.2217/rme.15.14] [PMID: 25933239]

[39] Liotta F, Angeli R, Cosmi L, *et al.* Toll-like receptors 3 and 4 are expressed by human bone marrow-derived mesenchymal stem cells and can inhibit their T-cell modulatory activity by impairing Notch signaling. Stem Cells 2008; 26(1): 279-89.
[http://dx.doi.org/10.1634/stemcells.2007-0454] [PMID: 17962701]

[40] Augello A, Tasso R, Negrini SM, *et al.* Bone marrow mesenchymal progenitor cells inhibit lymphocyte proliferation by activation of the programmed death 1 pathway. Eur J Immunol 2005; 35(5): 1482-90.
[http://dx.doi.org/10.1002/eji.200425405] [PMID: 15827960]

[41] Gonzalez-Rey E, Anderson P, González MA, Rico L, Büscher D, Delgado M. Human adult stem cells derived from adipose tissue protect against experimental colitis and sepsis. Gut 2009; 58(7): 929-39.
[http://dx.doi.org/10.1136/gut.2008.168534] [PMID: 19136511]

[42] Tögel F, Isaac J, Hu Z, Weiss K, Westenfelder C. Renal SDF-1 signals mobilization and homing of CXCR4-positive cells to the kidney after ischemic injury. Kidney Int 2005; 67(5): 1772-84.
[http://dx.doi.org/10.1111/j.1523-1755.2005.00275.x] [PMID: 15840024]

[43] Suga H, Eto H, Shigeura T, *et al.* IFATS collection: Fibroblast growth factor-2-induced hepatocyte growth factor secretion by adipose-derived stromal cells inhibits postinjury fibrogenesis through a c-Jun N-terminal kinase-dependent mechanism. Stem Cells 2009; 27(1): 238-49.
[http://dx.doi.org/10.1634/stemcells.2008-0261] [PMID: 18772314]

[44] Rehman J, Traktuev D, Li J, *et al.* Secretion of angiogenic and antiapoptotic factors by human adipose stromal cells. Circulation 2004; 109(10): 1292-8.
[http://dx.doi.org/10.1161/01.CIR.0000121425.42966.F1] [PMID: 14993122]

[45] da Silva Meirelles L, Chagastelles PC, Nardi NB. Mesenchymal stem cells reside in virtually all post-natal organs and tissues. J Cell Sci 2006; 119(Pt 11): 2204-13.
[http://dx.doi.org/10.1242/jcs.02932] [PMID: 16684817]

[46] Chang P, Qu Y, Liu Y, *et al.* Multi-therapeutic effects of human adipose-derived mesenchymal stem cells on radiation-induced intestinal injury. Cell Death Dis 2013; 4: e685.
[http://dx.doi.org/10.1038/cddis.2013.178] [PMID: 23788042]

[47] Aldridge V, Garg A, Davies N, *et al.* Human mesenchymal stem cells are recruited to injured liver in a β1-integrin and CD44 dependent manner. Hepatology 2012; 56(3): 1063-73.
[http://dx.doi.org/10.1002/hep.25716] [PMID: 22422467]

[48] Eggenhofer E, Benseler V, Kroemer A, *et al.* Mesenchymal stem cells are short-lived and do not

migrate beyond the lungs after intravenous infusion. Front Immunol 2012; 3(3): 297.
[PMID: 23056000]

[49] Ramírez M, Lucia A, Gómez-Gallego F, *et al.* Mobilisation of mesenchymal cells into blood in response to skeletal muscle injury. Br J Sports Med 2006; 40(8): 719-22.
[http://dx.doi.org/10.1136/bjsm.2006.028639] [PMID: 16807304]

[50] Rochefort GY, Delorme B, Lopez A, *et al.* Multipotential mesenchymal stem cells are mobilized into peripheral blood by hypoxia. Stem Cells 2006; 24(10): 2202-8.
[http://dx.doi.org/10.1634/stemcells.2006-0164] [PMID: 16778152]

[51] Wang Y, Johnsen HE, Mortensen S, *et al.* Changes in circulating mesenchymal stem cells, stem cell homing factor, and vascular growth factors in patients with acute ST elevation myocardial infarction treated with primary percutaneous coronary intervention. Heart 2006; 92(6): 768-74.
[http://dx.doi.org/10.1136/hrt.2005.069799] [PMID: 16251230]

[52] Lee RH, Pulin AA, Seo MJ, *et al.* Intravenous hMSCs improve myocardial infarction in mice because cells embolized in lung are activated to secrete the anti-inflammatory protein TSG-6. Cell Stem Cell 2009; 5(1): 54-63.
[http://dx.doi.org/10.1016/j.stem.2009.05.003] [PMID: 19570514]

[53] Wang N, Li Q, Zhang L, *et al.* Mesenchymal stem cells attenuate peritoneal injury through secretion of TSG-6. PLoS One 2012; 7(8): e43768.
[http://dx.doi.org/10.1371/journal.pone.0043768] [PMID: 22912904]

[54] Danchuk S, Ylostalo JH, Hossain F, *et al.* Human multipotent stromal cells attenuate lipopolysaccharide-induced acute lung injury in mice via secretion of tumor necrosis factor-α-induced protein 6. Stem Cell Res Ther 2011; 2(3): 27.
[http://dx.doi.org/10.1186/scrt68] [PMID: 21569482]

[55] Roddy GW, Oh JY, Lee RH, *et al.* Action at a distance: systemically administered adult stem/progenitor cells (MSCs) reduce inflammatory damage to the cornea without engraftment and primarily by secretion of TNF-α stimulated gene/protein 6. Stem Cells 2011; 29(10): 1572-9.
[http://dx.doi.org/10.1002/stem.708] [PMID: 21837654]

[56] Scruggs BA, Zhang X, Bowles AC, *et al.* Multipotent stromal cells alleviate inflammation, neuropathology, and symptoms associated with globoid cell leukodystrophy in the twitcher mouse. Stem Cells 2013; 31(8): 1523-34.
[http://dx.doi.org/10.1002/stem.1397] [PMID: 23606584]

[57] Moghimi SM, Hunter AC, Murray JC. Long-circulating and target-specific nanoparticles: theory to practice. Pharmacol Rev 2001; 53(2): 283-318.
[PMID: 11356986]

[58] Schrepfer S, Deuse T, Reichenspurner H, Fischbein MP, Robbins RC, Pelletier MP. Stem cell transplantation: the lung barrier. Transplant Proc 2007; 39(2): 573-6.
[http://dx.doi.org/10.1016/j.transproceed.2006.12.019] [PMID: 17362785]

[59] Chamberlain G, Fox J, Ashton B, Middleton J. Concise review: mesenchymal stem cells: their phenotype, differentiation capacity, immunological features, and potential for homing. Stem Cells 2007; 25(11): 2739-49.
[http://dx.doi.org/10.1634/stemcells.2007-0197] [PMID: 17656645]

[60] Choi H, Lee RH, Bazhanov N, Oh JY, Prockop DJ. Anti-inflammatory protein TSG-6 secreted by activated MSCs attenuates zymosan-induced mouse peritonitis by decreasing TLR2/NF-κB signaling in resident macrophages. Blood 2011; 118(2): 330-8.
[http://dx.doi.org/10.1182/blood-2010-12-327353] [PMID: 21551236]

[61] Walenda T, Bokermann G, Ventura Ferreira MS, *et al.* Synergistic effects of growth factors and mesenchymal stromal cells for expansion of hematopoietic stem and progenitor cells. Exp Hematol 2011; 39(6): 617-28.
[http://dx.doi.org/10.1016/j.exphem.2011.02.011] [PMID: 21356269]

[62] Németh K, Leelahavanichkul A, Yuen PS, *et al.* Bone marrow stromal cells attenuate sepsis *via* prostaglandin E(2)-dependent reprogramming of host macrophages to increase their interleukin-10 production. Nat Med 2009; 15(1): 42-9.
[http://dx.doi.org/10.1038/nm.1905] [PMID: 19098906]

[62] Shah K. Mesenchymal stem cells engineered for cancer therapy. Adv Drug Deliv Rev 2012; 64(8): 739-48.
[http://dx.doi.org/10.1016/j.addr.2011.06.010] [PMID: 21740940]

[64] Dai T, Yang E, Sun Y, *et al.* Preparation and drug release mechanism of CTS-TAX-NP-MSCs drug delivery system. Int J Pharm 2013; 456(1): 186-94.
[http://dx.doi.org/10.1016/j.ijpharm.2013.07.070] [PMID: 23933442]

[65] Cavarretta IT, Altanerova V, Matuskova M, Kucerova L, Culig Z, Altaner C. Adipose tissue-derived mesenchymal stem cells expressing prodrug-converting enzyme inhibit human prostate tumor growth. Mol Ther 2010; 18(1): 223-31.
[http://dx.doi.org/10.1038/mt.2009.237] [PMID: 19844197]

[66] Chen TS, Arslan F, Yin Y, *et al.* Enabling a robust scalable manufacturing process for therapeutic exosomes through oncogenic immortalization of human ESC-derived MSCs. J Transl Med 2011; 9: 47.
[http://dx.doi.org/10.1186/1479-5876-9-47] [PMID: 21513579]

[67] Yeo RW, Lai RC, Zhang B, *et al.* Mesenchymal stem cell: an efficient mass producer of exosomes for drug delivery. Adv Drug Deliv Rev 2013; 65(3): 336-41.
[http://dx.doi.org/10.1016/j.addr.2012.07.001] [PMID: 22780955]

[68] van Dommelen SM, Vader P, Lakhal S, *et al.* Microvesicles and exosomes: opportunities for cell-derived membrane vesicles in drug delivery. J Control Release 2012; 161(2): 635-44.
[http://dx.doi.org/10.1016/j.jconrel.2011.11.021] [PMID: 22138068]

[69] Mohit E, Rafati S. Biological delivery approaches for gene therapy: strategies to potentiate efficacy and enhance specificity. Mol Immunol 2013; 56(4): 599-611.
[http://dx.doi.org/10.1016/j.molimm.2013.06.005] [PMID: 23911418]

[70] Chapel A, Bertho JM, Bensidhoum M, *et al.* Mesenchymal stem cells home to injured tissues when co-infused with hematopoietic cells to treat a radiation-induced multi-organ failure syndrome. J Gene Med 2003; 5(12): 1028-38.
[http://dx.doi.org/10.1002/jgm.452] [PMID: 14661178]

[71] Spaeth E, Klopp A, Dembinski J, Andreeff M, Marini F. Inflammation and tumor microenvironments:

defining the migratory itinerary of mesenchymal stem cells. Gene Ther 2008; 15(10): 730-8. [http://dx.doi.org/10.1038/gt.2008.39] [PMID: 18401438]

[72] Somers P, Cornelissen R, Thierens H, Van Nooten G. An optimized growth factor cocktail for ovine mesenchymal stem cells. Growth Factors 2012; 30(1): 37-48. [http://dx.doi.org/10.3109/08977194.2011.634411] [PMID: 22077617]

[73] Lee JM, Kim BS, Lee H, Im GI. *In vivo* tracking of mesechymal stem cells using fluorescent nanoparticles in an osteochondral repair model. Mol Ther 2012; 20(7): 1434-42. [http://dx.doi.org/10.1038/mt.2012.60] [PMID: 22491215]

[74] Thomas T, Gori F, Spelsberg TC, Khosla S, Riggs BL, Conover CA. Response of bipotential human marrow stromal cells to insulin-like growth factors: effect on binding protein production, proliferation, and commitment to osteoblasts and adipocytes. Endocrinology 1999; 140(11): 5036-44. [PMID: 10537129]

[75] Ozaki Y, Nishimura M, Sekiya K, *et al.* Comprehensive analysis of chemotactic factors for bone marrow mesenchymal stem cells. Stem Cells Dev 2007; 16(1): 119-29. [http://dx.doi.org/10.1089/scd.2006.0032] [PMID: 17348810]

[76] Haider HKh, Jiang S, Idris NM, Ashraf M. IGF-1-overexpressing mesenchymal stem cells accelerate bone marrow stem cell mobilization *via* paracrine activation of SDF-1alpha/CXCR4 signaling to promote myocardial repair. Circ Res 2008; 103(11): 1300-8. [http://dx.doi.org/10.1161/CIRCRESAHA.108.186742] [PMID: 18948617]

[77] Lee CH, Cook JL, Mendelson A, Moioli EK, Yao H, Mao JJ. Regeneration of the articular surface of the rabbit synovial joint by cell homing: a proof of concept study. Lancet 2010; 376(9739): 440-8. [http://dx.doi.org/10.1016/S0140-6736(10)60668-X] [PMID: 20692530]

[78] Sieber C, Kopf J, Hiepen C, Knaus P. Recent advances in BMP receptor signaling. Cytokine Growth Factor Rev 2009; 20(5-6): 343-55. [http://dx.doi.org/10.1016/j.cytogfr.2009.10.007] [PMID: 19897402]

[79] Cho TJ, Gerstenfeld LC, Einhorn TA. Differential temporal expression of members of the transforming growth factor beta superfamily during murine fracture healing. J Bone Miner Res 2002; 17(3): 513-20. [http://dx.doi.org/10.1359/jbmr.2002.17.3.513] [PMID: 11874242]

[80] Montaner B, Pérez-Tomás R. Epidermal growth factor receptor (EGF-R) localization in the apical membrane of the enterocytes of rat duodenum. Cell Biol Int 1999; 23(7): 475-9. [http://dx.doi.org/10.1006/cbir.1999.0383] [PMID: 10728784]

[81] Ornitz DM, Xu J, Colvin JS, *et al.* Receptor specificity of the fibroblast growth factor family. J Biol Chem 1996; 271(25): 15292-7. [http://dx.doi.org/10.1074/jbc.271.25.15292] [PMID: 8663044]

[82] Ortega S, Ittmann M, Tsang SH, Ehrlich M, Basilico C. Neuronal defects and delayed wound healing in mice lacking fibroblast growth factor 2. Proc Natl Acad Sci USA 1998; 95(10): 5672-7. [http://dx.doi.org/10.1073/pnas.95.10.5672] [PMID: 9576942]

[83] Itoh N. The Fgf families in humans, mice, and zebrafish: their evolutional processes and roles in development, metabolism, and disease. Biol Pharm Bull 2007; 30(10): 1819-25.

[http://dx.doi.org/10.1248/bpb.30.1819] [PMID: 17917244]

[84] Matsumoto K, Nakamura T. Hepatocyte growth factor (HGF) as a tissue organizer for organogenesis and regeneration. Biochem Biophys Res Commun 1997; 239(3): 639-44. [http://dx.doi.org/10.1006/bbrc.1997.7517] [PMID: 9367820]

[85] Bevan D, Gherardi E, Fan TP, Edwards D, Warn R. Diverse and potent activities of HGF/SF in skin wound repair. J Pathol 2004; 203(3): 831-8. [http://dx.doi.org/10.1002/path.1578] [PMID: 15221943]

[86] Cross MJ, Dixelius J, Matsumoto T, Claesson-Welsh L. VEGF-receptor signal transduction. Trends Biochem Sci 2003; 28(9): 488-94. [http://dx.doi.org/10.1016/S0968-0004(03)00193-2] [PMID: 13678960]

[87] Zachary I, Gliki G. Signaling transduction mechanisms mediating biological actions of the vascular endothelial growth factor family. Cardiovasc Res 2001; 49(3): 568-81. [http://dx.doi.org/10.1016/S0008-6363(00)00268-6] [PMID: 11166270]

[88] Kim JY, Xin X, Moioli EK, *et al.* Regeneration of dental-pulp-like tissue by chemotaxis-induced cell homing. Tissue Eng Part A 2010; 16(10): 3023-31. [http://dx.doi.org/10.1089/ten.tea.2010.0181] [PMID: 20486799]

[89] Rezzoug F, Seelan RS, Bhattacherjee V, Greene RM, Pisano MM. Chemokine-mediated migration of mesencephalic neural crest cells. Cytokine 2011; 56(3): 760-8. [http://dx.doi.org/10.1016/j.cyto.2011.09.014] [PMID: 22015108]

[90] Gillitzer R, Goebeler M. Chemokines in cutaneous wound healing. J Leukoc Biol 2001; 69(4): 513-21. [PMID: 11310836]

[91] Kitaori T, Ito H, Schwarz EM, *et al.* Stromal cell-derived factor 1/CXCR4 signaling is critical for the recruitment of mesenchymal stem cells to the fracture site during skeletal repair in a mouse model. Arthritis Rheum 2009; 60(3): 813-23. [http://dx.doi.org/10.1002/art.24330] [PMID: 19248097]

[92] Tögel F, Isaac J, Hu Z, Weiss K, Westenfelder C. Renal SDF-1 signals mobilization and homing of CXCR4-positive cells to the kidney after ischemic injury. Kidney Int 2005; 67(5): 1772-84. [http://dx.doi.org/10.1111/j.1523-1755.2005.00275.x] [PMID: 15840024]

[93] Abbott JD, Huang Y, Liu D, Hickey R, Krause DS, Giordano FJ. Stromal cell-derived factor-1alpha plays a critical role in stem cell recruitment to the heart after myocardial infarction but is not sufficient to induce homing in the absence of injury. Circulation 2004; 110(21): 3300-5. [http://dx.doi.org/10.1161/01.CIR.0000147780.30124.CF] [PMID: 15533866]

[94] Sordi V, Malosio ML, Marchesi F, *et al.* Bone marrow mesenchymal stem cells express a restricted set of functionally active chemokine receptors capable of promoting migration to pancreatic islets. Blood 2005; 106(2): 419-27. [http://dx.doi.org/10.1182/blood-2004-09-3507] [PMID: 15784733]

[95] Thevenot PT, Nair AM, Shen J, Lotfi P, Ko CY, Tang L. The effect of incorporation of SDF-1alpha into PLGA scaffolds on stem cell recruitment and the inflammatory response. Biomaterials 2010; 31(14): 3997-4008. [http://dx.doi.org/10.1016/j.biomaterials.2010.01.144] [PMID: 20185171]

[96] Wang L, Li Y, Chen J, *et al.* Ischemic cerebral tissue and MCP-1 enhance rat bone marrow stromal cell migration in interface culture. Exp Hematol 2002; 30(7): 831-6.
[http://dx.doi.org/10.1016/S0301-472X(02)00829-9] [PMID: 12135683]

[97] Chen J, Li Y, Wang L, *et al.* Therapeutic benefit of intravenous administration of bone marrow stromal cells after cerebral ischemia in rats. Stroke 2001; 32(4): 1005-11.
[http://dx.doi.org/10.1161/01.STR.32.4.1005] [PMID: 11283404]

[98] Belema-Bedada F, Uchida S, Martire A, Kostin S, Braun T. Efficient homing of multipotent adult mesenchymal stem cells depends on FROUNT-mediated clustering of CCR2. Cell Stem Cell 2008; 2(6): 566-75.
[http://dx.doi.org/10.1016/j.stem.2008.03.003] [PMID: 18522849]

[99] Ishida Y, Kimura A, Kuninaka Y, *et al.* Pivotal role of the CCL5/CCR5 interaction for recruitment of endothelial progenitor cells in mouse wound healing. J Clin Invest 2012; 122(2): 711-21.
[http://dx.doi.org/10.1172/JCI43027] [PMID: 22214846]

[100] Stich S, Loch A, Leinhase I, *et al.* Human periosteum-derived progenitor cells express distinct chemokine receptors and migrate upon stimulation with CCL2, CCL25, CXCL8, CXCL12, and CXCL13. Eur J Cell Biol 2008; 87(6): 365-76.
[http://dx.doi.org/10.1016/j.ejcb.2008.03.009] [PMID: 18501472]

[101] D'Haese JG, Demir IE, Friess H, Ceyhan GO. Fractalkine/CX3CR1: why a single chemokine-receptor duo bears a major and unique therapeutic potential. Expert Opin Ther Targets 2010; 14(2): 207-19.
[http://dx.doi.org/10.1517/14728220903540265] [PMID: 20055718]

[102] Ji JF, He BP, Dheen ST, Tay SS. Interactions of chemokines and chemokine receptors mediate the migration of mesenchymal stem cells to the impaired site in the brain after hypoglossal nerve injury. Stem Cells 2004; 22(3): 415-27.
[http://dx.doi.org/10.1634/stemcells.22-3-415] [PMID: 15153618]

[103] Nagira M, Imai T, Hieshima K, *et al.* Molecular cloning of a novel human CC chemokine secondary lymphoid-tissue chemokine that is a potent chemoattractant for lymphocytes and mapped to chromosome 9p13. J Biol Chem 1997; 272(31): 19518-24.
[http://dx.doi.org/10.1074/jbc.272.31.19518] [PMID: 9235955]

[104] Saeki H, Tamaki K. Thymus and activation regulated chemokine (TARC)/CCL17 and skin diseases. J Dermatol Sci 2006; 43(2): 75-84.
[http://dx.doi.org/10.1016/j.jdermsci.2006.06.002] [PMID: 16859899]

[105] Pébay A, Bonder CS, Pitson SM. Stem cell regulation by lysophospholipids. Prostaglandins Other Lipid Mediat 2007; 84(3-4): 83-97.
[http://dx.doi.org/10.1016/j.prostaglandins.2007.08.004] [PMID: 17991611]

[106] Watterson KR, Lanning DA, Diegelmann RF, Spiegel S. Regulation of fibroblast functions by lysophospholipid mediators: potential roles in wound healing. Wound Repair Regen 2007; 15(5): 607-16.
[http://dx.doi.org/10.1111/j.1524-475X.2007.00292.x] [PMID: 17971005]

[107] Song HY, Lee MJ, Kim MY, *et al.* Lysophosphatidic acid mediates migration of human mesenchymal stem cells stimulated by synovial fluid of patients with rheumatoid arthritis. Biochim Biophys Acta

2010; 1801(1): 23-30.
[http://dx.doi.org/10.1016/j.bbalip.2009.08.011] [PMID: 19733258]

[108] Annabi B, Thibeault S, Lee YT, *et al.* Matrix metalloproteinase regulation of sphingosine--phosphate-induced angiogenic properties of bone marrow stromal cells. Exp Hematol 2003; 31(7): 640-9.
[http://dx.doi.org/10.1016/S0301-472X(03)00090-0] [PMID: 12842709]

[109] Chae SS, Paik JH, Furneaux H, Hla T. Requirement for sphingosine 1-phosphate receptor-1 in tumor angiogenesis demonstrated by *in vivo* RNA interference. J Clin Invest 2004; 114(8): 1082-9.
[http://dx.doi.org/10.1172/JCI200422716] [PMID: 15489955]

[110] Li C, Kong Y, Wang H, *et al.* Homing of bone marrow mesenchymal stem cells mediated by sphingosine 1-phosphate contributes to liver fibrosis. J Hepatol 2009; 50(6): 1174-83.
[http://dx.doi.org/10.1016/j.jhep.2009.01.028] [PMID: 19398237]

[111] Andersson U, Erlandsson-Harris H, Yang H, Tracey KJ. HMGB1 as a DNA-binding cytokine. J Leukoc Biol 2002; 72(6): 1084-91.
[PMID: 12488489]

[112] Ulloa L, Messmer D. High-mobility group box 1 (HMGB1) protein: friend and foe. Cytokine Growth Factor Rev 2006; 17(3): 189-201.
[http://dx.doi.org/10.1016/j.cytogfr.2006.01.003] [PMID: 16513409]

[113] Palumbo R, Bianchi ME. High mobility group box 1 protein, a cue for stem cell recruitment. Biochem Pharmacol 2004; 68(6): 1165-70.
[http://dx.doi.org/10.1016/j.bcp.2004.03.048] [PMID: 15313414]

[114] Meng E, Guo Z, Wang H, *et al.* High mobility group box 1 protein inhibits the proliferation of human mesenchymal stem cells and promotes their migration and differentiation along osteoblastic pathway. Stem Cells Dev 2008; 17(4): 805-13.
[http://dx.doi.org/10.1089/scd.2007.0276] [PMID: 18715162]

[115] Idriss HT, Naismith JH. TNF alpha and the TNF receptor superfamily: structure-function relationship(s). Microsc Res Tech 2000; 50(3): 184-95.
[http://dx.doi.org/10.1002/1097-0029(20000801)50:3<184::AID-JEMT2>3.0.CO;2-H] [PMID: 10891884]

[116] Baek SJ, Kang SK, Ra JC. *In vitro* migration capacity of human adipose tissue-derived mesenchymal stem cells reflects their expression of receptors for chemokines and growth factors. Exp Mol Med 2011; 43(10): 596-603.
[http://dx.doi.org/10.3858/emm.2011.43.10.069] [PMID: 21847008]

[117] Ponte AL, Marais E, Gallay N, *et al.* The *in vitro* migration capacity of human bone marrow mesenchymal stem cells: comparison of chemokine and growth factor chemotactic activities. Stem Cells 2007; 25(7): 1737-45.
[http://dx.doi.org/10.1634/stemcells.2007-0054] [PMID: 17395768]

[118] Zhang A, Wang Y, Ye Z, Xie H, Zhou L, Zheng S. Mechanism of TNF-α-induced migration and hepatocyte growth factor production in human mesenchymal stem cells. J Cell Biochem 2010; 111(2): 469-75.
[http://dx.doi.org/10.1002/jcb.22729] [PMID: 20533298]

[119] Wang M, Crisostomo PR, Herring C, Meldrum KK, Meldrum DR. Human progenitor cells from bone marrow or adipose tissue produce VEGF, HGF, and IGF-I in response to TNF by a p38 MAPK-dependent mechanism. Am J Physiol Regul Integr Comp Physiol 2006; 291(4): R880-4. [http://dx.doi.org/10.1152/ajpregu.00280.2006] [PMID: 16728464]

[120] Xiao Q, Wang SK, Tian H, *et al.* TNF-α increases bone marrow mesenchymal stem cell migration to ischemic tissues. Cell Biochem Biophys 2012; 62(3): 409-14. [http://dx.doi.org/10.1007/s12013-011-9317-y] [PMID: 22081437]

[121] West AP, Koblansky AA, Ghosh S. Recognition and signaling by toll-like receptors. Annu Rev Cell Dev Biol 2006; 22: 409-37. [http://dx.doi.org/10.1146/annurev.cellbio.21.122303.115827] [PMID: 16822173]

[122] Tomchuck SL, Zwezdaryk KJ, Coffelt SB, Waterman RS, Danka ES, Scandurro AB. Toll-like receptors on human mesenchymal stem cells drive their migration and immunomodulating responses. Stem Cells 2008; 26(1): 99-107. [http://dx.doi.org/10.1634/stemcells.2007-0563] [PMID: 17916800]

[123] Waterman RS, Tomchuck SL, Henkle SL, Betancourt AM. A new mesenchymal stem cell (MSC) paradigm: polarization into a pro-inflammatory MSC1 or an Immunosuppressive MSC2 phenotype. PLoS One 2010; 5(4): e10088.

[124] Pevsner-Fischer M, Morad V, Cohen-Sfady M, *et al.* Toll-like receptors and their ligands control mesenchymal stem cell functions. Blood 2007; 109(4): 1422-32. [http://dx.doi.org/10.1182/blood-2006-06-028704] [PMID: 17038530]

[125] Yuan L, Sakamoto N, Song G, Sato M. Low-level shear stress induces human mesenchymal stem cell migration through the SDF-1/CXCR4 axis *via* MAPK signaling pathways. Stem Cells Dev 2013; 22(17): 2384-93.

[126] Hong B, Li H, Zhang M, *et al.* p38 MAPK inhibits breast cancer metastasis through regulation of stromal expansion. Int J Cancer 2015; 136(1): 34-43. [http://dx.doi.org/10.1002/ijc.28958] [PMID: 24806617]

[127] Torsvik A, Bjerkvig R. Mesenchymal stem cell signaling in cancer progression. Cancer Treat Rev 2013; 39(2): 180-8. [http://dx.doi.org/10.1016/j.ctrv.2012.03.005] [PMID: 22494966]

[128] Chen J, Crawford R, Chen C, Xiao Y. The key regulatory roles of the PI3K/Akt signaling pathway in the functionalities of mesenchymal stem cells and applications in tissue regeneration. Tissue Eng Part B Rev 2013; 19(6): 516-28. [http://dx.doi.org/10.1089/ten.teb.2012.0672] [PMID: 23651329]

[129] Zha YH, He JF, Mei YW, Yin T, Mao L. Zinc-finger transcription factor snail accelerates survival, migration and expression of matrix metalloproteinase-2 in human bone mesenchymal stem cells. Cell Biol Int 2007; 31(10): 1089-96. [http://dx.doi.org/10.1016/j.cellbi.2007.03.023] [PMID: 17512761]

[130] Zgheib A, Pelletier-Bonnier É, Levros LC Jr, Annabi B. Selective JAK/STAT3 signalling regulates transcription of colony stimulating factor-2 and -3 in Concanavalin-A-activated mesenchymal stromal cells. Cytokine 2013; 63(2): 187-93.

[http://dx.doi.org/10.1016/j.cyto.2013.04.027] [PMID: 23688618]

[131] Silver DL, Geisbrecht ER, Montell DJ. Requirement for JAK/STAT signaling throughout border cell migration in Drosophila. Development 2005; 132(15): 3483-92.
[http://dx.doi.org/10.1242/dev.01910] [PMID: 16000386]

[132] Gao H, Priebe W, Glod J, Banerjee D. Activation of signal transducers and activators of transcription 3 and focal adhesion kinase by stromal cell-derived factor 1 is required for migration of human mesenchymal stem cells in response to tumor cell-conditioned medium. Stem Cells 2009; 27(4): 857-65.
[http://dx.doi.org/10.1002/stem.23] [PMID: 19350687]

CHAPTER 2

Three-dimensional Printing in Dentistry: An Advanced Technology for Craniofacial Regeneration

Maobin Yang[1,*], **Qian Zeng**[2], **Maria Pereira Vieira**[3] and **Bernardo Olsson**[4]

[1] *Temple University, Kornberg School of Dentistry, Philadelphia, USA*

[2] *Sun Yat-sen University, Guanghua School of Stomatology, Guangzhou, China*

[3] *Federal University of Mato Grosso do Sul, Campo Grande, Mato Grosso do Sul, Brazil*

[4] *Federal University of Santa Catarina, Florianópolis, Santa Catarina, Brazil*

Abstract: Since three-dimensional printing (3DP) technology was first invented in 1980s, it has been used in a range of fields, including architecture, manufacturing, engineering, education and medicine. In dentistry, 3DP provides a rapid and precise technology that helps diagnosis, treatment planning and treatment of various dental diseases. In craniofacial regeneration, 3DP helps to design and fabricate dental prosthesis and implants, and it also improves craniofacial tissue engineering. Not only scaffolds can be printed by 3DP, stem cells and growth factors can also be controlled and printed by 3DP. 3DP technology has become a major tool in craniofacial regeneration.

Keywords: Craniofacial Regeneration, Dentistry, Three-dimensional Printing.

2.1. INTRODUCTION

Three-dimensional printing (3DP) is a process in which three-dimensional objects are created by fusing or depositing materials layer by layer. The materials used for 3DP can be plastic, metal, ceramic, biomaterials or even living cells [1, 2]. 3DP is

* **Corresponding author Maobin Yang:** Department of Endodontology. Director, Regenerative Health Research Laboratory, Kornberg School of Dentistry, Temple University, Philadelphia, USA; Tel: 01-215-707-2810; Fax: 215-707-1482; E-mail: myang@temple.edu.

Jing Wang & YunFeng Lin (Eds.)

also known as additive manufacturing (AM), rapid prototyping (RP), or solid free-form technology (SFF) [1]. This new technology allows complex physical structures to be fabricated accurately and rapidly.

Since its first invention in 1984, 3DP has been developed into different modeling systems and its application has been expanded to various areas [1-8]. 3DP originally was only used to produce an object with designed shape, but now it can generate more functional products such as food, clothes, houses, auto parts, gun and aircraft. It is bringing radical changes to our daily life. In medical field, 3DP can be used in tissue and organ fabrication, in the creation of customized prostheses, implants, and anatomical models, and in pharmaceutical researches regarding drug dosage forms, drug delivery, and new drug discovery [1]. With the constantly evolving, 3DP has been brought to the dentistry field. It is making a difference in various dental specialty fields including endodontics [9-14], oral surgery [15-23], prosthodontics [24-27], periodontics [28, 29], orthodontics [30-34], and implant [35-45].

2.2. THE HISTORY OF 3DP

The concept of 3DP was first introduced by Chuck Hull in 1984. It was described as stereolithography (SLA), the printing of successive layers of materials to create a 3D object [2]. Later in 1988, Hull and the company 3D Systems developed the first 3D printer termed "Stereolithography Apparatus", and introduced the first commercially available 3D printer called SLA-250 [1, 46]. Subsequently, a new technique named "Fused Deposition Modeling" (FDM) was engineered by Scott Crump at Stratasys in 1990. In 1991, the Solid Ground Curing (SGC) technology by Cubital, and the Laminated Object Manufacturing technology (LOM) by Helisys were both commercialized [47].

In 1992, Selective Laser Sintering (SLS) system from DTM corporation and the Soliform Stereolithography system from Teijin Seiki were introduced to 3DP market [47]. In 1993, the first apparatus termed "3D printer" was patented by Massachusetts Institute of Technology (MIT) to attain plastic, metal and ceramic parts [48]. In 1996, Stratasys introduced a new 3DP machine named "Genisys" that used an extrusion process similar to FDM. In the same year, the company Z Corporation launched Z402, a concept modeling 3D printer based on MIT's inkjet printing technology. Also in 1996, three new 3DP systems emerged to the market: Personal Modeler 2100 from Ballistic Particle Manufacturing (BPM) technology, DuPont's Somos Stereolithography technology and Zippy Paper lamination systems [47].

In 2005, Z Corporation launched the first high definition 3D color printer named Spectrum Z510 [49], and in 2006 the RepRap, a desktop 3D printer for plastic objects printing, was introduced to market. It was a self-replicating 3D printer and provided an open source option to print the majority of its own parts [46, 49]. In 2007, Z Corporation introduced the ZPrinter 450, which could automatically remove and recycle loose powder material from printer. In 2009, the FDM patent was expired, and in 2014, SLS patent was expired. Since then various types of 3D printing system have been released. According to Wohlers 2014 Report, the worldwide 3D printing industry is now expected to grow from $3.07 billion in revenue in 2013 to $12.8 billion by 2018, and exceed $21 billion by 2020 [50].

The medical application of 3DP was dated back to 1996 when the idea of 3D-printed organs was proposed by Dr. Gabor Forgacs. In 1999, Dr. Atala's group at the Wake Forest Institute for Regenerative Medicine for the first time created artificial scaffolds in the shape of the desired bladder and coated it with cells taken from their patients, and successfully grew an artificial bladder [51]. This set the stage for true "bioprinting". According to the First International Workshop on Bioprinting and Biopatterning, bioprinting was defined as "the use of material transfer processes for patterning and assembling biologically relevant materials, molecules, cells, tissues, and biodegradable biomaterials with a prescribed organization to accomplish one or more biological functions" [52]. In the early 2000s, 3DP technology was applied in medical and dental fields to print dental implants and custom prosthesis [1]. At these early stages of bioprinting, only biomaterial scaffold was printed but not living cells. In the past decade, the 3D bioprinting technology was greatly evolved. It has been used to print not only the scaffolds for organs regeneration, but also print protein and DNA patterning, as well as multi-cellular assemblies and tissue patterning [52]. Though 3DP is still very far from being able to print organs, it is possible to print artificial scaffolds in the shape of an organ with living cells [5].

2.3. CATEGORY OF 3DP TECHNOLOGY

The principle of 3DP is that it generates the rapid prototyping models by a computer-aided design (CAD). The original design is drafted in a CAD program and then transformed to either Stereolithography file (STL) or Object file (OBJ) format. The STL file is established as the gold standard for data transfer between the CAD software and 3D printer. During printing, STL file is processed by software which converts the model into a series of thin layers and stacks them together to produce a 3D construct [46]. With the increased applications of 3DP in industry, medicine and research field, a variety of new methods and techniques have been developed, including fused deposition modeling, laminated object manufacturing, selective laser sintering, stereolithography, thermal inkjet, multijet modeling and polyjetpringting [5, 7, 53, 54]. Among these, laser sintering printer and deposition printer are the major types of printer [55].

2.3.1. Stereolithography (SLA)

As the first commercialized rapid prototyping system, SLA employs liquid photopolymer or epoxy resin in a vat and draws one layer at a time, then uses a low-power ultraviolet beam to cure the resin and solidify the layer. The second layer is then added, and the above process is repeated until the 3D model is fabricated [5, 54]. Currently, SLA is considered as the gold standard in producing 3D models. The resolution of SLA reaches up to 0.025mm. There are three drawbacks of SLA: 1) it is less cost-efficient than other 3DP techniques due to the high cost of raw materials and machine maintenance; 2) it is labor-intensive as it needs manual post-build handling; 3) it is time-consuming to produce a large object [16, 56, 57]. A modified SLA system, called "Continuous Liquid Interface Production (CLIP)" simplifies traditional SLA by creating a persistent liquid interface with an oxygen-permeable window below the ultraviolet image project plane, thus significantly increases the production speed [58].

2.3.2. Fused Deposition Modeling (FDM)

FDM was invented by Scott Crump in the late 1980s and it is currently the most commonly used 3DP technology [5, 16, 59]. In FDM system, melted thermoplastic is extruded from a nozzle and deposited on a moving x-y plane.

After a layer is solidified on build plate, the plate is lowered by 0.1 mm before the second layer starts to build [54]. Acrylonitrile butadiene styrene plastic and polylactic acid are two thermoplastics frequently used in FDM [5]. The major disadvantages of FDM printers include: the difficulty of manually cleaning support structures from complex anatomical models; and the limitation of printing multiple colors or materials. To overcome these disadvantages, dual-extruder technology is recently developed, which allows for two color print jobs, and this technology has been established in the mainstream of 3DP technology [54].

2.3.3. Selective Laser Sintering (SLS)

SLS was invented by Carl Dekard in the 1980s. This method applies a high-power laser onto powdered forms to sinter them in layers. A variety of materials can be used in SLS, including thermoplastic, polystyrene, ceramics, glass, nylon, steel, titanium, aluminum or silver [5, 54]. During SLS process, support structures can be avoided as SLS itself provides support, thus the SLS printed objected can attain a smoother surface. During the SLS printing, the unsintered powders need to be consistently removed from the models, and these powders can be reused [57]. Potential hazard may be produced by SLS printers because of the presence of lasers, pistons, and gas chambers with high temperature, therefore it requires expert handling [54]. The SLS printer has also relative high cost compared to other systems.

2.3.4. Inkjet Printing

As a non-contact printing system, inkjet printing uses thermal, piezoelectric, and electromagnetic technology to digitally deposit tiny droplets of “ink” onto a substrate, therefore it is also known as drop-on demand printer [53, 60]. The inkjet printing technique has been widely used to print electronic materials or complex integrated circuits in electronics and micro-engineering industries [61]. For bio-printing, inkjet printing is able to directly print DNA molecules for the fabrication of high-density DNA microarray [60, 62]. Inkjet printing offers advantages such as low cost, high resolution, high speed and great compatibility with many materials. It is able to control and alter drop size and density, hence potentially providing concentration gradients of biomaterials, including cells or growth factors. The limitation of inkjet printing is that it can only use liquid materials form droplet, and at this stage it is still difficult to achieve very high resolution of cell density [7].

2.3.5. Laminated Object Manufacturing (LOM)

LOM was developed by Helisys in 1991. It draws adhesive onto material sheets in order to create a bond between layers. The sheets can be made of different materials, such as paper, plastic film, or metal foil. One layer is first cut by laser cutter to form a specific pattern, and after the unused materials are removed, a second layer is added on the previous layer. This procedure is repeated until the whole object is printed [46]. LOM has advantage to produce relative large-size parts with the relative low cost, but its resolution is lower than SLS. In addition, the temperature needs accurate control during the procedure to avoid inefficient adhesive heating or structural damage. The choice of materials used for this process is also limited due to their plasticity into sheet form and ability of adhesive integration [5].

2.3.6. Polyjet Printing

Polyjet printing was patented by Sachs in 1994. During the process, jetting head deposits a single thin layer of photopolymer by sliding back and forth along the x-axis, followed by an immediate ultraviolet curing and hardening [63]. Similar to inkjet printing, polyjet printing can create a solid object from a liquid polymer by cutting the liquid with light. The difference is that polyjet prints layers of liquid and inkjet prints drops of liquid. In addition, polyjet printing applies ultraviolet light to harden liquid compound of reactive monomers [5, 64]. Polyjet process is one of the most accurate and fastest 3DP processes. It can produce a layer with thickness as thin as 16μm. The thin layers contribute to even and smooth surfaces of the final object [63]. In medical field, a variety of devices can be printed by Polyjet, including dental delivery trays, surgical orthopedic guides and hearing aids materials.

2.4. COMPARISON OF 3DP AND SUBTRACTIVE MANUFACTURING TECHNIQUE

3DP is quite different from the current manufacturing technology which is described as "subtractive". Subtractive manufacturing cuts and removes materials to achieve the desired geometry. The process is a machining process numerically controlled by computer, and performed by power-driven machine tools, such as saws, lathes, milling machines and drill presses. Subtractive manufacturing mechanically cuts the materials to achieve the desired geometry [8], and it is relatively easy to achieve accurate shape and precise dimension in less time than the conventional dental process. However, subtractive manufacturing usually wastes a considerable amount of raw materials. These materials must be removed during the process and unfortunately most of them can't be recycled [65]. Another limitation of subtractive process is that it is difficult to machine mass productions at a time [8]. In dentistry, the major CAD/CAM technology used in prosthetics is based on the subtractive manufacturing.

3DP technology is known as additive manufacturing. According to the American Society for Testing and Materials (ASTM), additive manufacturing is defined as "the process of joining materials to make objects from 3D model data, usually layer upon layer, as opposed to subtractive manufacturing methodologies" [66]. Compared to subtractive manufacturing, 3DP has several advantages: 1) it is cost effective as materials will not be wasted; 2) it allows for customized on-demand production; 3) it reduces the time to manufacture and makes localized production possible; 4) it can fabricate complex designs at a time [8, 67]. In medicine, 3DP is capable to customize and personalize medical products, equipment and drugs. It increases cost-effectiveness and productivity, and also offers great opportunities for sharing design and manufactures among researchers [1].

2.5. CURRENT APPLICATION OF 3DP IN DENTISTRY

The application of 3DP is explosively growing in the manufacturing industry and consumer market as it is possible to print various products, from clothing to gun, from car parts to designed jewelry [68]. In the past decade, 3DP has been introduced to public libraries, from where users are able to access 3D printers and scanners [5, 55]. The power of cloud computing is also used to access the personal 3DP library to print personal files and customize 3D products.

3DP has also emerged in the medical fields for diagnosis, treatment planning and treatment guidance. The current medical applications of 3DP can be organized into three categories: creating prostheses, implants, and anatomical models; tissue and organ fabrication; and pharmaceutical discovery, drug delivery, and dosage forms [68]. Currently, 3DP is capable to print bones, exoskeletons, trachea, ears, corneas, blood vessels, vascular networks, and various dental prostheses. In tissue engineering, scaffolds, growth factors and stem cells can also be printed separately or integratedly [2, 7, 55].

2.5.1. 3DP In Clinical Dentistry

In dentistry, various treatment options have been developed to preserve oral health and provide natural-like aesthetics with enhanced technology, leading to the reduction of treatment time, minimization of potential error with higher care quality assurance [6]. 3DP has showed potential success in diagnosis, treatment planning and clinical procedures, and it also greatly benefits dental education. The application of 3DP has been introduced into various dental specialties, including endodontics, oral surgery, periodontics, prosthodontics, orthodontics and implant.

2.5.1.1. Endodontics

Although the root canal anatomy and its variations have been well studied in normal teeth, the complex root canal system still remains challenging for root canal treatment regarding locating, accessing, instrumenting and cleaning root canals. It is important to understand root canal anatomy and its variations to achieve a successful endodontic therapy. 3DP works as a useful diagnostic aid in endodontics. Since 3DP provides a visualization of canal anatomy through digital reconstruction, it can guide endodontic treatment especially in teeth with abnormal canal morphologies. Lee *et al.* used computer-aided rapid prototyping (CARP) technology to document unusual root anatomy from a mandibular first molar, and successfully used 3DP to produce a tooth model which contained all roots [11]. The rapid prototyping tooth model can also be used to diagnose the exact location and size of apical lesion as well as root resorption. Kim *et al.*

reported a case of replicating multiple invasive root resorptions using rapid prototyping technology. The prototyping model was compared to the tooth after extraction, and proved to accurately simulate the lesions [10]. Byun *et al.* reported a case of endodontic treatment of a challenging tooth with a severely dilacerated root. They printed a comprehensive 3D tooth model including internal root canal structures. A custom-made guide jig was made based on the tooth model, and all canals were accessed and treated without complications [9].

Auto-transplantation is a treatment that transplants a donor tooth into an extraction site or a surgically prepared socket [69]. It has been reported that 3DP can facilitate the auto-transplantation [13, 14, 70-73]. In a case, the donor tooth model and jaw model were fabricated *via* 3DP technology, which helped clinicians better understand the root morphology with the surrounding anatomic structures, and make a better pre-operative treatment planning. This also helped minimize the tooth extra-oral time and avoided possible injury to transplanted tooth and adjacent tissues during procedure [14]. Park *et al.* demonstrated the clinical application of rapid prototyping tooth models for auto-transplantation [13]. In another case by Honda, a replica of graft tooth was produced by a stereo lithographic system. The replica was directly put into a socket to adjust the fitness before the transplantation of donor tooth. The outcome was very good as the donor tooth fitted into socket well. Both operation time and damage to the periodontal ligament were reduced due to the better pre-operative treatment plan [71].

In addition to the clinical applications, 3DP technique has also been used in the endodontic training and research. In a study, the efficiency of two rotary systems in preparation of the curved canals was compared using a simulated 3D printed tooth. The rapid prototyping mandibular molar replicas were produced by 3DP. The rotary systems were tested on molar replicas, and the results suggested that micro CT-based rapid prototyping resin tooth was a great tool to assess endodontic operations [12].

2.5.1.2. Oral Surgery

In the past decade, the application of 3DP has been widely used to improve the accuracy and quality of maxillofacial surgery. 3DP models are capable to simulate not only the static anatomical structures but also complex anatomical movements. The surgeons can better understand the visuospatial relationship within anatomical structures, which helps to shorten operative time, reduce intraoperative blood loss and promote wound healing [54]. Both oral surgeons and patients are benefiting greatly from 3DP technology.

3DP can be used to print replicas of surgery site as a preoperative guide for different types of maxillofacial surgeries. In mandibular reconstruction surgery, different 3DP approaches have been used to provide preoperative planning, including SLA, SLS and FDM [74-77]. A clinical study compared the 3D printed reconstruction plates with conventional reconstruction plates in mandibular reconstruction. The 3D manufactured reconstruction plate was based on medical rapid prototyping (MRP), which was designed with CAD/CAM technique and fabricated by a combined powder bed and inkjet 3D printer. It showed that the MRP-model-based plates achieved better mandibular contour [74]. In another study by Tsai *et al.*, SLA system was applied to design and generate customized mandible models, fibula models, cutting fixtures and registered fixtures, furnishing all-in-one process of mandible reconstruction. The process was proved to enhance the accuracy of surgery, shorten the operation duration, reduce the surgical risk, and result in a better mandible geometry [76]. Liu *et al.* proposed another 3DP solution for guided surgery to repair mandibular defect. In this case, surgical guide, original mandible model and the repair model were fabricated *via* SLS process. Based on prototyping models, grafts for defect repair were cut from fibular bone and shaped with high accuracy during surgery. The results showed that the SLS process saved 20% operating time compared to the conventional surgery [77]. 3DP was also reported to design and reconstruct 3D digital models of whole mandible [15].

3DP has been used in orthognathic surgery to customize treatment plan. With the orthognathic surgery guide generated *via* 3DP, the surgeon can precisely execute the procedure as planned in computer [78, 79]. Recently, a new computer-aided designed and manufactured template was used to guide the osteotomy and repositioning of maxillary. This CAD/CAM template provided a reliable method for transfer of maxillary surgical planning, which was a useful alternative to current intermediate splint technique. Its feasibility and validity were clearly illustrated in this study [22]. 3DP was also used to print wafers for maxillary movement during the bimaxillary orthognathic surgery. A study showed that the accuracy of the wafers by digital model surgery (DMS) was similar to those produced by manual model surgery. The rapid-prototyped inter-occlusal wafer produced with the aid of DMS can be an alternative to maxillary orthognathic surgery [21].

2.5.1.3. Prosthodontics

For the past decade, CAD/CAM milling system has become a major modality for rapid manufacture of dental prostheses for individual patient. However, there are some disadvantages of the systems, including limited accuracy, possible microscopic cracks, and a waste of material due to the subtractive process [65, 80]. 3DP overcomes these disadvantages and brings an advanced alternative in dental prosthetic manufacture. Up to date, different types of prostheses can be printed by 3DP, such as crowns, complete dentures, removable partial dentures and obturators [24, 81-85].

A case report by Cho described the fabrication of a mirror-image anterior single crown using a 3D printer. With the aid of computer and rapid prototyping technology, the shape of the contralateral central incisor was duplicated and reproduced. The results demonstrated that 3DP provided efficient and precise reproduction of central incisor with adequate anatomic contour [83].

Complete denture has been fabricated by 3DP. In a clinical study by Masanao, a 3D printed complete denture was compared with that made by the conventional method. During the manufacturing process, photopolymer materials were sprayed in subsequent ultra-thin layers onto a building tray, followed by ultraviolet curing for each layer to produce 3D printed dentures. However, the 3D printed dentures produced by this method showed inferior esthetics compared to conventional dentures, partly due to the monotonous color of 3DP material, suggesting that teeth and gum-colored materials need to be improved for 3D printed denture [85].

Williams *et al.* fabricated a removable partial denture (RPD) framework with a 3DP technique named Selective Laser Melting (SLM). The framework showed accurate fit that was comparable to those made by traditional casting methods [86]. Recently, a fully automatic RPD manufacture system was introduced. This system could print patient jaw models, the initial and optimized RPDs. It was also capable of shape optimization. The contact pressure between the denture base of RPD prototypes and the mandibular residual ridge model was tested *in vitro*. The results suggested that the customized optimal denture design minimized pain and discomfort, and reduced long-term residual ridge resorption [81].

3DP is also applied in the fabrication of obturators to restore maxillary deficiency. Jiao *et al.* combined CAD and rapid prototyping to manufacture obturators for individual patients after maxillary resection. SLA was used to produce resin positive mold. After clinical modification of border contours, the obturator fit the defect cavity very well [87].

In addition, 3DP is capable to build a device to treat sleep apnea. Mortadi *et al.* designed a device based on the 3D topography of a dental cast. The device was then printed using ProJet 3000 Plus 3D printing system. The printed device displayed an excellent fit on the teeth, palatal and gingival tissues, and effectively improved the sleep apnea in the patient [88].

2.5.1.4. Periodontics

Periodontal disease is a chronic inflammation which affects over 80% of adults worldwide. It may ultimately result in teeth loss due to the destruction of surrounding tissues [89, 90]. Periodontal regeneration requires reconstitution of the periodontium complex, including periodontal ligament (PDL), alveolar bone, and cementum [91]. Currently, a variety of 3D scaffolds have been used to guide periodontal regeneration [28, 29, 91 - 95].

Reconstructing the spatial orientation of PDL remains the most challenging obstacle in periodontal regeneration. Park *et al.* developed a customized biphasic construct with aligned micro-channels positioned along the periodontal compartment surface *via* 3DP. The fiber-guiding scaffold contributed to better periodontal fiber orientation than other amorphous scaffolds [92, 94]. The results showed that the regenerated PDL alignment was similar to the healthy, nature ligament tissues. In addition, the regenerated tissues restored tooth movement under biomechanical masticatory loading [92]. Later, Park *et al.* used the same construct to facilitate morphogenesis of bone-ligament complexes. Bone architecture and PDL interface architecture were designed using periodontal defect topography in STL surface images. The fiber-guiding scaffolds were then manufactured by a 3D printer. The result showed that the anatomically designed scaffolds facilitated the regeneration, arrangement, and organization of multilayer periodontal tissues. In addition, the scaffold architecture directed the integration of PDL into the mineralized tissue, resulting in functional tissue interfaces that mimicked natural anatomy [29].

Vaquette *et al.* also reported a design of biphasic scaffold consisting of a FDM compartment for bone formation and an electrospun membrane for periodontal regeneration. Osteoblasts were cultured on the bone compartment while multiple cell sheets of periodontal ligament (PDL) were cultured on the membrane. The results showed that higher mineralization density in the bone compartment and a thin mineralized cementum-like tissue deposited on the dentin surface. The scaffolds also demonstrated better attachment on the dentin than the ones without cells sheets [93]. Later, Costa *et al.* developed a new scaffold made from electrospun membrane with larger pore size, and calcium phosphate was deposited onto its FDM compartment. This modification significantly increased bone formation and allowed the attachment of functionally oriented periodontal ligament–like tissue [91].

Hierarchical scaffolds have been developed by 3DP in order to simulate architectural configuration of the periodontium. These scaffolds contain different compartments recapitulating the native structures of the periodontal complex, which contains multiple layers including gingiva, periodontal ligament, bone and cementum [96]. Recently, 3DP can solely print a tri-phasic scaffold in order to promote the formation of cementum-like tissue, periodontal ligament, and bone at the same time [28].

2.5.1.5. Orthodontics

The applications of 3DP in orthodontic practice are increasing along with the advancement of CAD/CAM. 3DP has showed a range of promising possibilities in the fabrication of orthodontic appliances. It has been used to fabricate removable orthodontic appliances. Martorelli *et al.* compared the orthodontic removable appliances fabricated by rapid prototyping system with those by computer controlled milling machine. It was concluded that both methods produced clear removable orthodontic appliance [33]. In a recent study, lingual fix orthodontic appliances were customized *via* 3DP. 3DP simplified the manufacturing process by eliminating soldering phase and the extra step of determining the lever arm on the lateral cephalograms, subsequently modified the design on the study mode, thus improved accuracy of orthodontic appliance design [32]. Retainers can also be designed and fabricated through a combination of cone-bean computed tomography (CBCT) and 3DP [34].

Physical models fabricated by 3DP techniques have been introduced in orthodontics. Hazeveld *et al.* investigated the accuracy and reproducibility of physical dental models that were reconstructed by three different rapid prototyping techniques: digital light, jetted photopolymer and powder based inkjet printing. It was concluded that dental models reconstructed by these three techniques were equally accurate and reproducible, suggesting that these replica models were appropriate for selected applications in orthodontics [31]. Furthermore, 3DP technology is able to display the mandible growth. Reynolds created a 3D physical model *via* Spectrum Z510 color-printing system to illustrate the growth condition of the mandible as orthodontic records. A growth map was printed by the color 3D printer, allowing tracking the mandible growth over a year. This process provided a relatively simple way to visualize and evaluate mandibular growth in patients [97].

3DP is also used to measure three dimensional forces. Three-dimensional force is the key factor determining teeth movement during orthodontic treatment, but it is difficult to measure intraorally due to the limited space of oral cavity. Liu *et al.* created a simulated oral model for measuring orthodontic force using 3DP. The models of teeth, maxilla, and positioning guide were fabricated layer by layer with a resin material [98].

2.5.1.6. *Implant*

Currently dental implant is the preferred option for substituting lost teeth that are caused by disease, trauma, surgery or other problems. Dental implants have demonstrated high success rate and showed successful integrity with surrounding tissues [43]. The use of 3DP for dental implants has been developed for various purposes, such as implant designs, assessment of porosity titanium, implant surgery guidance and osseointegration of printed implants [35-37, 39, 42 - 45, 99, 100].

Chen *et al.* used SLM technique to fabricate customized root-analog implants. Mechanical property tests showed that these customized implants had high strength and sufficient dimensional accuracy. The results proved that SLM was an efficient mean for printing fully dense implants [35].

3DP technology is capable to control internal and surface characters of implant. For example, it can manufacture titanium implants with dual porosity features, thus increase the augmented bone volume and bone height. El-Hajje *et al.* applied inkjet printing to build a porous titanium implant. Mechanical and physical properties were measured and compared to the theoretical values of bone. The data showed that 3D printed titanium had a porosity of 32.2-53.4% with a median pore diameter of 20-24μm, which facilitated cell attachment and reduced the elastic modulus [36]. Gagg *et al.* manufactured porous titanium implants using a combination of 3DP and sintering process. It was found that the sintering time and temperature determined implant morphology and mechanical properties. It suggested that a sintering period of 2 hours or more at a minimum temperature of 1300°C was suitable for titanium bonding and optimal mechanical properties [100].

3DP technology is helpful for implant surgery. In a study by Fluggea, implant drilling guide was printed with the integration of laser surface scanning, CBCT, computer-aided design and SLA printer. This process allowed data acquisition during the first consultation and avoided a loss of accuracy during the transfer of virtual planning to the drilling guide [37]. Kuhl *et al.* also used 3DP to design and fabricate implant surgery templates on cast models. A high accuracy was achieved using the printed templates for guided implant surgery [39].

Sufficient alveolar bone height and volume are critical to ensure implant stability and osseointegration. In a recent study, 3D printed synthetic onlay blocks were used in *in vivo* animal model to achieve vertical bone augmentation. The blocks were implanted to the calvaria of rabbits for 8 weeks before it was retrieved for histological analysis. The result showed that the printed blocks were strongly fused to the calvarial bone surface and infiltrated by newly formed bone. The printed monetite blocks promoted craniofacial bone augmentations more than 4mm [44]. Recently, Tamimmi *et al.* used 3DP to design and construct monetite onlays with different macro-pores that based on bone metabolic activity. These customized onlays were tested in rabbits to evaluate bone augmentation. It revealed the printed onlays promoted higher bone volume, less heterogeneity and more implant osseointegration. The results implied that osseointegration of dental implants with 3D printed synthetic onlays was feasible, and that macropore geometry design according to bone metabolic activity played an important role in bone volume augmentation [43].

2.5.2. 3DP in Bioengineering Research in Dentistry

Craniofacial tissues, including bones, teeth, muscles, cartilages, blood vessels and nerves, work as a functional unit to ensure structural support, protection, sensation and allow movement. Trauma, congenital malformations or surgeries may cause defect in craniofacial tissues. Due to the complexity of craniofacial structures, it is challenging to recover the functional craniofacial tissues. Tissue engineering, a new technology using a combination of stem cells, biomaterials and growth factors, has successfully reconstructed some craniofacial tissues in animal models. It provides a new possible treatment modality for craniofacial regeneration.

3DP technique, especially bioprinting, has become a vital tool for bioengineer tissues regeneration ranging from bone and teeth to vascular structure and nerves [46]. In regenerative medicine and dentistry, 3DP has been proposed to print organs and tissue analogs [7]. A customized 3D printed scaffold for tissue engineering allows for individualized treatment [68, 101]. It can generate scaffolds in which living cells are seeded before implanted into individual patient.

Though 3DP technology is becoming a major approach to fabricate scaffolds, its application in printing living cells, scaffold-free cells, or growth factors for the regenerative medicine is still at the early stage.

Currently, certain types of living cells can be printed with 3DP. In a study by Gruene, spots of human adipose-derived stem cells (ASCs) and endothelial colony-forming cells (ECFCs) were printed by 3DP. The results demonstrated that these cell spots could be arranged layer-by-layer in a three-dimensional array with any cell–cell ratio, cell quantity or cell-type combination. The spot spacing could also be controlled within this array [102]. In a study by Atala's group, multiple cell types were printed at pre-determined locations in a sodium alginate-collagen composite using a modified thermal inkjet printer. The *in vitro* results showed that all cells maintained viability, normal proliferation rate, phenotypic expression, and physiological functions. The bio-printed constructs were able to survive and mature into functional tissues if adequate vascularization occurred *in vivo*, demonstrating the feasibility of fabricating complex heterogeneous tissue containing multiple cell types using 3DP [103]. In a recent study, cell-printing and transferring technology were used to construct the cell-equipped amniotic membrane as a better cell delivery system for bone regeneration [104]. A valve-based cell printer was developed recently that could print human embryonic stem cells in uniform sized droplets. The printer was proved to be gentle enough to maintain stem cell viability and ability to differentiate into various cell types for tissue regeneration [105].

Cells can also be printed after mixing with biomaterial. For example, cell-laden hydrogels can be printed. Campos *et al.* encapsulated human mesenchymal stem cells and MG-63 cells into agarose hydrogels, and subsequently printed them in three-dimensional structures. The results showed that the printed cells remained viable and allowed for cell proliferation and production of extracellular matrix [106].

Growth factors have also been printed by 3DP. Miller *et al.* utilized inkjet printing for creating gradients of bone morphogenetic protein-2 (BMP-2) and insulin-like growth factor-II(IGF-II) on a fibrin-coated substrate. The growth factors gradient persisted up to 10 days. This printing technique was also used to create a combinatorial array containing BMP-2 and fibroblast growth factor-2 (FGF-2), which promoted cell proliferation and differentiation. In another study by Elmer, sub-micron fibers were patterned with FGF-2 or BMP-2 using inkjet printing. Mouse C2C12 myoblasts and C3H10T1/2 mesenchymal fibroblasts were seeded on the fibers. The printed FGF-2 and BMP-2 patterns inhibited myocyte differentiation, and promoted tenocyte and osteoblast differentiation along the fibers. It suggested that oriented sub-micron fibers with printed growth factors had a spatial control on cell survival and alignment [107]. The results of these studies are greatly contributing to regenerative medicine.

Up to date, the 3DP technology in directly printing cells or growth factors is lacking in literature of dentistry. Only one study by Xue *et al.* reported that they applied 3DP to print human dental pulp cells mixture with sodium alginate-gelatin hydrosol. The self-defined shape and 3D dimension were constructed, and cell survival rate reached 87% [108]. More researches are needed to promote the cell-based regeneration of craniofacial defects.

In conclusion, 3DP has been widely explored in different specialties, and it works as an important adjunct in diagnosis, treatment planning, prosthetic design, oral surgery and implant. It also greatly promotes dental education and tissue engineering researches. As technology development and more studies devoted in this field, the application of 3DP for craniofacial regeneration will be translated to clinic and greatly benefit dentists and patients.

CONFLICT OF INTEREST

The authors confirm that they have no conflict of interest to declare for this publication.

ACKNOWLEDGEMENTS

This work is financially supported by Temple University Kornberg School of Dentistry Research start-up fund.

REFERENCES

[1] Ventola CL. Medical applications for 3D printing: current and projected uses. P&T 2014; 39(10): 704-11.

[PMID: 25336867]

[2] Schubert C, van Langeveld MC, Donoso LA. Innovations in 3D printing: a 3D overview from optics to organs. Br J Ophthalmol 2014; 98(2): 159-61.
[PMID: 24288392]

[3] Suwanprateeb J. Strength improvement of critical-sized three dimensional printing parts by infiltration of solvent-free visible light-cured resin. J Mater Sci Mater Med 2006; 17(12): 1383-91.
[http://dx.doi.org/10.1007/s10856-006-0614-8] [PMID: 17143771]

[4] Ebert J, Ozkol E, Zeichner A, *et al.* Direct inkjet printing of dental prostheses made of zirconia. J Dent Res 2009; 88(7): 673-6.
[http://dx.doi.org/10.1177/0022034509339988] [PMID: 19641157]

[5] Prince JD. 3D Printing: An Industrial Revolution. J Electron Resour Med Libr 2014; 11(1): 39-45.
[http://dx.doi.org/10.1080/15424065.2014.877247]

[6] Bhambhani R, Bhattacharya J, Sen SK. Digitization and its futuristic approach in prosthodontics. J Indian Prosthodont Soc 2013; 13(3): 165-74.
[PMID: 24431730]

[7] Murphy SV, Atala A. 3D bioprinting of tissues and organs. Nat Biotechnol 2014; 32(8): 773-85.
[http://dx.doi.org/10.1038/nbt.2958] [PMID: 25093879]

[8] van Noort R. The future of dental devices is digital. Dent Mater 2012; 28(1): 3-12.
[http://dx.doi.org/10.1016/j.dental.2011.10.014] [PMID: 22119539]

[9] Byun C, Kim C, Cho S, *et al.* Endodontic treatment of an anomalous anterior tooth with the aid of a 3-dimensional printed physical tooth model. J Endod 2015; 41(6): 961-5.
[http://dx.doi.org/10.1016/j.joen.2015.01.016] [PMID: 25732403]

[10] Kim E, Kim KD, Roh BD, Cho YS, Lee SJ. Computed tomography as a diagnostic aid for extracanal invasive resorption. J Endod 2003; 29(7): 463-5.
[http://dx.doi.org/10.1097/00004770-200307000-00009] [PMID: 12877264]

[11] Lee SJ, Jang KH, Spangberg LS, *et al.* Three-dimensional visualization of a mandibular first molar with three distal roots using computer-aided rapid prototyping. Oral Surg Oral Med Oral Pathol Oral Radiol Endod 2006; 101(5): 668-74.
[http://dx.doi.org/10.1016/j.tripleo.2005.06.013] [PMID: 16632282]

[12] Ordinola-Zapata R, Bramante CM, Duarte MA, Cavenago BC, Jaramillo D, Versiani MA. Shaping ability of reciproc and TF adaptive systems in severely curved canals of rapid microCT-based prototyping molar replicas. J Appl Oral Sci 2014; 22(6): 509-15.
[http://dx.doi.org/10.1590/1678-775720130705] [PMID: 24918662]

[13] Park JM, Tatad JC, Landayan ME, Heo SJ, Kim SJ. Optimizing third molar autotransplantation: applications of reverse-engineered surgical templates and rapid prototyping of three-dimensional teeth. J Oral Maxillofac Surg 2014; 72(9): 1653-9.
[http://dx.doi.org/10.1016/j.joms.2014.04.012] [PMID: 25043146]

[14] Lee SJ, Jung IY, Lee CY, Choi SY, Kum KY. Clinical application of computer-aided rapid prototyping for tooth transplantation. Dent Traumatol 2001; 17(3): 114-9.
[http://dx.doi.org/10.1034/j.1600-9657.2001.017003114.x] [PMID: 11499760]

[15] Ciocca L, De Crescenzio F, Fantini M, Scotti R. CAD/CAM and rapid prototyped scaffold construction for bone regenerative medicine and surgical transfer of virtual planning: a pilot study. Comput Med Imaging Graph 2009; 33(1): 58-62. [http://dx.doi.org/10.1016/j.compmedimag.2008.10.005] [PMID: 19054651]

[16] Cohen A, Laviv A, Berman P, Nashef R, Abu-Tair J. Mandibular reconstruction using stereolithographic 3-dimensional printing modeling technology. Oral Surg Oral Med Oral Pathol Oral Radiol Endod 2009; 108(5): 661-6. [http://dx.doi.org/10.1016/j.tripleo.2009.05.023] [PMID: 19716728]

[17] Cui J, Chen L, Guan X, Ye L, Wang H, Liu L. Surgical planning, three-dimensional model surgery and preshaped implants in treatment of bilateral craniomaxillofacial post-traumatic deformities. J Oral Maxillofac Surg 2014; 72(6): 1138.e1-1138.e14. [http://dx.doi.org/10.1016/j.joms.2014.02.023] [PMID: 24679954]

[18] Fernandes da Silva AL, Borba AM, Simão NR, Pedro FL, Borges AH, Miloro M. Customized polymethyl methacrylate implants for the reconstruction of craniofacial osseous defects. Case Rep Surg 2014; 2014: 358569. [http://dx.doi.org/10.1155/2014/358569]

[19] Hong SB, Eliaz N, Leisk GG, Sach EM, Latanision RM, Allen SM. A new Ti-5Ag alloy for customized prostheses by three-dimensional printing (3DP). J Dent Res 2001; 80(3): 860-3. [http://dx.doi.org/10.1177/00220345010800030301] [PMID: 11379885]

[20] Kang SH, Kim MK, Kim BC, Lee SH. Orthognathic Y-splint: a CAD/CAM-engineered maxillary repositioning wafer assembly. Br J Oral Maxillofac Surg 2014; 52(7): 667-9. [http://dx.doi.org/10.1016/j.bjoms.2014.01.023] [PMID: 24957472]

[21] Kim BC, Lee CE, Park W, *et al.* Clinical experiences of digital model surgery and the rapid-prototyped wafer for maxillary orthognathic surgery. Oral Surg Oral Med Oral Pathol Oral Radiol Endod 2011; 111(3): 278-85.e1. [http://dx.doi.org/10.1016/j.tripleo.2010.04.038] [PMID: 20692187]

[22] Li B, Zhang L, Sun H, Yuan J, Shen SG, Wang X. A novel method of computer aided orthognathic surgery using individual CAD/CAM templates: a combination of osteotomy and repositioning guides. Br J Oral Maxillofac Surg 2013; 51(8): e239-44. [http://dx.doi.org/10.1016/j.bjoms.2013.03.007] [PMID: 23566536]

[23] Metzger MC, Hohlweg-Majert B, Schwarz U, Teschner M, Hammer B, Schmelzeisen R. Manufacturing splints for orthognathic surgery using a three-dimensional printer. Oral Surg Oral Med Oral Pathol Oral Radiol Endod 2008; 105(2): e1-7. [http://dx.doi.org/10.1016/j.tripleo.2007.07.040] [PMID: 18230371]

[24] Jiao T, Zhu C, Dong X, Gu X. Rehabilitation of maxillectomy defects with obturator prostheses fabricated using computer-aided design and rapid prototyping: a pilot study. Int J Prosthodont 2014; 27(5): 480-6. [http://dx.doi.org/10.11607/ijp.3733] [PMID: 25191895]

[25] Kim MS, Lee JY, Shin SW. Fabricating an obturator using rapid prototyping to design the framework: a case report. Int J Prosthodont 2014; 27(5): 439-41. [http://dx.doi.org/10.11607/ijp.3838] [PMID: 25191886]

[26] Lee JH. Accelerated techniques for a post and core and a crown restoration with intraoral digital scanners and CAD/CAM and rapid prototyping. J Prosthet Dent 2014; 112(5): 1024-9.
[http://dx.doi.org/10.1016/j.prosdent.2014.04.019] [PMID: 24951388]

[27] Lima JM, Anami LC, Araujo RM, Pavanelli CA. Removable partial dentures: use of rapid prototyping. J Prosthodont 2014; 23(7): 588-91.
[http://dx.doi.org/10.1111/jopr.12154] [PMID: 24750472]

[28] Lee CH, Hajibandeh J, Suzuki T, Fan A, Shang P, Mao JJ. Three-dimensional printed multiphase scaffolds for regeneration of periodontium complex. Tissue Eng Part A 2014; 20(7-8): 1342-51.
[http://dx.doi.org/10.1089/ten.tea.2013.0386] [PMID: 24295512]

[29] Park CH, Rios HF, Taut AD, *et al.* Image-based, fiber guiding scaffolds: a platform for regenerating tissue interfaces. Tissue Eng Part C Methods 2014; 20(7): 533-42.
[http://dx.doi.org/10.1089/ten.tec.2013.0619] [PMID: 24188695]

[30] Groth C, Kravitz ND, Jones PE, Graham JW, Redmond WR. Three-dimensional printing technology. J Clin Orthod 2014; 48(8): 475-85.
[PMID: 25226040]

[31] Hazeveld A, Huddleston Slater JJ, Ren Y. Accuracy and reproducibility of dental replica models reconstructed by different rapid prototyping techniques. Am J Orthod Dentofacial Orthop 2014; 145(1): 108-15.
[http://dx.doi.org/10.1016/j.ajodo.2013.05.011] [PMID: 24373661]

[32] Kwon SY, Kim Y, Ahn HW, Kim KB, Chung KR, Kim Sunny SH. Computer-aided designing and manufacturing of lingual fixed orthodontic appliance using 2D/3D registration software and rapid prototyping. Int J Dent 2014; 2014(2014): 164164.

[33] Martorelli M, Gerbino S, Giudice M, Ausiello P. A comparison between customized clear and removable orthodontic appliances manufactured using RP and CNC techniques. Dent Mater 2013; 29(2): e1-e10.
[http://dx.doi.org/10.1016/j.dental.2012.10.011] [PMID: 23140842]

[34] Nasef AA, El-Beialy AR, Mostafa YA. Virtual techniques for designing and fabricating a retainer. Am J Orthod Dentofacial Orthop 2014; 146(3): 394-8.
[http://dx.doi.org/10.1016/j.ajodo.2014.01.025] [PMID: 25172262]

[35] Chen J, Zhang Z, Chen X, Zhang C, Zhang G, Xu Z. Design and manufacture of customized dental implants by using reverse engineering and selective laser melting technology. J Prosthet Dent 2014; 112(5): 1088-95.e1.
[http://dx.doi.org/10.1016/j.prosdent.2014.04.026] [PMID: 24939253]

[36] El-Hajje A. Physical and mechanical characterisation of 3D-printed porous titanium for biomedical applications. J Mater Sci Mater Med 2014; 25(11): 471-80.
[PMID: 24243223]

[37] Flügge TV, Nelson K, Schmelzeisen R, Metzger MC. Three-dimensional plotting and printing of an implant drilling guide: simplifying guided implant surgery. J Oral Maxillofac Surg 2013; 71(8): 1340-6.
[http://dx.doi.org/10.1016/j.joms.2013.04.010] [PMID: 23866950]

[38] Gagg G, Ghassemieh E, Wiria FE. Analysis of the compressive behaviour of the three-dimensional printed porous titanium for dental implants using a modified cellular solid model. Proc Inst Mech Eng H 2013; 227(9): 1020-6.
[http://dx.doi.org/10.1177/0954411913489802] [PMID: 23804952]

[39] Kühl S, Payer M, Zitzmann NU, Lambrecht JT, Filippi A. Technical accuracy of printed surgical templates for guided implant surgery with the codiagnostiX™ software. Clin Implant Dent Relat Res 2015; 17 (Suppl. 1): e1177-82.
[http://dx.doi.org/10.1111/cid.12152]

[40] Murugesan K, Anandapandian PA, Sharma SK, Vasantha Kumar M. Comparative evaluation of dimension and surface detail accuracy of models produced by three different rapid prototype techniques. J Indian Prosthodont Soc 2012; 12(1): 16-20.
[http://dx.doi.org/10.1007/s13191-011-0103-8] [PMID: 23449946]

[41] Palmer R. Introduction to dental implants. Br Dent J 1999; 187(3): 127-32.
[PMID: 10481363]

[42] Park JM, Yi TK, Koak JY, Kim SK, Park EJ, Heo SJ. Comparison of five-axis milling and rapid prototyping for implant surgical templates. Int J Oral Maxillofac Implants 2014; 29(2): 374-83.
[http://dx.doi.org/10.11607/jomi.3265] [PMID: 24683563]

[43] Tamimi F, Torres J, Al-Abedalla K, *et al.* Osseointegration of dental implants in 3D-printed synthetic onlay grafts customized according to bone metabolic activity in recipient site. Biomaterials 2014; 35(21): 5436-45.
[http://dx.doi.org/10.1016/j.biomaterials.2014.03.050] [PMID: 24726538]

[44] Torres J, Tamimi F, Alkhraisat MH, *et al.* Vertical bone augmentation with 3D-synthetic monetite blocks in the rabbit calvaria. J Clin Periodontol 2011; 38(12): 1147-53.
[http://dx.doi.org/10.1111/j.1600-051X.2011.01787.x] [PMID: 22092695]

[45] Xiong Y, Qian C, Sun J. Fabrication of porous titanium implants by three-dimensional printing and sintering at different temperatures. Dent Mater J 2012; 31(5): 815-20.
[http://dx.doi.org/10.4012/dmj.2012-065] [PMID: 23037845]

[46] Gross BC, Erkal JL, Lockwood SY, Chen C, Spence DM. Evaluation of 3D printing and its potential impact on biotechnology and the chemical sciences. Anal Chem 2014; 86(7): 3240-53.
[http://dx.doi.org/10.1021/ac403397r] [PMID: 24432804]

[47] Wohlers T, Gornet T. History of additive manufacturing. Wohlers Report 2011; 2011: 1-34.

[48] Sachs EM, Haggerty JS, Cima MJ, Williams PA. Three-dimensional printing techniques. US patent 1993.

[49] Ishengoma FR, Mtaho AB. 3D printing : developing countries perspectives. Int J Comput Appl 2014; 104(11): 30-4.

[50] Collins F. Wohlers report 2014 uncovers annual growth of 349% for 3D printing and additive manufacturing industry 2014. Available at: www.wohlersassociates.com/press63.html

[51] Oberpenning F, Meng J, Yoo JJ, Atala A. *De novo* reconstitution of a functional mammalian urinary bladder by tissue engineering. Nat Biotechnol 1999; 17(2): 149-55.
[http://dx.doi.org/10.1038/6146] [PMID: 10052350]

[52] Mironov V, Reis N, Derby B. Review: bioprinting: a beginning. Tissue Eng 2006; 12(4): 631-4. [http://dx.doi.org/10.1089/ten.2006.12.631] [PMID: 16674278]

[53] Cui X, Boland T, D'Lima DD, Lotz MK. Thermal inkjet printing in tissue engineering and regenerative medicine. Recent Pat Drug Deliv Formul 2012; 6(2): 149-55. [http://dx.doi.org/10.2174/187221112800672949] [PMID: 22436025]

[54] Chae MP, Rozen WM, McMenamin PG, Findlay MW, Spychal RT, Hunter-Smith DJ. Emerging applications of bedside 3D printing in plastic surgery. Front Surg 2015; 2: 25. [http://dx.doi.org/10.3389/fsurg.2015.00025] [PMID: 26137465]

[55] Hoy MB. 3D printing: making things at the library. Med Ref Serv Q 2013; 32(1): 94-9. [http://dx.doi.org/10.1080/02763869.2013.749139] [PMID: 23394423]

[56] Ono I, Gunji H, Suda K, Kaneko F. Method for preparing an exact-size model using helical volume scan computed tomography. Plast Reconstr Surg 1994; 93(7): 1363-71. [http://dx.doi.org/10.1097/00006534-199406000-00005] [PMID: 8208802]

[57] Herlin C, Koppe M, Béziat JL, Gleizal A. Rapid prototyping in craniofacial surgery: using a positioning guide after zygomatic osteotomy - A case report. J Craniomaxillofac Surg 2011; 39(5): 376-9. [http://dx.doi.org/10.1016/j.jcms.2010.07.003] [PMID: 20813539]

[58] Tumbleston JR, Shirvanyants D, Ermoshkin N, *et al.* Additive manufacturing. Continuous liquid interface production of 3D objects. Science 2015; 347(6228): 1349-52. [http://dx.doi.org/10.1126/science.aaa2397] [PMID: 25780246]

[59] Watson RA. A low-cost surgical application of additive fabrication. J Surg Educ 2014; 71(1): 14-7. [http://dx.doi.org/10.1016/j.jsurg.2013.10.012] [PMID: 24411417]

[60] Okamoto T, Suzuki T, Yamamoto N. Microarray fabrication with covalent attachment of DNA using bubble jet technology. Nat Biotechnol 2000; 18(4): 438-41. [http://dx.doi.org/10.1038/74507] [PMID: 10748527]

[61] Sirringhaus H, Kawase T, Friend RH, *et al.* High-resolution inkjet printing of all-polymer transistor circuits. Science 2000; 290(5499): 2123-6. [http://dx.doi.org/10.1126/science.290.5499.2123] [PMID: 11118142]

[62] Goldmann T, Gonzalez JS. DNA-printing: utilization of a standard inkjet printer for the transfer of nucleic acids to solid supports. J Biochem Biophys Methods 2000; 42(3): 105-10. [http://dx.doi.org/10.1016/S0165-022X(99)00049-4] [PMID: 10737215]

[63] Singh R. Process capability study of polyjet printing for plastic components. J MechSciTechnol 2011; 25(4): 1011-5.

[64] Vaupotič B, Brezočnik M, Balic J. Use of PolyJet technology in manufacture of new product. J Achiev Mater ManufEng 2006; 18(1): 319-22.

[65] Bindl A, Mörmann WH. Fit of all-ceramic posterior fixed partial denture frameworks *in vitro.* Int J Periodontics Restorative Dent 2007; 27(6): 567-75. [PMID: 18092451]

[66] Nadkarni RA. Title: Modern instrumental methods of elemental analysis of petroleum products and

lubricants. Philadelphia, PA, United States: ASTM 1991.
[http://dx.doi.org/10.1520/STP1109-EB]

[67] Tracy JW. A revolution in the making. Fam Pract Manag 2007; 14(8): 11.
[PMID: 17915363]

[68] Klein GT, Lu Y, Wang MY. 3D printing and neurosurgery--ready for prime time? World Neurosurg 2013; 80(3-4): 233-5.
[http://dx.doi.org/10.1016/j.wneu.2013.07.009] [PMID: 23871811]

[69] Natiella JR, Armitage JE, Greene GW. The replantation and transplantation of teeth. A review. Oral Surg Oral Med Oral Pathol 1970; 29(3): 397-419.
[http://dx.doi.org/10.1016/0030-4220(70)90143-X] [PMID: 4983973]

[70] Lee SJ, Kim E. Minimizing the extra-oral time in autogenous tooth transplantation: use of computer-aided rapid prototyping (CARP) as a duplicate model tooth. Restor Dent Endod 2012; 37(3): 136-41.
[http://dx.doi.org/10.5395/rde.2012.37.3.136] [PMID: 23431083]

[71] Honda M, Uehara H, Uehara T, *et al.* Use of a replica graft tooth for evaluation before autotransplantation of a tooth. A CAD/CAM model produced using dental-cone-beam computed tomography. Int J Oral Maxillofac Surg 2010; 39(10): 1016-9.
[http://dx.doi.org/10.1016/j.ijom.2010.06.002] [PMID: 20630704]

[72] Shahbazian M, Jacobs R, Wyatt J, *et al.* Accuracy and surgical feasibility of a CBCT-based stereolithographic surgical guide aiding autotransplantation of teeth: *in vitro* validation. J Oral Rehabil 2010; 37(11): 854-9.
[http://dx.doi.org/10.1111/j.1365-2842.2010.02107.x] [PMID: 20626574]

[73] Keightley AJ, Cross DL, McKerlie RA, Brocklebank L. Autotransplantation of an immature premolar, with the aid of cone beam CT and computer-aided prototyping: a case report. Dent Traumatol 2010; 26(2): 195-9.
[http://dx.doi.org/10.1111/j.1600-9657.2009.00851.x] [PMID: 20089073]

[74] Azuma M, Yanagawa T, Ishibashi-Kanno N, *et al.* Mandibular reconstruction using plates prebent to fit rapid prototyping 3-dimensional printing models ameliorates contour deformity. Head Face Med 2014; 10: 45.
[http://dx.doi.org/10.1186/1746-160X-10-45] [PMID: 25338640]

[75] Engel M, Hoffmann J, Castrillon-Oberndorfer G, Freudlsperger C. The value of three-dimensional printing modelling for surgical correction of orbital hypertelorism. Oral Maxillofac Surg 2015; 19(1): 91-5.
[http://dx.doi.org/10.1007/s10006-014-0466-1] [PMID: 25249178]

[76] Tsai MJ, Wu CT. Study of mandible reconstruction using a fibula flap with application of additive manufacturing technology. Biomed Eng Online 2014; 13: 57.
[http://dx.doi.org/10.1186/1475-925X-13-57] [PMID: 24885749]

[77] Liu YF, Xu LW, Zhu HY, Liu SS. Technical procedures for template-guided surgery for mandibular reconstruction based on digital design and manufacturing. Biomed Eng Online 2014; 13(8): 63.
[http://dx.doi.org/10.1186/1475-925X-13-63] [PMID: 24886431]

[78] Hirsch DL, Garfein ES, Christensen AM, Weimer KA, Saddeh PB, Levine JP. Use of computer-aided

design and computer-aided manufacturing to produce orthognathically ideal surgical outcomes: a paradigm shift in head and neck reconstruction. J Oral Maxillofac Surg 2009; 67(10): 2115-22. [http://dx.doi.org/10.1016/j.joms.2009.02.007] [PMID: 19761905]

[79] Gateno J, Teichgraeber JF, Xia JJ. Three-dimensional surgical planning for maxillary and midface distraction osteogenesis. J Craniofac Surg 2003; 14(6): 833-9. [http://dx.doi.org/10.1097/00001665-200311000-00004] [PMID: 14600624]

[80] Wang H, Aboushelib MN, Feilzer AJ. Strength influencing variables on CAD/CAM zirconia frameworks. Dent Mater 2008; 24(5): 633-8. [http://dx.doi.org/10.1016/j.dental.2007.06.030] [PMID: 17765301]

[81] Chen J, Ahmad R, Suenaga H, *et al.* Shape optimization for additive manufacturing of removable partial dentures - anew paradigm for prosthetic CAD/CAM. PLoS One 2015; 10(7): e0132552. [http://dx.doi.org/10.1371/journal.pone.0132552] [PMID: 26161878]

[82] Katase H, Kanazawa M, Inokoshi M, Minakuchi S. Face simulation system for complete dentures by applying rapid prototyping. J Prosthet Dent 2013; 109(6): 353-60. [http://dx.doi.org/10.1016/S0022-3913(13)60316-9] [PMID: 23763778]

[83] Cho SH, Chang WG. Mirror-image anterior crown fabrication with computer-aided design and rapid prototyping technology: a clinical report. J Prosthet Dent 2013; 109(2): 75-8. [http://dx.doi.org/10.1016/S0022-3913(13)60018-9] [PMID: 23395331]

[84] Atzeni E, Iuliano L, Minetola P, Salmi A. Proposal of an innovative benchmark for accuracy evaluation of dental crown manufacturing. Comput Biol Med 2012; 42(5): 548-55. [http://dx.doi.org/10.1016/j.compbiomed.2012.01.009] [PMID: 22364825]

[85] Inokoshi M, Kanazawa M, Minakuchi S. Evaluation of a complete denture trial method applying rapid prototyping. Dent Mater J 2012; 31(1): 40-6. [http://dx.doi.org/10.4012/dmj.2011-113] [PMID: 22277604]

[86] Williams RJ, Bibb R, Eggbeer D, Collis J. Use of CAD/CAM technology to fabricate a removable partial denture framework. J Prosthet Dent 2006; 96(2): 96-9. [http://dx.doi.org/10.1016/j.prosdent.2006.05.029] [PMID: 16911885]

[87] Jiao T, Zhu C, Dong X, Gu X. Rehabilitation of maxillectomy defects with obturator prostheses fabricated using computer-aided design and rapid prototyping: a pilot study. Int J Prosthodont 2014; 27(5): 480-6. [http://dx.doi.org/10.11607/ijp.3733] [PMID: 25191895]

[88] Al Mortadi N, Eggbeer D, Lewis J, Williams RJ. Design and fabrication of a sleep apnea device using computer-aided design/additive manufacture technologies. Proc Inst Mech Eng H 2013; 227(4): 350-5. [http://dx.doi.org/10.1177/0954411912474741] [PMID: 23637210]

[89] Pihlstrom BL, Michalowicz BS, Johnson NW. Periodontal diseases. Lancet 2005; 366(9499): 1809-20. [http://dx.doi.org/10.1016/S0140-6736(05)67728-8] [PMID: 16298220]

[90] Armitage GC. Development of a classification system for periodontal diseases and conditions. Ann Periodontol 1999; 4(1): 1-6. [http://dx.doi.org/10.1902/annals.1999.4.1.1] [PMID: 10863370]

[91] Costa PF, Vaquette C, Zhang Q, Reis RL, Ivanovski S, Hutmacher DW. Advanced tissue engineering

scaffold design for regeneration of the complex hierarchical periodontal structure. J Clin Periodontol 2014; 41(3): 283-94.
[http://dx.doi.org/10.1111/jcpe.12214] [PMID: 24304192]

[92] Park CH, Rios HF, Jin Q, *et al.* Biomimetic hybrid scaffolds for engineering human tooth-ligament interfaces. Biomaterials 2010; 31(23): 5945-52.
[http://dx.doi.org/10.1016/j.biomaterials.2010.04.027] [PMID: 20471083]

[93] Vaquette C, Fan W, Xiao Y, Hamlet S, Hutmacher DW, Ivanovski S. A biphasic scaffold design combined with cell sheet technology for simultaneous regeneration of alveolar bone/periodontal ligament complex. Biomaterials 2012; 33(22): 5560-73.
[http://dx.doi.org/10.1016/j.biomaterials.2012.04.038] [PMID: 22575832]

[94] Park CH, Rios HF, Jin Q, *et al.* Tissue engineering bone-ligament complexes using fiber-guiding scaffolds. Biomaterials 2012; 33(1): 137-45.
[http://dx.doi.org/10.1016/j.biomaterials.2011.09.057] [PMID: 21993234]

[95] Park CH, Kim KH, Rios HF, Lee YM, Giannobile WV, Seol YJ. Spatiotemporally controlled microchannels of periodontal mimic scaffolds. J Dent Res 2014; 93(12): 1304-12.
[http://dx.doi.org/10.1177/0022034514550716] [PMID: 25216511]

[96] Ivanovski S, Vaquette C, Gronthos S, Hutmacher DW, Bartold PM. Multiphasic scaffolds for periodontal tissue engineering. J Dent Res 2014; 93(12): 1212-21.
[http://dx.doi.org/10.1177/0022034514544301] [PMID: 25139362]

[97] Reynolds M, Reynolds M, Adeeb S, El-Bialy T. 3-d volumetric evaluation of human mandibular growth. Open Biomed Eng J 2011; 5: 83-9.
[http://dx.doi.org/10.2174/1874120701105010083] [PMID: 22046201]

[98] Liu YF, Zhang PY, Zhang QF, Zhang JX, Chen J. Digital design and fabrication of simulation model for measuring orthodontic force. Biomed Mater Eng 2014; 24(6): 2265-71.
[PMID: 25226926]

[99] Reyes A, Turkyilmaz I, Prihoda TJ. Accuracy of surgical guides made from conventional and a combination of digital scanning and rapid prototyping techniques. J Prosthet Dent 2015; 113(4): 295-303.
[http://dx.doi.org/10.1016/j.prosdent.2014.09.018] [PMID: 25681352]

[100] Gagg G, Ghassemieh E, Wiria FE. Effects of sintering temperature on morphology and mechanical characteristics of 3D printed porous titanium used as dental implant. Mater Sci Eng C Mater Biol Appl 2013; 33(7): 3858-64.
[http://dx.doi.org/10.1016/j.msec.2013.05.021] [PMID: 23910288]

[101] Rengier F, Mehndiratta A, von Tengg-Kobligk H, *et al.* 3D printing based on imaging data: review of medical applications. Int J CARS 2010; 5(4): 335-41.
[http://dx.doi.org/10.1007/s11548-010-0476-x] [PMID: 20467825]

[102] Gruene M, Pflaum M, Hess C, *et al.* Laser printing of three-dimensional multicellular arrays for studies of cell-cell and cell-environment interactions. Tissue Eng Part C Methods 2011; 17(10): 973-82.
[http://dx.doi.org/10.1089/ten.tec.2011.0185] [PMID: 21585313]

[103] Xu T, Zhao W, Zhu JM, Albanna MZ, Yoo JJ, Atala A. Complex heterogeneous tissue constructs containing multiple cell types prepared by inkjet printing technology. Biomaterials 2013; 34(1): 130-9. [http://dx.doi.org/10.1016/j.biomaterials.2012.09.035] [PMID: 23063369]

[104] Tsugawa J, Komaki M, Yoshida T, Nakahama K, Amagasa T, Morita I. Cell-printing and transfer technology applications for bone defects in mice. J Tissue Eng Regen Med 2011; 5(9): 695-703. [http://dx.doi.org/10.1002/term.366] [PMID: 21953867]

[105] Faulkner-Jones A, Greenhough S, King JA, Gardner J, Courtney A, Shu W. Development of a valve-based cell printer for the formation of human embryonic stem cell spheroid aggregates. Biofabrication 2013; 5(1): 015013. [http://dx.doi.org/10.1088/1758-5082/5/1/015013] [PMID: 23380571]

[106] Duarte Campos DF, Blaeser A, Weber M, *et al.* Three-dimensional printing of stem cell-laden hydrogels submerged in a hydrophobic high-density fluid. Biofabrication 2013; 5(1): 015003. [http://dx.doi.org/10.1088/1758-5082/5/1/015003] [PMID: 23172592]

[107] Ker ED, Nain AS, Weiss LE, *et al.* Bioprinting of growth factors onto aligned sub-micron fibrous scaffolds for simultaneous control of cell differentiation and alignment. Biomaterials 2011; 32(32): 8097-107. [http://dx.doi.org/10.1016/j.biomaterials.2011.07.025] [PMID: 21820736]

[108] Xue SH, Lv PJ, Wang Y, Zhao Y, Zhang T. [Three dimensional bioprinting technology of human dental pulp cells mixtures]. Beijing Da Xue Xue Bao 2013; 45(1): 105-8. [PMID: 23411530]

CHAPTER 3

Induced Pluripotent Stem Cells: Proliferation, Migration, MicroRNA, Signaling Molecules

Yuanzhi Xu[1], Feiyu Wang[2], Jing Wang[1,*], Yun Lu[3], Yun Shen[1], Songtao Wu[4], Zhaozhao Chen[4], Runyi Mao[2] and **Raorao Wang[1]**

[1] *Department of Stomatology, Shanghai Tenth People's Hospital, Tongji University School of Medicine, Shanghai, China*

[2] *Department of Oral Surgery, Ninth People's Hospital, Shanghai Jiao Tong University School of Medicine, Shanghai, China*

[3] *Department of Orthodontics, College of Stomatology, Dalian Medical University, Dalian, China*

[4] *State Key Laboratory of Oral Diseases, West China School of Stomatology, Sichuan University, Chengdu, China*

Abstract: Since it was first demonstrated that induced pluripotent stem cells (iPS cells) could be derived from mature cells, significant progress has been made in the field of acquisition, characteristics, identification and application of iPS cells. Until now, diverse means have been proven to generate iPS cells successfully in many biological species and more cell types. Meanwhile, researchers continue to target the efficiency of induction. To identify the characteristics of induced pluripotent stem cells and attest to their pluripotency, one must verify the expression of new derived stem cell genes and proteins, doubling times, methylation patterns, teratoma formation, embryoid body formation, viable chimera formation and capacity to differentiate into all cell types. In other words, induced pluripotent stem cells are theoretically similar, or even same, to natural pluripotent stem cells, for instance embryonic stem (ES) cells. Furthermore, iPS cells have the potential to take the place of ES cells eventually, from which numerous ethical difficulties arise for the treatment of a mass of diseases, for use in therapeutics for drug discovery, disease modeling and regenerative medicine, *etc*. However, many problems exist as barriers to clinical transformation, including risk for induced oncogenesis and the stability of reprogramming. Here we summarize the current

[*] **Corresponding author Jing Wang:** Department of Stomatology, Shanghai Tenth People's Hospital, Tongji University School of Medicine, No. 301, Middle Yanchang Road, Shanghai 200072, P.R. China; Tel: 86-21-66301726; Fax: 86-21-66301725; E-mail: dentist_tracy@hotmail.com

Jing Wang & YunFeng Lin (Eds.)

acquaintance of iPS generation and evaluate the advantages and disadvantages of their applications in clinical medicine.

Keywords: Biological safety, Biometric identification, Cellular biological properties, Cell differentiation, Cellular reprogramming, Induced pluripotent stem cell, Investigation of drugs, Pluripotency, Regenerative medicine.

INTRODUCTION

During normal development, cells are generally endowed to increasingly differentiated fates. But according to research achievements for cellular reprogramming, the process of differentiation is reversible. iPS cells are a type of pluripotent stem cell that is generated from somatic cells, contrary to natural course of cell fate. The iPS cell technology was first introduced by the Japanese investigator Shinya Yamanaka [1] who transfected four genes (containing Oct3/4, Sox2, c-Myc and Klf4, OSKM) into mouse embryonic fibroblasts (MEFs) *via* retroviral transduction and converted somatic cells into pluripotent stem cells. The Nobel Prize of 2012 was granted to him for his contribution in discovering that mature cells can be reversibly reprogrammed to become pluripotent status.

Among the pluripotent stem cells, the embryonic stem cell is the best-known, because, as the beginning and origin of life, they can differentiate into all kinds of cell lines. However, since the usage of embryonic stem cells relate to impairing the embryo [2], significant ethical opposition has been raised. Furthermore, the source is limited, and patient-matched embryonic stem cell lines cannot currently be obtained. There are too many obstacles to the clinical production of embryonic stem cells.

In contrast, iPS cells are derived from differentiated cells or mature tissue, reprogrammed to an embryonic-like state. Reprogrammed human induced pluripotent stem cells (hiPS) are similar to human embryonic stem (hES) cells in many fields, such as morphology, epigenetic status of pluripotent genes, surface antigen expression, telomerase activity and proliferation rate. Beyond that, it has been demonstrated that hiPS cells have the capacity to differentiate into cell types of all three germ layers *in vitro* and *in vivo*. The cell source is extensive, and they can be used to give each individual his or her own pluripotent stem cell line.

Autologous donors also reduce the likelihood of immune rejection of transplantation. Thus, hiPS cell is a promising cell classification as a replacement of embryonic stem cells in the field of regenerative medicine, thus mitigating ethical risks. However, more and more evidence indicated that hiPS cells and hES cells are not identical at many aspects. There are also verified differences in inheritance among different iPS cell lines that may show that differences in the somatic cell source or even genetic variability among similar cells lead to discrepant pluripotent stem cell.

In recent studies, it has been demonstrated that more than one outcome of iPS cell states exists from the usage of a secondary reprogramming means, in which the initial differentiated cells are mouse fibroblasts continued derived by OSKM containing doxycycline. A novel category of steady-state pluripotent cells was emerged, as given F-class status after the fuzzy presentation of cell colonies in culture. It is a new condition that deserves to be studied.

Although various methods have been established to generate iPS cells, with some success in terms of degree of pluripotency, each means has advantages and disadvantages with respect to future clinical applications.

While iPS cell technology has developed more rapidly than ever, it has not advanced to a stage where the related techniques of cell therapy or organ synthesis are safe and reliable. Carcinogenicity is one of the significant risks blocking the usage of iPS cells in human while there are challenges like low efficiency, incomplete, genomic insertion posing serious risks that limit the use in clinic. Thus, the task of iPS cell application continues to be both challenging and full of opportunities.

2. IPS CELLS: METHODS OF ACQUISITION, PROMOTING GROWTH AND INCREASING INDUCED EFFICIENCY

2.1. Selection of Proper Recipient Cells

Since the successful derivation, in 2006, of iPS cells from adult mouse fibroblasts through the ectopic co-expression of four related genes, fibroblasts have been the most common cell type to reprogram, likely because these cells are readily available or can be safely obtained by skin biopsy and are easy to culture in the laboratory. In 2007, another milestone was achieved when two independent research teams created iPS cells from somatic human cells [3]. With the same derived principle used in earlier mouse models, Yamanaka move forward a single step to successfully reprogrammed human fibroblasts to generate derived pluripotent stem cells *via* the same four genes with a retroviral system. In contrast, the team led by Thomson used OCT4, SOX2, LIN28 and NANOG with a lentiviral system. Subsequently, these historic contributions inspired an astonishing host of follow-up researches. Numerous different cell types, such as keratinocytes, neural stem cells, keratinocytes, stomach and liver cells, melanocytes, amniotic cells, bone marrow stromal cells and blood cells, have also been reprogrammed into iPS cells by a variety of different reprogramming methods.

2.2. Alternative Vector Systems

Numerous approaches to the delivery of reprogramming factors have been conducted. However, the iPS cell technique is still considered to have low efficiency, slow kinetics, incomplete reprogramming and potential indeterminism from exogenous gene insertion. No approach is perfect, and different alternatives are known to have specific advantages. Nevertheless, while scientists have attempted to eliminate or alleviate these problems, alternative approaches have their own limitations. The following list of iPS cell technologies is aimed to provide a quick and comprehensive reference.

Retroviral vectors played an initial role in the history of the development of iPS cell technology. To date, they are the most widely used for transgene delivery in cells cultured *in vitro* and in animal models. A segment of gene was introduced into a specific site of the plasmid to generate a plasmid-based retroviral vector. Mutations in specific genes were introduced by polymerase PCR-based site-directed mutagenesis. It is better to amplify the coding regions of candidate genes by real-time PCR, clone these sequences, and recombine the resulting plasmids.

However, chimeras and progeny mice cultivated from iPS cells frequently develop into tumors or cancer eventually, which may be due to the oncogenicity of transcription factor c-Myc. It is crucial to generate iPS cells without this potential oncogenic gene for the safety of future clinical treatment. In a later study [4] published in 2008, Yamanaka introduced a method for creating iPS cells without c-Myc. The course takes longer and has even lower efficiency, but the resulting chimeras did not develop cancer. In the meantime, other key strategies for avoiding such tumorigenesis have involved the use of alternative vectors: plasmids, adenovirus, protein compounds or naked DNA. In 2008, Hochedlinger *et al.* [5] used certain system to create adenoviruses that express human Oct4, Sox2, Klf4 and c-Myc. The requisite four transcription factors were cloned into sites of cytomegalovirus vector. Homologous recombination was performed in bacterial cells. The adenoviruses were embed in HEK293 cells and purified to high titer. There is another advantage for the use of adenoviruses, which is that they have to be exposed for only a brief span of time for the enough effective reprogramming process to occur. Also in 2008, Yamanaka's team [6] delivered the four necessary genes with a plasmid. They used cDNA encoding Oct3/4, Klf4 and Sox2 to connect with peptide for inserting into the plasmid. In addition, another plasmid expressing c-Myc was constructed, after which transfected plasmids were incorporated into the host genome. Although the method of using plasmid avoids retrovirus, they still require c-Myc and KLF4, the two cancer-promoting genes, to accomplish reprogramming. Another serious problem with this method is that the lower efficiency and the transgene expression levels were observed at low level more with plasmid transfection than with retroviruses. Furthermore, the risk of insertional mutagenesis still exists by the use of gene insertion. To reverse the low efficiency levels, the technique titled the piggyBac transposon system can deliver the key reprogramming factors more effectively without leaving any risk of footprint mutagenesis in the host cell genome.

2.3. Induction Genes

Some genes are capable of induction, especially certain crucial genes are indispensible for the generation of iPS cells.

The earliest teams reporting the generation of iPS cells, Yamanaka [1] and Thomson[3], used a reprogramming cocktail consisting of Oct-3/4 and some of the Sox gene family (Sox1, Sox2, Sox3 and Sox15). It is now believed that the two kinds of genes are crucial transcriptional regulators involved in the reprogramming process whose absence makes induction not go on. Otherwise, additional genes, including members of the Myc family (c-myc, L-myc and N-myc), the Klf family (Klf1, Klf2, Klf4 and Klf5), LIN28 and Nanog, can be used to take the place of each other. They have also been reported to increase induction efficiency.

Oct-3/4 is one of the families known as octamer transcription factors. An earlier study [7] reported that it plays a crucial role in the maintenance of pluripotency in both ES and early embryos cells. The absence of Oct-3/4 in pluripotent stem cells leads to spontaneous differentiation, and the expression of Oct-3/4 thus retains the pluripotency and differentiation potential of natural pluripotent stem cells. Numerous other genes in the Oct family failed to show the ability of induction, which demonstrates the specificity of Oct-3/4 in the reprogramming process.

The Sox family: The genes of Sox family are also related to maintaining pluripotency, which is similar to the function of Oct-3/4 [8]. Meanwhile, compared with Oct-3/4, It is also related to the character of multipotent and unipotent, which is exclusively expressed in pluripotent stem cells. Thus, Sox2 became the initial gene used for induction by both Yamanaka and Thomson. Other genes in the Sox family, such as Sox1, Sox3, Sox15 and Sox18, also verified to take effect in the induction process and generate iPS cells with comparatively low efficiency.

The Klf family: This family, comprised of Klf4, has been testified to make an inevitable contribution to the long-term maintenance of the ES cell phenotype and the fast proliferation of ES cells in culture system [9]. It was initially used by Yamanaka *et al.* as a transcription factor to generate iPS cells. However, the research team of Thomson reported that Klf4 was not necessary for the generation of human iPS cells subsequently. Klf1, Klf2, Klf4 and Klf5 were found to be factors capable of generating iPS cells, although with reduced efficiency.

The Myc family: The genes of Myc family are often up-regulate the incidence of tumors. They are proto-oncogenes implicated in the genesis of cancer. A study by Yamanaka indicated that c-myc is an initiator in the generation of iPS cells.

However, usage of the "myc" family of genes for the induction of iPS cells resulted in 20% of the chimeric mice developing cancer. The risk of oncogenicity caused widespread controversy and cast a shadow over future clinical applications. Fortunately, c-Myc can be dispensed with for the direct induction of mouse fibroblasts [10]. To remedy the low efficiency induced by the lack of c-Myc, L-myc and N-myc have been identified to have the ability of inducibility, with efficiency similar to that of c-myc.

Nanog: Nanog was known to be beneficial in clone recovery from human ES-cell -derived mesenchymal cells. At the same time, it was not obliged to be required in the initial appearance of such clones [3]. However, when Nanog was overexpressed in mouse ES cells, the increase in reprogramming efficiency was 200-fold [11]. The expression of Nanog also promotes the cloning efficiency of human ES cells [12] and could further increase the survival and success rate of early reprogramming process. Therefore, it was adopted by Thomson [3], who discovered that Nanog, as one of the transcription factors, can generate iPS cells.

LIN28: LIN28 is an mRNA binding protein, which expressed in embryonic carcinoma cells and embryonic stem cells that associated with differentiation and proliferation. In the field of reprogrammed mesenchymal cell clone recovery, LIN28 had not only consistent but more modest effect. Thomson demonstrated [3] that it is a factor in iPS cell generation.

Glis1: As an alternative to C-myc, Glis1 has been selected to collaborate with the other three transcription factors Oct-3/4, Klf4 and Sox2 to induce pluripotency of certain cells, with the advantage of reduced tumor risk [13].

UTF1: When UTF1 along with OSKM was expressed in human primary fibroblasts, more iPS colonies were obtained. It has been demonstrated [14] that UTF can accelerate the speed of generation of iPS cells.

SALL4: When SALL4 was co-expressed with OSK in human adult fibroblasts from dermis, a ten-fold increase rate could be observed [15].

2.4. Acquisition of iPS Cells

It has been demonstrated that there exists not the only one kind of induced pluripotent stem cell. Another novel variety of steady-state pluripotent cells, named F-class state, was found with the fuzzy appearance in cell colonies. Secondary reprogramming systems resolve the low efficiency to generate sufficient numbers of cells. Researchers used doxycycline-inducible reprogramming transgenes along with iPS-derived differentiated cells.

Sustained high-doxycycline through the course of induce mouse embryonic fibroblast (MEF) to become iPS cells with the F-class state. However, lowering the doxycycline concentration showed different outcome, leading to the ESC-like state.

The primary technique for the induction of iPS cells is as follows (Fig. **1**):

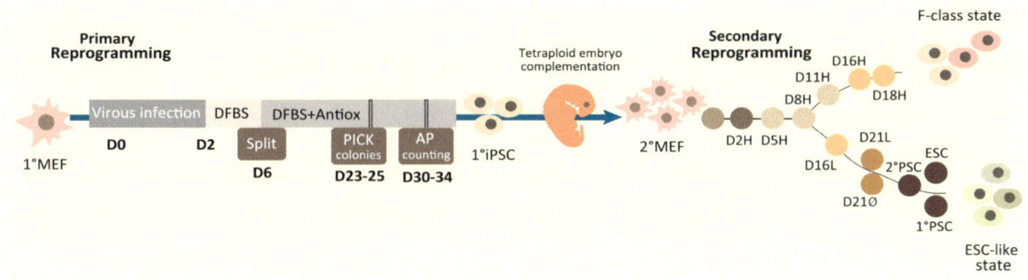

Fig. (1). Acquisition of iPS cells.

1. Harvest embryos from 13.5-day-pregnant mice, divorced from the mother and wiping off the uterine horn. Clean with sterile phosphate-buffered saline (PBS), cutting away the surrounding uterine membranes and envelope.
2. Continue to remove the head, tail, limbs and internal organs of embryo. Wash the remaining bodies with PBS, minced, then transfer into trypsin with EDTA solution, 3 mL for per embryo, and incubate at 37°C for 5 min. After incubation, end solution by adding an equal amount of medium; further mechanically dissociate the remaining tissue by gentle pipetting. Discard the supernatant, transfer the remaining mixture to a 10-cm dish and add 10 mL of MEF media.
3. After 24-48 hrs, the plate would be fully covered with MEFs. Gently wash the dish containing MEFs with PBS to remove traces of media and then use 3 mL trypsin/EDTA solution for trypsinization, incubate for 3 min at 37°C and add an equal amount of medium to stop digestive reaction. Collect cells by

centrifugation (1,500 rpm for 5 min) and resuspend them in fresh medium. According to the density of $1*10^6$ cells (passage 1) per 10-cm dish, MEFs were cultured at the circumstance of 37°C with 5% CO_2.

4. Retroviral infection: The packaging cells can be obtained in a variety of cell lines, *e.g.* Plat-E cells and HEK293T cells. Before transduction, seed HEK 293T cells at a settled density of $8*10^6$ cells per 10-cm dish one day in advance. On the next day, introduce retroviral vectors into packaging cells *via* transfection reagent in accordance with the instructions. For a 10-cm culture dish, there is, in addition OCT4, SOX2, KLF4and c-MYC. pMXs-GFP can monitor efficiency as a separate control factor. Add the transfection and plasmid DNA to the mixture for incubation. Add the reagent mixture onto packaging cells drop by drop. Incubate cells overnight at an incubator settled 37°C with 5% CO_2. In the next moment, just use a fluorescence microscope to check the signal emerged by the reaction of pMXs-GFP and affirm that all HEK 293T cells incubated have been transfected.
5. After twenty-four hours for transduction, replace fresh medium. Seed MEFs at a density of $8*10^5$ cells per 10 cm dish on antibiotic-treated STO feeders. Another twenty-four hours later, filter the supernatants that contains virus derived from cultures through a cellulose acetate filter and add polybrene for standby. Incubate the prepared mixture to overnight for target cells. After then, replant the acquired cells in fresh medium. Add G418 at a final concentration after three days. Finally, wanted clones can be found and collected for 2 to 3 weeks.
6. Use the microscope to check the infected cells daily. The morphology of cells would be strongly changed with infected by the four exogenous genes. If the morphology change and proliferation of the infected urinary cells are substantial, then use plate with irradiated MEFs as feeder cells for UiPS cell induction. After 2 days, replace the old medium by the fresh human ESC medium. Thus, continue to incubate the cells for 7 days, replace the medium daily with fresh human ESC medium. 16 days after infection, replace the old medium daily by fresh mTesR1 medium.

2.5. Culture of Recipient and Induced Cells

For isolating embryonic fibroblasts, mature tissue was minced for 1mm*1mm in

HEPE-buffered medium and digested in collagenase. The cells were harvested from multiple solution times in certain culture media containing DMEM (high glucose), 10% fetal bovine serum (FBS) and 1% double antibiotics (penicillin/ streptomycin). Samples were centrifuged at 1,500 rpm for 5 min and the pellets suspended in embryonic fibroblast medium. Only early passages were used for iPS generation since use within three passages is more likely to avoid replicative senescence.

An essential bioengineering problem in the research of iPS cell is the limit of technologies that cannot expand the iPS cells in large scale. How to develop large-scale expansion of primitive iPS cells as well as expansion of their differentiated later generation is an urgent problem researchers face now. Achievement of this technology is essential not only for future further research and potential issue need to be solved. With the breakthrough, we would not only conduct rapid and parallel experiments, not worrying about whether cell quantity is enough, but also apply eventual translation products derived from iPS cells into clinical practice. Actually, the issue also can benefit the culture of embryonic and adult stem cells. Recently, Lei *et al.* published a scalable, hydrogel-based 3D culture method that could be free of animal- and human-derived factors [16]. According to paper published, both ES and iPS cells were tested for these means, manifesting versatility.

2.6. Increased Induced Efficiency

It is a common view that with defined transcription factors, we can reprogram somatic cells into iPS cells. However, the low efficiency of iPS cells acquisition and length of time required for finishing the reprogramming procedure have limited the application of this technology in more research field. For this reason, researchers have arisen numerous induction systems to make an advance in the efficiency of iPS cell generation and speed of the process when they expend effort to study the mechanism of reprogramming in depth.

Esteban and co-workers reported [17] that Vitamin C has the capability of alleviating roadblocks to reprogramming, along with reducing cell senescence. Vitamin C, known as the function of antioxidant in compound mix, was added with a proper concentration to promote reprogramming processes and efficiency, resulting in the improvement of iPS cell generation and the conversion of pre-iPS cells into iPS cells. The reprogramming efficiency observed is up to as high as 6.2%. With human fibroblast-like cells (FLCs) used as feeder layers, human foreskin fibroblasts not only reprogrammed into a state of pluripotency by OSKM, but also showed a reprogramming efficiency superior to that obtained on MEF feeder layers, with more effectiveness in supporting the growth of human pluripotent stem cells [18]. In addition, extensive demethylation appeared in the promoter regions were shown in the adult mouse-derived fibroblasts that were cultured in the surrounding of collagen complexes. It would lead to increased proliferation along with decreased senescence. The efficiency of reprogramming fibroblasts into iPS cells became significantly higher than that without collagen complexes. The above facts demonstrated that collagen complexes not only stimulate proliferation but also inhibit cell death and promote the growth out of the fibroblasts-state. It was a promising approach to improving the efficiency of reprogramming [19].

As to the problem that some cells may stopped in partially reprogrammed states because of incomplete expression of defined transcription factors, it has been demonstrated [20] that the speed of reprogramming could be accelerated by RNA inhibition of transcription factors, and more fully reprogrammed cells can be obtained with RNA inhibition to improve the overall efficiency of the reprogramming, along with DNA methyltransferase inhibitors called 5-azacytidine. Small-molecule compounds also played a big role in improving inductive efficiency [21]. In particular, the increasing reprogramming efficiency will be more than 100-fold, comparing valproic acid (VPA), an HDAC inhibitor, with Oct4-GFP. The use of VPA also makes sure that pluripotent stem cells can be efficiently induced without the effect of the oncogene c-Myc.

3. IPS CELLS: METHODS FOR IDENTITY AND SAFETY FOR REGENERATIVE MEDICINE

3.1. Cellular Biological Properties

Theoretically, the generated iPS cells derived from somatic cells were particularly similar to those natural pluripotent stem cells, such as ES cells. However, more and more published paper show evidence that indicates [22] that the biological properties, such as epigenetic profile, differ between ES and iPS cell lines. Following the difference, concern about unpredictable variations was raised for the reason that different cell lines will lead to potentially inconsistent result in iPS-based disease research.

Morphology: iPS cells were similar to ESCs. According to reports [23], human iPS cells are similar to hESCs, forming flat, sharp-edged and tightly packed colonies. Meanwhile colonies formed by mouse iPS cells are similar to mESCs, comprising less flat and more aggregated than human ESCs. However, there are various differences between mouse and human iPS cells in appearance. They are mainly reflected in which colonies of mouse iPS cell presenting more dome-like and refractive than colonies of human iPS cells and colonies of human iPS cells are flatter than those of mice.

Cell markers: According to current literature, it should be expected that fully reprogrammed cells would be changed with low levels of fibroblast-typical cell markers and high levels of ESC-typical cell markers. Reprogramming to pluripotency involves the sequential activation of the alkaline phosphatase (AP), stage-specific embryonic antigen 1 (SSEA1) and Nanog [24]. Furthermore, human iPS cells also expressed ESC markers such as Oct4, Nanog, SSEA-3, SSEA-4, TRA-1-60, TRA-1-81, DNA methyltransferase 3β (DNMT3β) and REX1.

When one follows the trail of the expression of markers during MEF reprogramming derived by OKSM, it is evident that, by day 3, down-regulation of the fibroblast-special expression signal can be seen. At day 6, major branches express the ESC markers CD326, CD54 and SSEA1. Fractions express high levels of the partly reprogrammed cell (PRC) markers CD73, CD49d and CD200. Till day 9, CD326 and SSEA1 expression converged in a subpopulation and persisted on days 12 and 16.

Telomere and telomerase activity: The premature death of the sheep Dolly first led to the emphasis on a shorter telomere. Regaining telomere length is considered an important feature that can reveal the completeness of reprogramming iPS cells. In the first set of established human iPS cells studied by Yamanaka's team, high levels of telomerase activity were detected. However, whether iPS cells regain telomere length equal to that in ES cells after reprogramming is unclear. Although Marion *et al.* [25] demonstrated that, during reprogramming, a high level of telomerase activity is not necessary to generate iPS cells, the efficiency of iPS cells derived from telomerase-deficient mice shows a dramatic decrease. Therefore, using iPS cells for transplantation therapies may also lead to premature aging if iPS cells do not have sufficient telomere length.

Gene-expression profile: Different mouse iPS cell clones have internal differences as well as differences between mouse ESCs and mouse iPS cells. Certain gene expressions were commonly up-regulated in ESCs and iPS cells, including *Myb*, *Kit*, *Gdf3*, *Zic3, Dppa3, Dppa4, Dppa5, Nanog, Sox2, Esrrb and Rex1,* which indicated the comparability of the two stem cells. Some genes up-regulate more significantly in ESCs than in iPS cells, including *Dnmt3a*, *Dnmt3b*, *Dnmt3l*, *Utf1*, *Tcl1* and the *LIF receptor* gene. Subsequent analysis confirms that iPS cells are limited similar, but not absolutely identical, to ES cells [1].

Epigenetic profile: There is more and more DNA methylation accumulated during the process of reprogramming gradually, and only cells of the ESC-like state achieved loss. During reprogramming process, pluripotent stem cells of ESC-like state can be distinguished *via* the extent of demethylation compared with the other failure, because activated factors prohibit demethylation. It has also been observed that genes with CpG-rich promoters demonstrate stable low methylation and strong engagement of histone markers, whereas other genes with CpG-poor promoters are stable *via* methylation. Some epigenetic changes are equipped with function specifically to erase somatic status in the epigenome so as to transform into an ESC-like status. The changes mentioned above include chromatin reorganization, DNA demethylation of promoter regions of defined genes, resetting of histone modifications and reactivation of X chromosome silenced in somatic cells.

Once transcription factors are transfected into cells, changes in histone modifications occur immediately, suggesting that changes in histone markers occur early signifying the initiation of the reprogramming process. Subsequently, a peak of H3K4me2 accumulates in the promoters of many pluripotency genes, which are enriched in binding sites of Oct4 and Sox2. These pluripotent genes consist of Sall4, Fgf4 and so on. This stage is also associated with promoter hypomethylation in certain regions that are essential for conversion. In parallel, a large number of promoters and enhancers of somatic genes begin to lose H3K4me2, consistent with early down-regulation of MEF markers such as Thy1 and Postn. Above all, the epigenetic modifications of MEF-specific factors and changes of defined pluripotency genes result in changes in their expression, which marked the first but important step in the conversion of somatic cells transformed into pluripotent stem cells [26].

RNA expression: The retained miRNA pools of reprogramming cells are similar to total RNA content in size, suggesting the importance of precise regulation of the gross amount of miRNA. iPS cells with the ESC-like state were detected the increasing expression of miR-290/95, 302/367 and 106a/363 during reprogramming. Specifically, the miR-302/367 is known as the capacity to stimulate iPS cell proliferation, survival, MET and DNA demethylation. miR-290/295 is known as the function of stimulating cell proliferation and survival, demonstrating that the activities of certain miRNAs are the signature of the reprogramming progress.

3.2. Pluripotency

The derived iPS cell lines are isolated and sent to test the molecular features of whether they are fully reprogrammed cells. They are routinely assessed in functional assays such as teratoma formation, tetraploid complementation and differentiated cells. Although a series of strict assay is available to be used for testifying human iPS cells, it is still not sufficiently powerful and efficacious to lead to the conclusion that iPS cells can differentiate into all kinds of the cell types of human. Thus, it also does not have the capacity to assess the mechanism of iPS cells developing into the germ line. Existing research cautions that the forthcoming standards for iPS cells still suffer intense debate, especially when they are potentially used for treatment. A set of unified and accepted standards that can be recognized worldwide is critical for the research and applications moving forward.

Teratoma formation: The highest-rigorous test for mouse iPS cells is tetraploid complementation, which injecting iPS cells into tetraploid blastocysts to observe whether they are capable of developing directly into an entire organism. embryoid bodies formed from iPS cells have been discovered at 7 days in suspension cultures. Neural cells appeared after 8 days differentiation in adherent culture. Immunofluorescent staining revealed that human iPS cells have the ability to differentiate into all three germs layers-mesoderm, endoderm and ectoderm. Furthermore, after iPS cells injected into NOD/SCID mice 5-6 weeks, large teratomas formed. Histology tests showed a wide range of tissue types, including neural tissue, gut-like epithelium, cartilage and muscle.

Various types of differentiated cells: *In vitro*, it has been realized to guide ES cell differentiation into various cell types. Theoretically, iPS cells can differentiate into all types of cells. Until now, iPS cell-derived differentiated cells including neurocyte, Hepatocyte, Hematopoietic cell, skeletal muscle cell, endothelial cell *et al.* Detail view will be discussed in section 3.4.

Tetraploid complementation: Chimeric mice of tetraploid complementation is part of a precise trial that can prove the practicality of iPS cells for the characterization of cellular reprogramming and developmental potency as useful tools. This study confirms that iPS cells can obtain true pluripotency that is similar to that of ES cells. Based on a previous study [24], it has attested the possibility that iPS cells have the ability to generate chimeras, which are competent for germ-line transmission. One or two cell lines from each of the experimental runs were chosen, and injected into normal blastocysts, which were transferred to pseudopregnant recipient females. According to forming live iPS 4N-comp pups, researchers found that the rate of iPS 4N-comp animal production were similar to that of normal ES cells in our laboratory. The outcomes showed definitive evidence that iPS cells are truly pluripotent. Next, the embryonic development of offspring from iPS 4N-comp mice was studied. The male mouse was mated with the iPS mouse. Without developmental delay, the iPS mouse was able to fertilize, and more important thing was they produced viable preimplantation-stage embryos with normal morphology. The offspring further demonstrated the viability and practicability of the iPS 4N-comp mice, in other words, reflecting the true pluripotency of the iPS lines, which are critical for important applications in therapeutic interventions [28].

3.3. The Restrictions and Safety Issues for Regenerative Medicine

The procedure of creating iPS cells is initially based on the vectors of lentiviruses or retroviruses to transfer defined genes. However, integrating into host chromosomes due to transfection through virus can cause insertional mutagenesis and potentially malignant transformations, which is not perfect for clinical studies. At present, iPS cells' propensity to form tumors is the major concern about the potential clinical application of iPS cells. Consequently, use a different vector for input to avoid genomic insertion is an economically alternative way. Although Plasmids, adenoviruses and transposon vectors have lower generating efficiency, they have all been successfully applied [5, 6].

Similar to ESCs, iPS cells readily form teratomas when injected into immune deficient mice. Teratoma formation is considered by FDA as a major obstacle for application of stem cell in regenerative medicine. Recent study showed that after human-induced pluripotent stem cells were transplanted into the mice which were on motor functional recovery after spinal cord injuries, the cells differentiated into three neural lineages in the spinal cord. The cells had the ability to stimulate regrowth of the injured spinal cord, maintain myelination and form synapses. These positive outcomes were observed for over 112 days after the injury, without tumor formation.

Since, at this time, iPS cells can be created with high efficiency only with modifications, which predicted to be less safe and more tumorigenic than hESCs. All the genes that have been shown to promote iPS cell formation have also been related to cancer in one way or another. Some of the genes are known as oncogenes, including the members of the Myc family. While omitting Myc allows for iPS cell fabrication, the efficiency is weaken up to 100-fold. Thus, new technique to generate iPS cells has been rapidly and continuously evolving. Utilizing Plasmids to synthesize RNAs and proteins have been used to induce the pluripotent of somatic cell types. All of these methods are likely to be more acceptable for clinical studies, relative to viral transduction.

The strategy also appears to provide a way about the ethical complexities of using stem cells derived from human embryos. However, concerns about possible side effects remained. Specifically bad news came from a study showing that when iPS cells were injected into the mice from which they had been derived, they provoked immune systems casting doubt on one of the key advantages of the cells. Zhao *et al.* [29] conducted a study involving a teratoma formation assay demonstrated that iPS cells produced an immune response which was strong enough to cause rejection of the cells. They found teratomas when a similar process was performed on genetically equivalent ES cells, which demonstrated the cells were tolerated by the immune system. In 2013, Araki *et al.* [30] tried to reproduce the outcomes obtained by Zhao *et al.* by utilizing another trial. Cells were taken from a chimera which had been grown from iPS cell clones and a mouse embryo, and then transplanted into syngeneic mice. They performed a similar procedure using ES cells instead of iPS cell clones. Outcomes indicated that there were no significant difference in the immunogenic responses produced by the iPS cells and the ES cells. Furthermore, Araki *et al.* reported little or no immunogenic response for both cell lines. Thus, Araki *et al.* were unable to reach the same conclusions as Zhao *et al.*

Based on what we have learned, iPS cells are as safe as ES cells. But the continued threaten of such cells have not been controlled and inducing genetic damage and malignant cell growth is ever-present, and the potential and fate of these cells *in vivo* are under intense investigation. Although many issues stand between the lab and the clinic, iPS cells are just as promising as ES cells for cell transplantation. The differences between these two kinds of stem cells are peripheral compared with those in how individual cell lines grow and differentiate in culture.

These findings support another positive study on iPS cells, which was published last year. The study found that the reprogramming process induced fewer mutations than previous thought. Flora Vaccarino and her colleagues [31] utilized high-resolution DNA analysis to compare the genomes of iPS cells and the adult cells from which they were derived. Most of the DNA mutations in the iPS cells were founded not arise during reprogramming but had been present in the parent cells.

4. IPS CELLS: METHODS OF INDUCTION TOWARD DIFFERENT CELL TYPES, FACTORS INFLUENCING THE POTENTIAL OF DIFFERENTIATION, AND MECHANISMS DETERMINING THE FATE OF IPS CELL SAFTER INDUCTION

iPS cells can be differentiated into ectodermal, mesodermal and endodermal origin by mainly three methods, as well as with some developmental techniques according to different cells. Since endothelial and hematopoietic cells differentiated from iPS cells have undergone more detailed studies on the differentiating procedure, the following paragraphs provide more introductions about the factors influencing the ability of differentiation and mechanisms determining the fate of iPS cell induction.

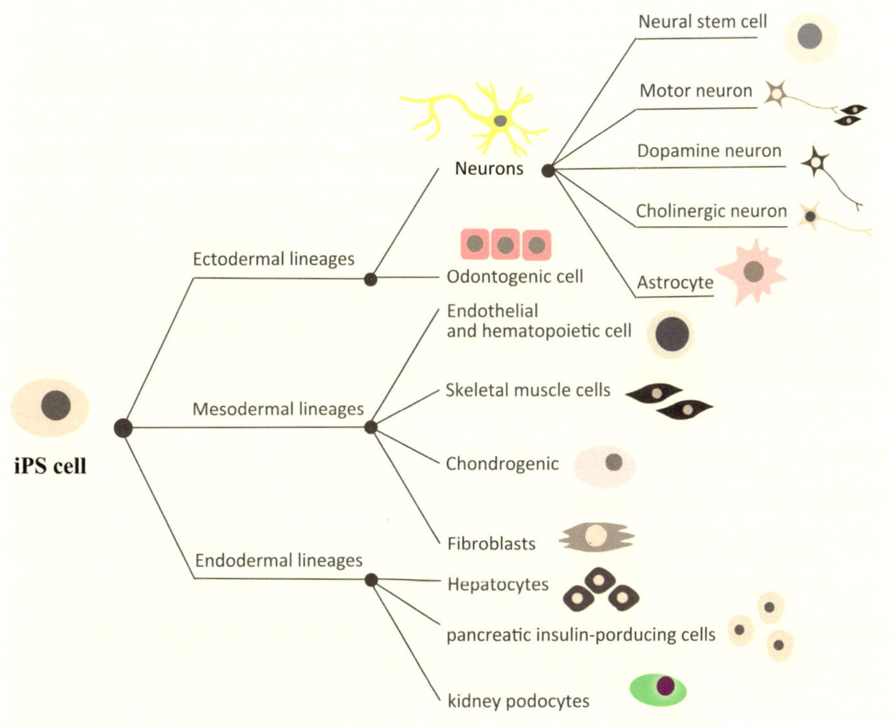

Fig. 2. Cell types of iPS cell differentiation.

Following the emergence of iPS cells, various cell types can be derived (Fig. **2**). The methods for ES cell differentiation are being tested on iPS cells, but whether these protocols work the same way for iPS cells requires more researches.

Generally, there are three primary methods to induce iPS cell differentiation according to endothelial cells induced from iPS cells [32]. First, iPS cells can be differentiated into embryoid bodies (EB) in a suspension culture under promoting self-aggregation conditions. Large cell lineages will emerge within EBs, which spontaneously undergo differentiation into ectoderm, endoderm and mesoderm germ layers, and promoted by numerous growth factors. Second, some types of stromal cells are co-cultured with differentiating iPS cells to promote ectoderm differentiating into cell lineages. Finally, tissue plate cultures submerged with protein substrates (fibronectin, gelatin, matrigel or other proteins) and together with specific culture media recombinant growth factors are used to differentiate iPS cells. Until now, several types of cells have been achieved from iPS cell differentiation, including ectodermal, mesodermal and endodermal lineages.

Neurons: Parkinson's disease happens common chronic progressive neurodegenerative disorders, which is characterized mainly by losing nigrostriatal dopaminergic neurons. Wernig *et al.* compared multiple hiPS cell lines and demonstrated that iPS cells can differentiate into neural precursor cells efficiently from a mouse model. After being transplanted into the fetal mouse brain, the cells migrated into various brain regions and differentiated into neurons and glia [33]. Soldner *et al.* demonstrated that iPS cells, being reprogrammed from the DFs of five Parkinson's disease patients, can be guided to differentiate into dopaminergic neurons by EB formation in hiPS cell lines [34]. Further, Rhee *et al.* compared multiple hiPS cell lines and demonstrated that the protein-based iPS cells were highly expandable without senescence [35]. A Cre-recombinase excisable viral vector system was used to remove the viral vectors carrying the reprogramming factors in those iPS cells. Cai *et al.* showed that hiPS cells could differentiate into dopaminergic neurons and survive in the 6-OHDA rat brain expressing midbrain dopaminergic neuron precursor cell markers through a rat model including Parkinson's disease included, though no significant recovery from induced Parkinson's disease symptoms was observed [36]. A study by Hu *et al.* concluded that hiPS cell lines carrying the transgenes or without the transgenes (reprogrammed by Yamanaka's or Thomson's four factors), derived from dermal or lung fibroblasts from donors of various ages, were converted to functional neurons, neuroepithelial cells, glia, or neuronal subtypes (motor neurons), and the results showed that all iPS cell lines tested followed the same developmental principles as hES cells but with variable potency. The differentiation of iPS cells appeared to follow normal development of motor neuron formation [37].

Odontogenic cells: Though until now there has been no definitive method for inducing iPS cell differentiation into odontogenic cells, Seki *et al.* described a novel method using gene transfection. First, they generated iPS-cell-derived neural-crest-like cells (iNCLCs) and then differentiated them into odontoblast-like cells by co-transfecting Pax9 and Bmp4 expression plasmids. An interaction between exogenous Pax9- and Bmp4-induced signaling modulated Dmp1 and Dspp expression. Their findings may suggest that iNCLCs differentiate into odontoblast-like cells [38].

Endothelial and hematopoietic cells: Protocols used to guide hES cell differentiation into hematopoietic and endothelial cell lineages were tested on iPS cells. An embryoid-body-mediated approach used to guide the differentiation of hES cells into a hemangioblastic lineage—a bipotential progenitor of endothelial and hematopoietic cell lineages—was tested on multiple lines of hiPS cells derived by Thomson's and Yamanaka's four factors. Although hiPS cells could be able to generate hemangioblasts/blast cells, hematopoietic cells, and endothelial cells, the efficiency was dramatically lower. Meanwhile, some conditions, such as severely limited growth, significantly increased apoptosis and expansion capability, as well as early senescence of endothelial cells and a substantially download the ability of hematopoietic colony-forming [39].

Initially, Flk1 expressed in Brachyury (BRY)—which was the key step to the induction of murine endothelial cells, then activating BMP4 signaling, subsequently, additional growth factors such as fibroblast growth factor 2 (FGF2), insulin-like growth factor-1 (IGF-1), placental growth factor (PIGF) and epidermal growth factor (EGF)—will promote endothelial cell to proliferate, differentiate, and eventual be maturate in *in vitro* culture. Some other vital transcription factors promoting endothelial cell emergence have been identified recently. *Er71/Etv2*, belonged to the E-26 (ETS) transcription factor family, directly binds to genes which are vital for endothelial cell lineage and blood developing, such as *Flk1*, limb domain only 2 *(Lmo2)* and stem cell leukemia *(Scl)* [40, 41]. *Tcea3* is highly expressed in mouse ESCs and modulates Activin A/Nodal signaling negatively, whose transcripts diminished may acquire more endothelial cells with enhanced tube formation in *in vitro* cell culture [42]. Ovo-like zinc finger 2 (*OVOL2*) as a novel protein interacting with *Er71/Etv2 was reported* [43]. Both co-localized in hematopoietic cell nucleus, which can increase endothelial and hematopoietic cell production in differentiating murine ES cells. *OVOL2* can significantly enhance the *ER71*-mediated activation to *Flk1* promoter, and when *OVOL2* transcripts were diminished, ES cell differentiation into endothelial and blood cells was lost to various degrees. *GATA2*, another co-expression with *ER71* binding the *Spi1* gene, has an effect similar to that of *OVOL2* [44]. Transcription factors, as *Lmo2, Gata2, Pitx2, Mycn, Tal1*and*Sox17,* screened by means of a doxycycline-inducible cassette in fetal liver cells, murine fibroblast cells and/or ES cells, were shown to restrict the transfected cells in a proliferative state with hemangioblastic potential. All results demonstrated that it may be some possibility to identify and control the condition of many progenitor cell lineages [45].

The iPS cells differentiated into endothelial cell lineage were also regulated by differential micro-RNAs (miR). MiR-21, which was overexpressed during murine ES cell differentiation into endothelial cells, may activate the Akt signaling pathway [46]. MiR-199 may be increased when murine ES cells differentiate into endothelial cells through targeting and suppressing the Notch ligand Jassed 1 (JAG1) expressed, but may inhibit differentiation of vascular smooth-muscle cells [47].

Skeletal muscle cells: miPS cells transplanted into Duchenne muscular dystrophy mice model, could lead to inhibited myogenic lineage differentiation in injured muscles, underlying long-lived muscle stem cell support, proved by Mizuno *et al.*, who demonstrated that miPS cells have the potential to develop into skeletal muscle stem/progenitor cells *in vitro* [48]. These findings suggest that iPS cells could be used in muscular dystrophies treatment compared with adult ES cells.

Chondrocytes: Current approaches for chondrogenic differentiation from iPS cells are inefficient and complicated. Nejadnik *et al.* developed a new and efficient approach that avoids embryoid formation. They directly differentiated hiPS cells into mesenchymal stem/stromal cells (MSCs) and chondrocytes. hiPS cell-MSC-derived chondrocytes showed GAG,Col2A1and SOX9 gene expression to be significantly increased compared with hiPS cell-MSCs. When hiPS cells and hiPS-MSC-derived chondrocytes were transplanted into osteochondral defects of rat arthritic joints, hyaline cartilage matrix production were recorded by magnetic resonance imaging studies, which may improve cartilage regeneration outcomes in arthritic joints [49].

Hepatocytes: Recent evidence has revealed that iPS cells could become sufficient functional endodermal-derived hepatocytes, demonstrating the real potential of iPS cells for the field of hepatology. The iPS-cell-derived hepatic endoderm or hepatocytes exhibit hepatic morphology, express essential hepatic markers and produce/secrete proteins (*e.g.*, α-fetoprotein), characteristics of functional hepatocytes [50]. Si-Tayeb *et al.* generated hepatocyte-like cells derived from hiPS cells expressing high efficiency. At initiation, monolayers of human embryonic stem cells (huES) were cultured in Roswell Park Memorial Institute (RPMI) media containing 100 ng/mL activin A and B27 supplements, and 5 days later, with the addition of BMP4/FGF2, the specification of hepatic cells was robust, with more than 80% of cells expressing HNF4a. The investigators added hepatocyte growth factor during the following 5 days, and during the final differentiation, the hepatocyte culture medium was used, which supplemented with Oncostatin M (20 ng/mL) at 5% CO_2/O_2 for another 5 days. Followed these, 80% of cells were shown to highly expression albumin, which could be identified by immunocytochemistry [51].

Pancreatic insulin-producing cells: Zhang *et al.* developed a highly efficient differentiation method to make iPS cells into mature pancreatic insulin-producing cells in a chemically defined culture system, and calculated that EGF facilitates the expansion of PDX1-positive pancreatic progenitors [52].

Kidney podocytes: Song and his colleagues first studied iPS cells differentiating directly to generate kidney cells with podocyte features. They used suspension culture containing BMP-7, activin A, retinoic acid, *etc*., and transferred it into a 0.1% gelatin-coated monolayer from days4 to 10, finally obtaining iPS cells with the podocyte phenotype [53].

Thus, it appears that further studies of large numbers of individual iPS cell lines or clones are needed to understand the extent of variation in differentiation potential that is related to the effects of cell origin or reprogramming methods, especially with regard to any clinical applications. Additionally, iPS-cell-derived differentiated cells must be able to fully mature into differentiated cells and have functions equivalent to those of the naturally formed tissue cells. Fine-tuning of the differentiation protocols is needed to achieve these desired goals.

5. APPLICATION OF INDUCED IPS CELLS IN TISSUE ENGINEERING: CELL SOURCE, UNDERSTANDING OF CELL ULARPLURIPOTENCY AND DIFFERENTIATION

As in tissue engineering, iPS cells may have significant applications in regenerative medicine, which can repair injured or degenerated tissue by the regeneration of these tissues from iPS cells then transplanting them into the injured or degenerated parts. Until now, only cells, tissues or organs could be translated into a patient from a healthy individual and a well-matched physiological profile of the patient. As we all know, numerous tests are conducted before transplanting, to minimize immune rejection. iPS cells offer a useful approach to the challenge of rejection, since the cells translated into the patient's body will be differentiated into the repaired iPS cells excluding immunogenicity[30]. Many injuries caused by accidents, activities or natural disasters can be treated by gene therapy with iPS cells [54 - 56], and in addition, occasional injuries and disease can also be treated with the help of iPS cells [57, 58].

Central nervous system regeneration: After iPS cells were successfully derived from NS/PCs, Tsuji *et al.* achieved the restoration of motor function by transplanting those cells into a spinal cord injury (SCI) mouse model in 2010 [59]. In the following year, human iPS-cell-derived stem cells were transplanted into a mouse SCI model to restore motor function successfully [60]. Moreover, human iPS-cell-derived NS/PCs could restore motor function when being transplanted into a marmoset SCI model, and long-term motor function was recovered without tumor formation observed [61]. Though these findings were from preclinical research, they provide a proof of concept that could potentially lead to a treatment method.

Liver regeneration: iPS cells can also be used to differentiate into hepatocytes for various liver problems. Liu and his colleagues applied human iPS cells derived from hepatic cells at differentiation stages, which can successfully repopulate the liver tissue of mice with liver cirrhosis, and produce human-specific liver protein, all of what demonstrated the engraftment and liver-regenerative capabilities of human iPS-cell-derived multistage hepatic cells *in vivo* [54].

Blood cell regeneration: Recently, iPS cells were utilized for *ex vivo* various blood components expansion. If possible, investigators could generate red blood cells that could be applied for blood generation, which is required worldwide for the purposes of treating various damage or diseases [62].

Retinitis pigmentosa epithelium (RPE) regeneration: Various degenerative diseases can also depend on gene therapy by iPS cells. Age-related Macular Degeneration (AMD) can cause impaired vision due to RPE degeneration. iPS cells derived from an RP patient were shown to differentiate into rod photoreceptor cell sheets [63]. Kamao *et al.* generated human iPS cells meeting clinical use requirements, and demonstrated that these cell sheets show phagocytic ability and gene-expression patterns similar to those of native RPE. Importantly, there was no tumor formation or immune rejection upon transplantation of autologous non-human primate iPS cell sheets [64].

Primary immunodeficiency cell regeneration: Until now, the specific cells generated from iPS cells are expected to be applied to treat various primary immunodeficiencies, such as Wiskott-Aldrich syndrome (WAS) [65] and Chronic Granulomatous disorder (CGD) [66]. WAS is caused by WASP gene mutation. With transplanting WASP-deficient (WKO) HSCs with WASP-expressing retroviruses, mature B and T cells recovered to normal cells [65]. CGD, an inherited disorder of phagocytes, affects the function of neutrophils. Jiang *et al.*

acquired iPS cells derived from three unrelated patients' fibroblasts, and these CGD-patient-specific iPS cell lines showed normal phagocytic properties without reactive oxygen species production correlative with clinical diagnosis of CGD [66]. Together, these results indicated that those patient-specific iPS cell lines represent an important tool for special disease treatment.

Altogether, analysis of these data showed wide applications for iPS cells in tissue engineering. However, it is still a long way to their utility in wide clinical applications, and further research is required.

CONFLICT OF INTEREST

The authors declare that they have no financial or other relationships with other people or organizations that might lead to a conflict of interest.

ACKNOWLEDGEMENTS

We thank Dr. Ling Wu for valuable comments and advices. This work is financially supported by the Natural Science Foundation of Shanghai (15ZR1432700) and the Natural Science Foundation of China (81500808).

REFERENCES

[1] Takahashi K, Yamanaka S. Induction of pluripotent stem cells from mouse embryonic and adult fibroblast cultures by defined factors. Cell 2006; 126(4): 663-76. [http://dx.doi.org/10.1016/j.cell.2006.07.024] [PMID: 16904174]

[2] Klimanskaya I, Chung Y, Becker S, Lu SJ, Lanza R. Human embryonic stem cell lines derived from single blastomeres. Nature 2006; 444(7118): 481-5. [http://dx.doi.org/10.1038/nature05142] [PMID: 16929302]

[3] Yu J, Vodyanik MA, Smuga-Otto K, *et al.* Induced pluripotent stem cell lines derived from human somatic cells. Science 2007; 318(5858): 1917-20. [http://dx.doi.org/10.1126/science.1151526] [PMID: 18029452]

[4] Nakagawa M, Koyanagi M, Tanabe K, *et al.* Generation of induced pluripotent stem cells without Myc from mouse and human fibroblasts. Nat Biotechnol 2008; 26(1): 101-6. [http://dx.doi.org/10.1038/nbt1374] [PMID: 18059259]

[5] Stadtfeld M, Nagaya M, Utikal J, Weir G, Hochedlinger K. Induced pluripotent stem cells generated without viral integration. Science 2008; 322(5903): 945-9. [http://dx.doi.org/10.1126/science.1162494] [PMID: 18818365]

[6] Okita K, Nakagawa M, Hyenjong H, Ichisaka T, Yamanaka S. Generation of mouse induced pluripotent stem cells without viral vectors. Science 2008; 322(5903): 949-53.

[http://dx.doi.org/10.1126/science.1164270] [PMID: 18845712]

[7] Nichols J, Zevnik B, Anastassiadis K, *et al.* Formation of pluripotent stem cells in the mammalian embryo depends on the POU transcription factor Oct4. Cell 1998; 95(3): 379-91. [http://dx.doi.org/10.1016/S0092-8674(00)81769-9] [PMID: 9814708]

[8] Avilion AA, Nicolis SK, Pevny LH, Perez L, Vivian N, Lovell-Badge R. Multipotent cell lineages in early mouse development depend on SOX2 function. Genes Dev 2003; 17(1): 126-40. [http://dx.doi.org/10.1101/gad.224503] [PMID: 12514105]

[9] Li Y, McClintick J, Zhong L, Edenberg HJ, Yoder MC, Chan RJ. Murine embryonic stem cell differentiation is promoted by SOCS-3 and inhibited by the zinc finger transcription factor Klf4. Blood 2005; 105(2): 635-7. [http://dx.doi.org/10.1182/blood-2004-07-2681] [PMID: 15358627]

[10] Wernig M, Meissner A, Cassady JP, Jaenisch R. c-Myc is dispensable for direct reprogramming of mouse fibroblasts. Cell Stem Cell 2008; 2(1): 10-2. [http://dx.doi.org/10.1016/j.stem.2007.12.001] [PMID: 18371415]

[11] Silva J, Chambers I, Pollard S, Smith A. Nanog promotes transfer of pluripotency after cell fusion. Nature 2006; 441(7096): 997-1001. [http://dx.doi.org/10.1038/nature04914] [PMID: 16791199]

[12] Darr H, Mayshar Y, Benvenisty N. Overexpression of NANOG in human ES cells enables feeder-free growth while inducing primitive ectoderm features. Development 2006; 133(6): 1193-201. [http://dx.doi.org/10.1242/dev.02286] [PMID: 16501172]

[13] Maekawa M, Yamaguchi K, Nakamura T, *et al.* Direct reprogramming of somatic cells is promoted by maternal transcription factor Glis1. Nature 2011; 474(7350): 225-9. [http://dx.doi.org/10.1038/nature10106] [PMID: 21654807]

[14] Zhao R, Daley GQ. From fibroblasts to iPS cells: induced pluripotency by defined factors. J Cell Biochem 2008; 105(4): 949-55. [http://dx.doi.org/10.1002/jcb.21871] [PMID: 18668528]

[15] Tsubooka N, Ichisaka T, Okita K, Takahashi K, Nakagawa M, Yamanaka S. Roles of Sall4 in the generation of pluripotent stem cells from blastocysts and fibroblasts. Genes Cells 2009; 14(6): 683-94. [http://dx.doi.org/10.1111/j.1365-2443.2009.01301.x] [PMID: 19476507]

[16] Lei Y, Schaffer DV. A fully defined and scalable 3D culture system for human pluripotent stem cell expansion and differentiation. Proc Natl Acad Sci USA 2013; 110(52): E5039-48. [http://dx.doi.org/10.1073/pnas.1309408110] [PMID: 24248365]

[17] Esteban MA, Wang T, Qin B, *et al.* Vitamin C enhances the generation of mouse and human induced pluripotent stem cells. Cell Stem Cell 2010; 6(1): 71-9. [http://dx.doi.org/10.1016/j.stem.2009.12.001] [PMID: 20036631]

[18] Du SH, Tay JC, Chen C, *et al.* Human iPS cell-derived fibroblast-like cells as feeder layers for iPS cell derivation and expansion. J Biosci Bioeng 2015; 120(2): 210-7. [http://dx.doi.org/10.1016/j.jbiosc.2014.12.009] [PMID: 25622768]

[19] Chang BS, Choi YJ, Kim JH. Collagen complexes increase the efficiency of iPS cells generated using fibroblasts from adult mice. J Reprod Dev 2015; 61(2): 145-53.

[http://dx.doi.org/10.1262/jrd.2014-081] [PMID: 25740096]

[20] Mikkelsen TS, Hanna J, Zhang X, *et al.* Dissecting direct reprogramming through integrative genomic analysis. Nature 2008; 454(7200): 49-55. [http://dx.doi.org/10.1038/nature07056] [PMID: 18509334]

[21] Huangfu D, Maehr R, Guo W, *et al.* Induction of pluripotent stem cells by defined factors is greatly improved by small-molecule compounds. Nat Biotechnol 2008; 26(7): 795-7. [http://dx.doi.org/10.1038/nbt1418] [PMID: 18568017]

[22] Maherali N, Sridharan R, Xie W, *et al.* Directly reprogrammed fibroblasts show global epigenetic remodeling and widespread tissue contribution. Cell Stem Cell 2007; 1(1): 55-70. [http://dx.doi.org/10.1016/j.stem.2007.05.014] [PMID: 18371336]

[23] Wakao S, Kitada M, Kuroda Y, *et al.* Morphologic and gene expression criteria for identifying human induced pluripotent stem cells. PLoS One 2012; 7(12): e48677. [http://dx.doi.org/10.1371/journal.pone.0048677] [PMID: 23272044]

[24] Wernig M, Meissner A, Foreman R, *et al.* *In vitro* reprogramming of fibroblasts into a pluripotent ES-cell-like state. Nature 2007; 448(7151): 318-24. [http://dx.doi.org/10.1038/nature05944] [PMID: 17554336]

[25] Marión RM, Blasco MA. Telomere rejuvenation during nuclear reprogramming. Curr Opin Genet Dev 2010; 20(2): 190-6. [http://dx.doi.org/10.1016/j.gde.2010.01.005] [PMID: 20176474]

[26] Buganim Y, Faddah DA, Jaenisch R. Mechanisms and models of somatic cell reprogramming. Nat Rev Genet 2013; 14(6): 427-39. [http://dx.doi.org/10.1038/nrg3473] [PMID: 23681063]

[27] Zhou W, Freed CR. Adenoviral gene delivery can reprogram human fibroblasts to induced pluripotent stem cells. Stem Cells 2009; 27(11): 2667-74. [http://dx.doi.org/10.1002/stem.201] [PMID: 19697349]

[28] Zhao XY, Li W, Lv Z, *et al.* iPS cells produce viable mice through tetraploid complementation. Nature 2009; 461(7260): 86-90. [http://dx.doi.org/10.1038/nature08267] [PMID: 19672241]

[29] Zhao T, Zhang ZN, Rong Z, Xu Y. Immunogenicity of induced pluripotent stem cells. Nature 2011; 474(7350): 212-5. [http://dx.doi.org/10.1038/nature10135] [PMID: 21572395]

[30] Araki R, Uda M, Hoki Y, *et al.* Negligible immunogenicity of terminally differentiated cells derived from induced pluripotent or embryonic stem cells. Nature 2013; 494(7435): 100-4. [http://dx.doi.org/10.1038/nature11807] [PMID: 23302801]

[31] Stevens HE, Mariani J, Coppola G, Vaccarino FM. Neurobiology meets genomic science: the promise of human-induced pluripotent stem cells. Dev Psychopathol 2012; 24(4): 1443-51. [http://dx.doi.org/10.1017/S095457941200082X] [PMID: 23062309]

[32] Wilson HK, Canfield SG, Shusta EV, Palecek SP. Concise review: tissue-specific microvascular endothelial cells derived from human pluripotent stem cells. Stem Cells 2014; 32(12): 3037-45. [http://dx.doi.org/10.1002/stem.1797] [PMID: 25070152]

[33] Wernig M, Zhao JP, Pruszak J, *et al.* Neurons derived from reprogrammed fibroblasts functionally integrate into the fetal brain and improve symptoms of rats with Parkinson's disease. Proc Natl Acad Sci USA 2008; 105(15): 5856-61.
[http://dx.doi.org/10.1073/pnas.0801677105] [PMID: 18391196]

[34] Soldner F, Hockemeyer D, Beard C, *et al.* Parkinson's disease patient-derived induced pluripotent stem cells free of viral reprogramming factors. Cell 2009; 136(5): 964-77.
[http://dx.doi.org/10.1016/j.cell.2009.02.013] [PMID: 19269371]

[35] Rhee YH, Ko JY, Chang MY, *et al.* Protein-based human iPS cells efficiently generate functional dopamine neurons and can treat a rat model of Parkinson disease. J Clin Invest 2011; 121(6): 2326-35.
[http://dx.doi.org/10.1172/JCI45794] [PMID: 21576821]

[36] Cai J, Yang M, Poremsky E, Kidd S, Schneider JS, Iacovitti L. Dopaminergic neurons derived from human induced pluripotent stem cells survive and integrate into 6-OHDA-lesioned rats. Stem Cells Dev 2010; 19(7): 1017-23.
[http://dx.doi.org/10.1089/scd.2009.0319] [PMID: 19824823]

[37] Karumbayaram S, Novitch BG, Patterson M, *et al.* Directed differentiation of human-induced pluripotent stem cells generates active motor neurons. Stem Cells 2009; 27(4): 806-11.
[http://dx.doi.org/10.1002/stem.31] [PMID: 19350680]

[38] Seki D, Takeshita N, Oyanagi T, *et al.* Differentiation of Odontoblast-Like Cells From Mouse Induced Pluripotent Stem Cells by Pax9 and Bmp4 Transfection. Stem Cells Transl Med 2015; 4(9): 993-7.
[http://dx.doi.org/10.5966/sctm.2014-0292] [PMID: 26136503]

[39] Feng Q, Lu SJ, Klimanskaya I, *et al.* Hemangioblastic derivatives from human induced pluripotent stem cells exhibit limited expansion and early senescence. Stem Cells 2010; 28(4): 704-12.
[http://dx.doi.org/10.1002/stem.321] [PMID: 20155819]

[40] Lee D, Park C, Lee H, *et al.* ER71 acts downstream of BMP, Notch, and Wnt signaling in blood and vessel progenitor specification. Cell Stem Cell 2008; 2(5): 497-507.
[http://dx.doi.org/10.1016/j.stem.2008.03.008] [PMID: 18462699]

[41] Liu F, Kang I, Park C, *et al.* ER71 specifies Flk-1+ hemangiogenic mesoderm by inhibiting cardiac mesoderm and Wnt signaling. Blood 2012; 119(14): 3295-305.
[http://dx.doi.org/10.1182/blood-2012-01-403766] [PMID: 22343916]

[42] Cha Y, Heo SH, Ahn HJ, *et al.* Tcea3 regulates the vascular differentiation potential of mouse embryonic stem cells. Gene Expr 2013; 16(1): 25-30.
[http://dx.doi.org/10.3727/105221613X13776146743343] [PMID: 24397209]

[43] Kim JY, Lee RH, Kim TM, *et al.* OVOL2 is a critical regulator of ER71/ETV2 in generating FLK1+, hematopoietic, and endothelial cells from embryonic stem cells. Blood 2014; 124(19): 2948-52.
[http://dx.doi.org/10.1182/blood-2014-03-556332] [PMID: 25267199]

[44] Shi X, Richard J, Zirbes KM, *et al.* Cooperative interaction of Etv2 and Gata2 regulates the development of endothelial and hematopoietic lineages. Dev Biol 2014; 389(2): 208-18.
[http://dx.doi.org/10.1016/j.ydbio.2014.02.018] [PMID: 24583263]

[45] Vereide DT, Vickerman V, Swanson SA, Chu LF, McIntosh BE, Thomson JA. An expandable, inducible hemangioblast state regulated by fibroblast growth factor. Stem Cell Rep 2014; 3(6): 1043-

57.
[http://dx.doi.org/10.1016/j.stemcr.2014.10.003] [PMID: 25458896]

[46] Di Bernardini E, Campagnolo P, Margariti A, *et al.* Endothelial lineage differentiation from induced pluripotent stem cells is regulated by microRNA-21 and transforming growth factor β2 (TGF-β2) pathways. J Biol Chem 2014; 289(6): 3383-93.
[http://dx.doi.org/10.1074/jbc.M113.495531] [PMID: 24356956]

[47] Chen T, Margariti A, Kelaini S, *et al.* MicroRNA-199b Modulates Vascular Cell Fate During iPS Cell Differentiation by Targeting the Notch Ligand Jagged1 and Enhancing VEGF Signaling. Stem Cells 2015; 33(5): 1405-18.
[http://dx.doi.org/10.1002/stem.1930] [PMID: 25535084]

[48] Mizuno Y, Chang H, Umeda K, *et al.* Generation of skeletal muscle stem/progenitor cells from murine induced pluripotent stem cells. FASEB J 2010; 24(7): 2245-53.
[http://dx.doi.org/10.1096/fj.09-137174] [PMID: 20181939]

[49] Nejadnik H, Diecke S, Lenkov OD, *et al.* Improved approach for chondrogenic differentiation of human induced pluripotent stem cells. Stem Cell Rev 2015; 11(2): 242-53.
[http://dx.doi.org/10.1007/s12015-014-9581-5] [PMID: 25578634]

[50] Sullivan GJ, Hay DC, Park IH, *et al.* Generation of functional human hepatic endoderm from human induced pluripotent stem cells. Hepatology 2010; 51(1): 329-35.
[http://dx.doi.org/10.1002/hep.23335] [PMID: 19877180]

[51] Si-Tayeb K, Noto FK, Nagaoka M, *et al.* Highly efficient generation of human hepatocyte-like cells from induced pluripotent stem cells. Hepatology 2010; 51(1): 297-305.
[http://dx.doi.org/10.1002/hep.23354] [PMID: 19998274]

[52] Zhang D, Jiang W, Liu M, *et al.* Highly efficient differentiation of human ES cells and iPS cells into mature pancreatic insulin-producing cells. Cell Res 2009; 19(4): 429-38.
[http://dx.doi.org/10.1038/cr.2009.28] [PMID: 19255591]

[53] Song B, Smink AM, Jones CV, *et al.* The directed differentiation of human iPS cells into kidney podocytes. PLoS One 2012; 7(9): e46453.
[http://dx.doi.org/10.1371/journal.pone.0046453] [PMID: 23029522]

[54] Liu H, Kim Y, Sharkis S, Marchionni L, Jang YY. *In vivo* liver regeneration potential of human induced pluripotent stem cells from diverse origins. Sci Transl Med 2011; 3(82): 82ra39.
[http://dx.doi.org/10.1126/scitranslmed.3002376] [PMID: 21562231]

[55] Tan Q, Lui PP, Rui YF, Wong YM. Comparison of potentials of stem cells isolated from tendon and bone marrow for musculoskeletal tissue engineering. Tissue Eng Part A 2012; 18(7-8): 840-51.
[http://dx.doi.org/10.1089/ten.tea.2011.0362] [PMID: 22011320]

[56] Suzuki N, Yamazaki S, Yamaguchi T, *et al.* Generation of engraftable hematopoietic stem cells from induced pluripotent stem cells by way of teratoma formation. Mol Ther 2013; 21(7): 1424-31.
[http://dx.doi.org/10.1038/mt.2013.71] [PMID: 23670574]

[57] Kazuki Y, Hiratsuka M, Takiguchi M, *et al.* Complete genetic correction of ips cells from Duchenne muscular dystrophy. Mol Ther 2010; 18(2): 386-93.
[http://dx.doi.org/10.1038/mt.2009.274] [PMID: 19997091]

[58] Doherty AM, Fisher EM. Microcell-mediated chromosome transfer (MMCT): small cells with huge potential. Mamm Genome 2003; 14(9): 583-92.
[http://dx.doi.org/10.1007/s00335-003-4002-0] [PMID: 14629108]

[59] Tsuji O, Miura K, Okada Y, *et al.* Therapeutic potential of appropriately evaluated safe-induced pluripotent stem cells for spinal cord injury. Proc Natl Acad Sci USA 2010; 107(28): 12704-9.
[http://dx.doi.org/10.1073/pnas.0910106107] [PMID: 20615974]

[60] Nori S, Okada Y, Yasuda A, *et al.* Grafted human-induced pluripotent stem-cell-derived neurospheres promote motor functional recovery after spinal cord injury in mice. Proc Natl Acad Sci USA 2011; 108(40): 16825-30.
[http://dx.doi.org/10.1073/pnas.1108077108] [PMID: 21949375]

[61] Kobayashi Y, Okada Y, Itakura G, *et al.* Pre-evaluated safe human iPSC-derived neural stem cells promote functional recovery after spinal cord injury in common marmoset without tumorigenicity. PLoS One 2012; 7(12): e52787.
[http://dx.doi.org/10.1371/journal.pone.0052787] [PMID: 23300777]

[62] Singh VK, Kalsan M, Kumar N, Saini A, Chandra R. Induced pluripotent stem cells: applications in regenerative medicine, disease modeling, and drug discovery. Front Cell Dev Biol 2015; 3: 2.
[http://dx.doi.org/10.3389/fcell.2015.00002] [PMID: 25699255]

[63] Yoshida T, Ozawa Y, Suzuki K, *et al.* The use of induced pluripotent stem cells to reveal pathogenic gene mutations and explore treatments for retinitis pigmentosa. Mol Brain 2014; 7(1): 45.
[http://dx.doi.org/10.1186/1756-6606-7-45] [PMID: 24935155]

[64] Kamao H, Mandai M, Okamoto S, *et al.* Characterization of human induced pluripotent stem cell-derived retinal pigment epithelium cell sheets aiming for clinical application. Stem Cell Rep 2014; 2(2): 205-18.
[http://dx.doi.org/10.1016/j.stemcr.2013.12.007] [PMID: 24527394]

[65] Klein C, Nguyen D, Liu CH, *et al.* Gene therapy for Wiskott-Aldrich syndrome: rescue of T-cell signaling and amelioration of colitis upon transplantation of retrovirally transduced hematopoietic stem cells in mice. Blood 2003; 101(6): 2159-66.
[http://dx.doi.org/10.1182/blood-2002-05-1423] [PMID: 12433691]

[66] Jiang Y, Cowley SA, Siler U, *et al.* Derivation and functional analysis of patient-specific induced pluripotent stem cells as an *in vitro* model of chronic granulomatous disease. Stem Cells 2012; 30(4): 599-611.
[http://dx.doi.org/10.1002/stem.1053] [PMID: 22311747]

CHAPTER 4

Craniofacial Defects and their Regeneration: Destruction and Regeneration of the Periodontium, Craniofacial Tumors, Trauma, and Congenital Defects

Jingang Xiao[1,2,*], **Lin Liu**[1], **Lei Wang**[2], **Kui Huang**[1], **Yong Li**[1] and **Zhihao Yao**[1]

[1] *Department of Oral and Maxillofacial Surgery, Hospital of Stomatology, Southwest Medical University, Luzhou, Sichuan Province, China*

[2] *Orofacial Reconstruction and Regeneration Laboratory, Hospital of Stomatology, Southwest Medical University, Luzhou, Sichuan Province, China*

Abstract: Injuries in the cranial and maxillofacial region are clinically important because this region contains the most exposed parts of the body and is important for expression of social etiquette. Currently, the most common methods used to repair cranial and maxillofacial defects are autologous tissue transplantation and artificial composite material reparation. However, these methods have limitations. In recent years, tissue engineering has been developed as an alternative method of craniofacial reparation and reconstruction. Stem cell therapy is an emerging strategy that may potentially be used for the reconstruction of craniofacial defects.

Keywords: Adipose tissue regeneration, Bone defects, Bone regeneration, Cartilage regeneration, Congenital defects, Craniofacial defects, Cartilage defects, Periodontal tissue regeneration, Periodontium defects, Regenerative medicine, Reparation and reconstruction, Scaffolds, Soft tissue defects, Soft tissue regeneration, Stem cells, Tissue engineering, Tooth loss, Tooth regeneration, Trauma, Tumors.

* **Corresponding author Jingang Xiao:** Department of Oral and Maxillofacial Surgery, Hospital of Stomatology, Southwest Medical University, No.2, Jiang Yang Nan Road, Luzhou 646000, P.R.China; Tel: 86-0830-3165521; Fax: 86-0830-3165520; E-mail: drxiaojingang@163.com.

4.1. INTRODUCTION

The cranial and maxillofacial region contains the most significantly exposed parts of the human body and important organs for expression of social etiquette. Cranial and maxillofacial defects commonly occur secondary to congenital malformation, infection, cancer, and trauma, but the clinical management of these defects remains challenging [1]. Other than a small number of congenital malformations, most such craniofacial defects are caused by acquired factors and are thus often called acquired defects. The most frequent cause of these defects in the first half of the 20th century was infectious diseases, especially sequelae of gangrenous stomatitis. However, oral and maxillofacial defects caused by traffic accidents and tumor resection have steadily increased with developments of the social economy and surgical treatment of craniofacial tumors [2 - 5].

Rapid development of craniofacial reconstruction and reconstruction surgery has been promoted with increasing emphasis on quality of life and formal establishment of the unity of form and function recovery. Surgical repair and reconstruction involves the correction of deformities and reconstruction of defects of various tissues and organs of the whole body, and its basic technology can be used in various clinical disciplines. Cranial and maxillofacial reparative and reconstructive surgery involves correction of deformities and reconstruction of defects of the oral and maxillofacial region. It not only involves oral and maxillofacial surgery but is also related to the clinical disciplines of craniofacial surgery, head and neck surgery, plastic surgery, otolaryngology, and other fields.

A patient's quality of life will substantially decline by severely disfiguring facial injuries. The craniofacial region has various specific functions, such as protection of the brain and optic tracts, mastication, speech, breathing, and hearing. The craniofacial region is also important for self-esteem and social recognition [6, 7]. Therefore, cranial and maxillofacial defects cause not only great physical damage but also psychological damage.

The three basic principles of reparation and reconstruction in any body region are repair of the defect, improvement in shape, and restoration of function. This is also true for reconstruction of oral and maxillofacial defects. To achieve success, a surgeon must be mindful of the functional and aesthetic requirements in planning for reconstruction [8]. In clinical practice, the repair method differs among different degrees of craniofacial defects. The development of surgical materials has led to the establishment of autologous tissue transplantation and artificial composite material reparation as routine methods for reparation of craniofacial defects. Autograft surgery for craniofacial defects has been widely used; most such procedures involve the use of a free flap and autologous bone graft. Hydroxyapatite (HA) and implants comprising artificial composite materials are also widely used in the clinical setting.

The cranial and maxillofacial region comprises an intricate bony/cartilaginous structural framework, secretory organs, muscle, and a sophisticated skin/subcutaneous structure, all of which are attached through a vast neurovascular network and numerous ligaments. Because of the limited source of autologous tissue, even the most sophisticated autologous reconstructive technology is insufficient to restore a wide range of maxillofacial defects. Therefore, massive craniomaxillofacial defects caused by trauma, tumor resection, or congenital malformations are still a unique challenge for reconstructive surgeons [9].

In the past 10 years, experimental studies involving facial vascularized allografts have increased the possibility of clinical use of these allografts in patients with disease, trauma, and congenital deformities. This technique may now be considered a viable option for restoration of complex craniofacial defects, the treatment outcomes of which are still not ideal. Vascularized allotransplantation allows for optimal anatomical reconstruction and provides desired functional, aesthetic, and psychological results superior to those obtained by traditional methods. Patients' functional status often recovers, allowing them to make facial expressions such as smiling and to perform various functions such as smelling, eating, drinking, and speaking [6]. With recent progress in oral maxillofacial surgery and immunotherapy, composite tissue allotransplantation now allows for the performance of tissue reconstruction with composites made of all necessary components. However, craniomaxillofacial allotransplantation has some shortcomings, such as the need for lifelong immunosuppression and mismatch between the patient's and donor's skeletal framework and defects, providing suboptimal results [9].

Tissue engineering as an alternative to autologous transplantation provides new hope for restoration and reconstruction of missing tissue and avoids immunosuppression. It has been used to successfully create many tissue subtypes, including bone and cartilage. The aim of regenerative medicine is to restore and replace tissue that has been damaged or lost during tissue engineering and biomimetic strategies [9, 10]. Tissue engineering is an emerging strategy that can potentially be used to reconstruct craniofacial defects [11]. Langer and Vacanti [12] define tissue engineering as "an interdisciplinary field that applies the principles of engineering and the life sciences toward the development of biological substitutes that restore, maintain or improve tissue function."

Tissue engineering has been widely used in craniofacial surgery and regenerative medicine during the past 10 years [13]. Tissue engineering is an interdisciplinary technique that creates a biological substitute through application of the principles of engineering to biology and medicine to restore, maintain, and improve normal function. The emerging discipline of regenerative medicine and tissue engineering is committed to a rational approach to maintain and preserve the micro-environment, which is based on growth factors for tissue induction, stem cells, and scaffolds [14]. Regenerative medicine has been suggested as a promising strategy with which to regenerate tissue using the triad principle of applying combinations of three building blocks: functionally active cells to (re)generate biologically functional tissues, supporting scaffolds, and growth factors [15].

Stem cell-based tissue engineering approaches have the capacity to regenerate damaged and diseased tissues. Craniofacial defects often arise from trauma, tumor resection, and congenital deformities that require restoration and reconstruction [2 - 5]. Studies have shown that stem cells delivered *via* scaffolds can regenerate tissue with excellent results [16, 17]. Human bone marrow-derived mesenchymal stem cells (BMSCs) are multipotent and have the capacity to differentiate into osteoblasts, chondrocytes, neurons, myoblasts, adipocytes, and fibroblasts [18]. Human BMSCs can be isolated and cultured from bone marrow, induced to differentiate *in vitro*, and combined with scaffolds for tissue defect restoration.

Growth factors are naturally occurring proteins induced by cell surface receptor-mediated chemokines, mitosis, and angiogenesis. They participate in the growth, development, repair, reconstruction, and regeneration of tissues and organs [19, 20].

The scaffold is a key factor in tissue regeneration. It provides a desired three-dimensional (3D) space that allows stem cells to adhere, proliferate, migrate, and differentiate. After transplantation *in vivo*, a cell-seeded scaffold can further organize; the scaffold gradually degrades and vessels gradually form [21]. The importance of biomimetic design has recently been recognized simply because the best way to achieve tissue regeneration should involve following the natural laws of this process [22].

This chapter discusses the classification and treatment of craniofacial defects. The focus of this chapter is to discuss the application of tissue engineering in regeneration of craniofacial defects and analyze and summarize *in vivo* and *in vitro* studies of tissue engineering for regeneration of craniofacial defects to gain new insight into craniofacial regeneration medicine.

4.2. CRANIOFACIAL DEFECTS: PERIODONTIUM, CRANIOFACIAL TUMORS, TRAUMA, AND CONGENITAL DEFECTS

The craniofacial area comprises the brow facial point below the horizontal line, from the face of a third and a third in two parts, including the mandible, maxilla, lips, cheeks, tongue, and other important anatomic structures [23 - 28]. Factors such as tumor resection [27 - 34], trauma [26, 34 - 40], and congenital deformities [23 - 25, 41 - 43] often lead to the lack of important anatomical structures and impaired physiological functions such as breathing, speaking, chewing, and swallowing. Such factors may also result in facial deformity, seriously affecting the patient's physical and mental health and quality of life [24, 26 - 28, 30 - 34, 36 - 43]. However, because of the particular location and complex anatomy of structures in the craniofacial region, different craniofacial defects require different methods of reparation and reconstruction. Thus, a classification of methods according to specific craniofacial defects would be indispensable for clinical and research work. Although many different methods of classification of craniofacial defects have been introduced in the past few decades, a simple, clear, understandable, and memorable classification with guiding significance for clinical work is lacking. This article addresses classification of repair methods involving the four aspects of craniofacial defects: the periodontium, craniofacial tumors, trauma, and congenital defects.

4.2.1. Periodontium Defects

The periodontium includes the gingiva, periodontal ligament, alveolar bone, and cementum [44] (Because this article addresses craniofacial defects, dental pulp is not discussed; part of the cementum can be found in the dental pulp). The periodontium constitutes a functional periodontal support system that is firmly attached to the alveolar bone, and makes the oral mucosa and the teeth in a good state at the same time when biting. When factors such as inflammation, disease, and trauma cause destruction and loss of the periodontal tissue, the teeth loosen and shift. These changes may cause temporomandibular disorders and dentofacial anomalies, seriously affecting patients' normal chewing ability and quality of life [44]. Information on periodontium defects and their etiology is lacking. Periodontium defects are herein classified with reference to several types of common clinical periodontal diseases.

Class 1: Gingival diseases that mainly affect the gingiva, including chronic gingivitis and gingival hyperplasia. Reparation and reconstruction of the periodontium can be achieved through clinical periodontal initial therapy and gingivectomy.

Class 2: Destructive periodontal disease that mainly causes gingival inflammation, loss of the periodontal ligament, and alveolar bone resorption [44]. Most of these periodontium defects are irreversible. In recent years, repair and reconstruction of the periodontal ligament and alveolar bone have been achieved through tissue transplants, periodontal tissue growth factor, and guided tissue regeneration. At the same time, remarkable progress in gene therapy and tissue engineering has been made, further promoting reparation and reconstruction of periodontium defects.

Class 3: Factors such as nutritional deficiencies, diseases, and injuries caused by periodontal deformation, trauma, shrinking, *etc.* Varying degrees of class 3 defects also cause changes in the periodontal membrane and alveolar bone absorption [44].

4.2.2. Craniofacial Tumors

Tumors are common diseases that seriously harm human beings [27]. They can be classified as benign or malignant. For most benign tumors, conservative treatment or limited resection of the primary tumor will not cause large tissue defects [32, 33]. For malignant tumors, surgery is the most important and effective treatment method [28].

Malignant oral tumors are common among head and neck cancers and often affect the tongue, gums, mouth, and palate. The prevalence of oral cancer is about 1 in 10 million people, accounting for 1.45% to 5.18% of all cancers. Most malignant oral tumors are squamous cell carcinoma, and the peak age at which they develop is 40 to 60 years. Treatment of oral cancer includes expanded resection of the primary tumor and regional lymph node dissection. However, this treatment leads to severe defects in the oral and maxillofacial tissue, resulting in functional disorders of chewing, swallowing, and speech as well as facial destruction, seriously affecting patients' quality of life after surgery [27, 28].

Malignant tumors of the tongue, gingiva, buccal mucosa, mouth, and palate as well as expansion of the primary lesion result in severe tissue defects. Expanded resection of the primary tumor and application of large muscle/skin flaps and free iliac bone grafts are used to repair defects caused by malignant tumors of the jaw bone and other structures [29].

The following description addresses the tumor location, the surgical method, and the defect repair method for soft tissue-only defects and in maxillofacial combined resection (soft tissue and jaw bone defects) [29].

4.2.2.1. Soft Tissue Defects (Lingual Defect, Penetrating Defect of Cheek, Palatal Defect, Tongue and Lip Defect, Parotid Gland and Duct Defect, Facial Nerve Defect)

1. Tongue defect

   I. Tongue body defect: Tongue defect 1/4, 1/2, 3/4, no functional areas

II. Tongue root defect: Tongue defect 1/4, 1/2, 3/4, no functional areas
III. Whole tongue defect

2. Lip defect: Upper and lower lip defects involving 1/4, 1/2, 3/4, or the whole lip
3. Mouth floor defect: Anterior or posterior to the floor of the mouth
4. Soft palate defect: Involving either ≤1/2 or >1/2 of the soft palate
5. Pharyngeal defect: Nasopharyngeal lateral defect, posterior pharyngeal defect

4.2.2.2. Maxillofacial Combined Resection (Soft Tissue and Jaw Bone Defects)

The maxillary central surface is an important bone structure supporting the facial contours. Because of the complexity of the anatomical structure of the maxilla, maxillary defects are more difficult to repair than mandibular defects [30]. The most commonly used classification system for maxillary defects is that described by James Brown, a professor in the UK [31]. In 2010, Dr. Brown introduced the following maxillary defect classification method (Fig. **1**).

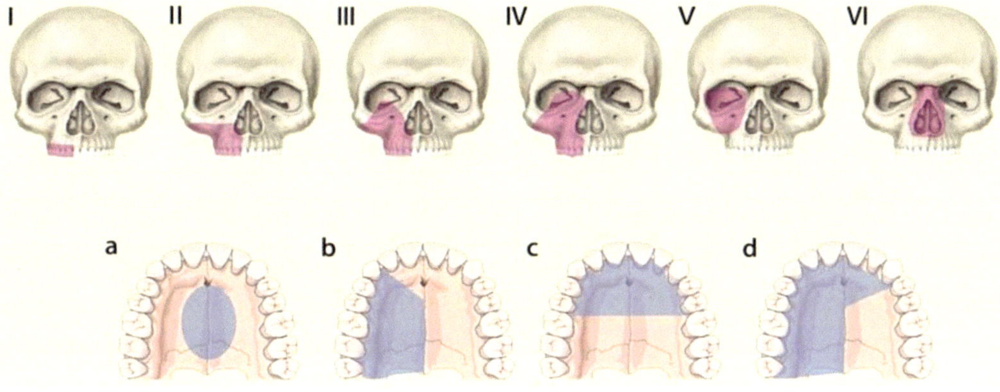

Fig. (1). Classification of vertical and horizontal maxillectomy and midface defects.

Vertical component (Fig. **1**, I–VI)

Class I: Maxillectomy with no oroantral fistula [31]. This type of defect only involves the alveolar bone. Small alveolar bone defects do not require any methods of reparation and reconstruction. However, large alveolar bone defects require skin flaps, pedicled skin flaps, free flaps (*e.g.*, forearm flaps), and similar techniques for repair and reconstruction.

Class II: Low maxillectomy [31]. Including alveolar bone and sinus wall defect, orbital floor and orbital weeks remain. Such defects often result in a maxillary sinus fistula to the mouth or nose, a mouth fistula, larger-volume maxillary defects, and defects of the dentition. Therefore, the purpose of reparation and reconstruction is to close the fistula, provide support for denture, and restore the contours of the patients. Clinically, pedicled flaps and free tissue flaps are more commonly used to repair and reconstruct low maxillectomy defects. However, larger defects require bone composite flaps for repair and reconstruction. Forearm skin flaps and shoulder blade and fibula grafts can also be used to repair and reconstruct low maxillectomy defects.

Class III: High maxillectomy [31]. This involves orbital floor defects, periorbital defects, and skull base defects. Because of the involvement of the skull base and periorbital tissue, it is difficult to make complex craniofacial reparation. Therefore, maxillofacial reconstruction is the best choice. When designing this type of defect repair and reconstruction in clinical practice, orbital floor reconstruction to restore to palpation, transplant patients autologous bone to recover surface, and solve the problem of face collapse. Free vascularized bone transplantation is currently the most commonly used method.

Class IV: Radical maxillectomy [31]. Class IV defects involve a lack of orbital contents, which will lead to a poor prognosis. This should be taken into account when determining the optimal reconstructive option. The most common reparative and reconstructive method for class IV defects involves broad myocutaneous flaps. The purpose of reparation and reconstruction of these defects is to eliminate the fistula interlinking the oral and nasal cavities and reconstruct the shape of the nose and eyes [45]. Importantly, repair and reconstruction of the dura can be accomplished by making use of vascularized muscle. This will prevent cerebrospinal fluid leakage.

Class V: Orbitomaxillary defects [31]. Such defects mainly involve the orbit around the upper jaw; they generally do not involve the alveolar processes or defects in the jaws. Pedicled flaps are most widely used in the clinical setting for repair and reconstruction of class V defects; such flaps include temporalis, temporoparietal, and buccal fat pad flaps [45]. However, anterolateral thigh or radial forearm flaps may be used for larger areas of skin loss.

Class VI: Nasomaxillary defects [31]. Such defects are similar to class IV defects; they only involve the maxillary region around the nose and generally do not involve the alveolar processes or defects in the jaws. Reconstruction is not required when the defect does not involve the palate or nasal bones. However, an osteocutaneous radial forearm flap may be used when a large area of dura is lacking and a high risk of cerebrospinal fluid leakage is present.

Horizontal component (Fig. **1**, a–d) [31]

Class a: palatal defect only; no involvement of dental alveolus

Class b: ≤1/2 unilateral

Class c: ≤1/2 bilateral or transverse anterior

Class d: >1/2 maxillectomy

Lower jaw injury after tumor resection can affect the patient's appearance and lead to unilateral muscle damage caused by malocclusion and other dysfunction [34]. An optimal method must be chosen to reconstruct a normal appearance, and restore function. A variety of surgical methods may be used. Because of the complexity of many mandibular defects, a simple, practical, widely accepted classification system is needed. The following simplified classification of mandibular defects draws from American professor Schultz [34]. The classification focuses on areas of vascular integrity to facilitate accurate microsurgical repair of complex mandibular defects. This mandibular defect classification system is illustrated in Fig. (**2**).

Class 1: Unilateral dentoalveolar defect not crossing the midline and not extending into the angle of the mandible [34].

Class 2: Unilateral defect extending beyond the angle of the mandible [34].

Class 3: Bilateral defect not involving the angles of the mandible [34].

Class 4: Bilateral defect with extension into at least one angle of the mandible [34].

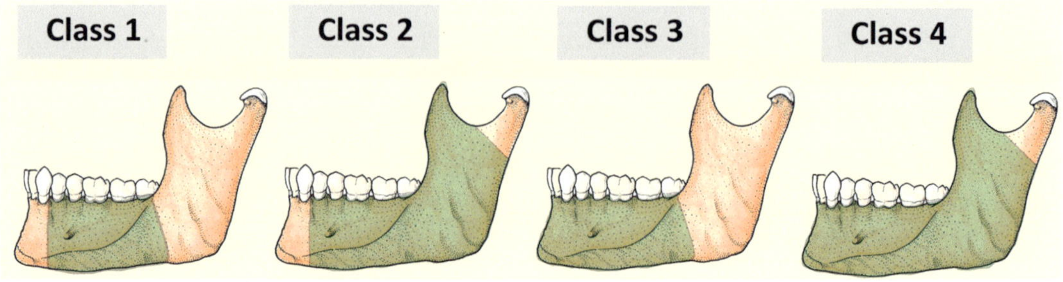

Fig. (2). Classification of mandibular defects.

4.2.3. Craniofacial Trauma

Trauma, especially maxillofacial trauma, remains a major threat to public health [26, 40]. Maxillofacial trauma can be divided into two types: maxillofacial soft tissue injury and maxillofacial fractures. Soft tissue injuries in the oral and maxillofacial region can occur either alone or with simultaneous bone fracture. Because of the complicated anatomic relationships among structures in the maxillofacial region, many emergency doctors fail to correctly assess maxillofacial injuries, potentially leading to missed or incorrect diagnoses.

The rapid development of automobiles and traffic systems throughout the 20th century has caused traffic accidents to become the main cause of oral and maxillofacial injury, accounting for 30% to 40% of such injuries. A gradual upward trend in traffic accidents as the cause of maxillofacial injuries has been seen in recent years. The most common causes of soft tissue injury in the oral and maxillofacial regions include abrasion, contusion, incisions, punctures, lacerations, bite wounds, and other such injuries. Different clinical symptoms and treatment methods are required for these various etiologies. The most common causes of maxillofacial fractures can be divided into three groups: motor vehicle accidents and assaults, which result in about 80% of such fractures; athletic and industrial accidents; and alcohol intoxication-induced drunk driving [39].

Oral and maxillofacial soft tissue injuries, jaw bone fractures, and some common facial fractures are described below.

4.2.3.1. Soft Tissue Injuries

Oral and maxillofacial soft tissue injuries can occur either alone or with simultaneous jaw fractures. Statistically, the incidence of pure maxillofacial soft tissue injuries accounts for about 65% of maxillofacial injuries. Clinically, it is common that injuries, burns, tumors, and so forth lead to oral and maxillofacial soft tissue defects [46]. Abrasions, contusions, incisions, punctures, lacerations, bite wounds, and others [38] can be characterized according to the etiology and severity of the injury. Craniofacial soft tissue injuries may cause severe deformities, which will lead to loss of the patient's normal appearance, affecting his or her social life. Oral and maxillofacial soft tissue defects can cause difficulties of eye closed, mouth closed and ventilation dysfunction, and also cause progressive damage to the eyes, mouth, nose, and other structures [38]. Additionally, oral and maxillofacial soft tissue injuries in infants, young children, and teenagers may affect the growth of bones and organs and result in secondary deformities. Therefore, establishment of maxillofacial soft tissue defect classification standards will facilitate agreement regarding maxillofacial restoration and reconstruction techniques among doctors. Various technical methods of maxillofacial soft tissue defect repair may be used to achieve an optimal therapeutic effect and improve patient satisfaction. A maxillofacial soft tissue defect classification system is described below.

The classification of maxillofacial soft tissue injuries is based on the concept of aesthetic zoning (including five regions namely forehead, cheek, nose, perioral and periocular), and a combination of common clinical situation. So maxillofacial soft tissue defects can be classified into the following four types according to different defect scopes: Class I: defects limited to portion of one region; Class II: defects limited to one entire region; Class III: defects involving any two regions; Class IV: defects involving three or more regions.

4.2.3.2. Fractures of the Jaws (Mandible and Maxilla)

4.2.3.2.1. Mandibular Fractures

The left and right sides of the mandible each comprise approximately 1/3 of the entire U-shaped structure, and the hemimandibular unite at the midline symphysis

[37, 40]. This anatomy makes the mandible susceptible to damage, and the occurrence rate of mandibular fracture is high. The mandibular condyle is an important part of the temporomandibular joint (TMJ), because the mandible is the only movable bone among craniofacial bones, and involved in the movement of the opening and closing. When the mandibular symphysis sustains a crush fracture, due to the pulling of attached muscles, the front part of the mandible shifts backward, potentially causing glossocoma (tongue falling), airway obstruction, or asphyxia and endangering the patient's life.

Symptoms of mandibular fractures include pain, the sensation that the bite is "off," numbness of the chin or lower lip, and abnormal lower jaw movement [40]. Jaw occlusion is the most important indicator of a mandibular fracture during emergency assessments.

The characteristics of mandibular fractures are related to the anatomic structure of the mandible. Fracture sites include the condyle (36%), body (21%), angle (20%), symphysis (14%), ramus (3%), alveolar ridge (3%) and coronoid (2%) [37] (Fig.**3**). Thus, the condyle, body (mental foramen), angle, and symphysis are the most common sites of mandibular fractures.

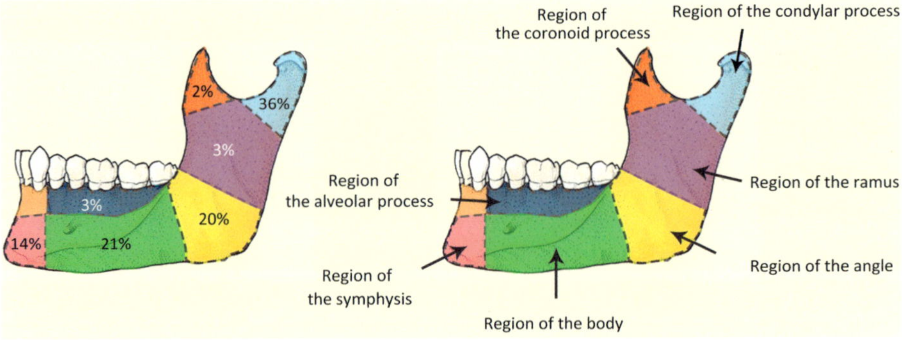

Fig. (3). Regions of the mandible and correlating percentages of fractures occurring in each region.

Mandibular fractures can be classified by anatomical location and whether they are crushed, displaced, or "green sticked" [37]. Common types include fractures of the symphysis, mental foramen [36], mandibular angle, and condyle. Mandibular fractures can also be classified as favorable or unfavorable according to their configuration and location. Favorable fractures usually have no obvious displacement; even if a shift in the mandible is present, it does not cause breathing difficulties or asphyxia. Examples of favorable fractures include the middle of the chnin fractures, mandibular angle fractures, and condylar fractures. Unfavorable fractures, in contrast, include fractures on the both sides of the middle joint and comminuted fractures, fractures of the mental foramen [36], because the pulling of attached muscle, the fracture fragments in the case of high level condylar fracture will shift backwards, resulting in glossocoma, causing breathing difficulty or asphyxia and potentially endangering the patient's life.

4.2.3.2.2. Maxillary Fractures

The upper jaw mainly located in the facial middle third is the largest bone in the middle of the face [35]. Maxillary fractures are serious because they can affect many important adjacent structures: the nose, eyes, mouth, head, and others. Therefore, such fractures often occlude or otherwise adversely affect the nose and other structures, negatively affect the patient's appearance, and may be seriously complicated by brain injury and skull base fractures.

Symptoms of maxillary fractures include nasopharyngeal bleeding, periorbital bruising, double vision, numbness over the ipsilateral cheek and maxillary dentition, and malocclusion [39]. Jaw occlusion is the most important indicator of a mandibular fracture during emergency assessments.

Maxillary fractures were divided into three types by the French physician René Le Fort according to the location of the fracture lines: Le Fort I, also termed infraversion or horizontal maxillary fractures; Le Fort II, also termed mesomaxillary or pyramidal fractures; and Le Fort III, also termed high maxillary fractures or craniofacial disjunction [39]. These fractures can occur alone or in combination.

4.2.3.3. Other Maxillofacial Fractures

4.2.3.3.1. Zygomatic Fractures

The malar bone and zygomatic arch constitute the surface of the most prominent part of the maxillofacial region and are vulnerable to fractures. The zygomatic arch which is thin and narrow is constituted by zygomatic process and processus temporalis, the fracture in this site may appear alone and can also occur simultaneously with other bones.

The clinical manifestations of fractures of the zygoma and maxilla are similar. The main difference is zygomatic depression deformity. Symptoms of zygomatic fractures include limited mouth opening, diplopia, infraorbital nerve injury, and midfacial ecchymosis [39].

In 1992, Zingg divided zygomatic fractures into three types. Type I is characterized by an incomplete fracture of the bone. Type II is characterized by a completely single zygomatic fractures. Type III involves the surrounding bone. About 10% of tripod fractures are not displaced and require only monitoring; 31% are displaced without rotation and can be elevated back into place [39].

4.2.3.3.2. Fractures of the Orbital Floor

Orbital fractures involve the orbital rim and orbital wall. Simple orbital fractures are rare, accounting for only 4% to 15% of all orbital fractures [39]. Clinically, blow-out orbital fractures are the most common; that is, an upward strike force causes a sharp increase in orbital pressure, resulting in collapse of the orbital wall down to the maxillary sinus. A characteristic orbital floor fracture then develops.

Symptoms of orbital fractures include subconjunctival hemorrhage, chemosis, and globe injuries. Serious blow-out fractures cause enophthalmos [39].

4.2.3.3.3. Nasal Fractures

Nasal fractures may occur in isolation secondary to traumatic injury or violence; they can also occur simultaneously with other maxillofacial fractures. An accurate diagnosis and correct treatment can achieve satisfactory results.

Symptoms of nasal fractures include asymmetry, ecchymosis, swelling, and tenderness of the nose. The most serious injuries may result in cerebrospinal rhinorrhea [39].

Nasal fractures have been classified according to the nature, direction, and size of the strike force. Lateral forces are the most common and have great destructive power [39].

4.2.4. Congenital Disease

A facial cleft is usually caused by improper or failed fusion of two adjacent facial processes [41]. The first 6 to 7 weeks of embryonic development is an important period, involving the integration and joint of two adjacent facial processes. In this period, many teratogenic factors can influence the process of surface growth and development. When such growth stops or slows, facial deformities such as cleft lip and palate may occur.

The exact cause and pathogenesis of fusion failure during embryonic palatal development remain incompletely understood. The results of many studies indicate that such fusion failure may be influenced by a variety of factors rather than a single factor. These factors may be generally categorized as genetic and environmental. Examples include malnutrition, infection, injury, endocrine disorders, drugs, physics, and alcohol and tobacco exposure.

Cleft lip and palate is the most common congenital deformity in the oral and maxillofacial region. And it is often associated with the development of other facial deformity resulting in physical defects, physiological dysfunction, and even serious psychological and behavioral problems. Plastic and reconstructive surgery is currently the only treatment method for cleft lip and palate. However, simple closed fracture treatment is unsatisfactory in such patients. Therefore, the proposed treatment of cleft lip and palate is multidisciplinary; specific treatment of the corresponding morphological, functional, and psychological defects at each developmental stage from birth to adulthood should be implemented. Separate discussions of cleft lip and cleft palate are presented below from the viewpoint of multidisciplinary sequential treatment.

4.2.4.1. Classification of Cleft Lip and Palate

4.2.4.1.1. Cleft Lip

Cleft lip is the most common congenital malformation of the oral and maxillofacial region and often occurs with cleft palate. Cleft lip may be part of a specific developmental facial syndrome, which accounts for about 10% of all cases of cleft lip. However, most cases of cleft lip occur independently of a specific syndrome. Cleft lip which is caused by failed fusion of globular process and maxillary process occurred in the lateral side of the upper lip and can be classified into unilateral or bilateral cleft lip. Cleft lip can be categorized as complete or incomplete according to the degree of pathological change.

1. Unilateral cleft lip

Type I cleft lip: Limited to a split lip.

Type II cleft lip: The lip is split, but the nasal floor is complete.

Type III cleft lip: The upper lip to the nasal floor is entirely incomplete.

2. Bilateral cleft lip

The classification method of bilateral cleft lip is same as what of the unilateral cleft lip, for examples bilateral cleft lip III degrees, bilateral cleft lip II degrees, the left III degrees and the right II degrees, *etc*. Clinically, cases of recessive cleft lip may also be seen. In these patients, the skin and mucous membrane are not split, but the tissues in the bottom of which are not united, causing shallow trench-like depressions appeared in lips.

4.2.4.1.2. Cleft Palate

Cleft palate is also a common congenital malformation of the oral and maxillofacial region. It can occur alone or simultaneously with cleft lip. The palate develops at about 8 weeks of embryonic development, and cleft palate is caused by failed fusion of bilateral lateral palatal process and medial nasal process. Various degrees of tissue defects and deformities may occur along with cleft palate. Which has bad influences on suck, diet and can cause physiological disorders; in severe cases, congenital heart disease can endanger the patient's life. Children with cleft palate may also have bone defects that disrupt jaw growth, leading to midface depression deformity, occlusal disorders, soft tissue defects, congenital missing teeth, and other problems, defective teeth, and other problems.

These disorders may result in serious psychological and behavioral problems. Therefore, comprehensive sequential therapy in patients with cleft palate is particularly important. In view of this, establishment of a cleft palate classification system method would help clinical doctors to develop effective treatments and improve patient satisfaction. The clinical classification methods of cleft palate are as follows:

1. Cleft of the soft palate
2. Cleft of the soft and hard palate
3. Unilateral cleft of the palate
4. Bilateral cleft of the palate associated with bilateral cleft lip and alveolus

4.2.4.2. Craniofacial Dysplasia

Facial development, including the combination and fusion of the facial processes, is closely related to differentiation of branchial arch and development of the embryo [41, 43]. The third week of embryonic development, a process appeared on the lower end of the forebrain called frontonasal process. Proliferation and migration of neural crest cells at this time result in differentiation of the mandibular process and maxillary protrusion in the lower part of the frontal sinus. At 28 days of embryonic development, three processes have been formed by this differentiation: one nasal process in the middle and two lateral nasal processes on the side. In the fifth week of development, the middle nasal process grow rapidly, and two globular processes appear at its distal end. At this point, the entire process of facial development is complete [23, 24, 43]. In the first 6 to 7 weeks, the above-described surface processes join and fuse, and the organs of the maxillofacial region are formed. A cleft is caused by improper or failed fusion of two adjacent facial processes under the influence of various factors. Some authors have divided clefts into the following four types of maxillofacial deformities. Type I cleft: between the lateronasal and maxillary process (so-called oro-nas- -ocular cleft), type II cleft: between the medionasal and maxillary process (usually termed a cleft lip), type III cleft: between the maxillary processes (better known as cleft palate), and type IV cleft: between the maxillary and mandibular process (known as macrostomia) [41].

In 1971, sőmmering [41] first proposed that the direction of a maxillofacial fracture follows the constant axes along the face; *i.e.*, through the midline open portion of the nose, nostrils, maxilla, and eyebrows. In 1819, Couronne [42] first proposed the concept of maxillofacial fracture. Over the next several years, various researchers have proposed a variety of classification systems for maxillofacial deformities [23, 24].

In 1976, Tessier [47] proposed a system of facial cleft classification using labels from 0 to 30, and the relationship between soft tissue and bone was emphasized for the first time in his classification system [25, 41, 42]. In general, a soft tissue cleft corresponds to a fissure of the underlying bony structure and *vice versa* [41]. Establishment of such a maxillofacial deformity classification system is simple, feasible, and easy to remember. In the Tessier classification system, the positions of maxillofacial deformities around the eyebrows, maxilla, nose, and lips are labeled along constant axes or lines and assigned a number 0 to 14, where 0 passes through the vertical centerline of the open face [25, 41, 47]. The ossification centers, numbered 1 to 13, are scattered along a spiral route that extends from the lateral aspect of the superior edge of the external orbital pillar to the mandibular midline. In the upper third of the skull, the ossification centers numbered 1 to 3 correspond to the formation of the anterior wall of the skull; those in the middle third, numbered 4 to 11, correspond to the orbital, nose and auditory meatus at a lower position; and those numbered 12 to 13 correspond to the lateral and anterior walls of the oral cavity [25, 41].

Various maxillofacial deformities are herein discussed with respect to the classic Tessier maxillofacial deformity classification system and drawing from the Mazzola classification system [47, 48].

4.2.4.2.1. Internasal and Nasal Dysplasia

Internasal and nasal dysplasias are known as Tessier type 0 and 1 clefts [41]. According to the Tessier 0-14 partition theory, the nose passes through the vertical midline open region of the face. Internasal and nasal dysplasia results from failure of fusion of the two nasal halves.

Clinical symptoms of internasal and nasal dysplasia include complete separation of the nasal halves and abnormal proliferation of unilateral nasal mucosa, which cause partial tissue defects. However, absence of half of the nose may also be found inpatients with some serious nasal dysplasias [48]. These patients may develop a tube similar to the nasal cavity at the site of the defect; this tube is called a proboscis lateralis. Complications of internasal dysplasia include median cleft lip [49, 50] and duplication of the frenulum labiorum.

4.2.4.2.2. Maxillary and Midface Dysplasia

Maxillary development originates from the first branchial arch, which can be involved in the formation of most of the midfacial soft and hard tissue through embryonic palatal fusion. These structures include the bottom of the orbit, the zygoma, the soft and hard palate, the maxillary alveolar bone, and part of the cheek. Thus, the three main aspects of maxillary and midface dysplasia are nasomaxillary dysplasia (or naso-ocular cleft), medial maxillary dysplasia, and lateral maxillary dysplasia [41, 47].

Nasomaxillary dysplasia (or naso-ocular cleft) is known as a Tessier type 1 or 3 cleft [41]. In such cases, barriers to fusion during embryonic development are divided into two types according to the developmental stage. The first type is failure of maxillary and lateronasal fusion, resulting in a naso-ocular cleft [41, 47]. Naso-ocular clefts are characterized by midface dysplasia with normal lip development. The second type is development of a complete cleft, called an oro-naso-ocular cleft [41, 47]. Such clefts occur due to failed fusion of the medionasal and maxillary regions and are characterized by a cleft lip between the central and lateral incisors. These clefts may cause severe alveolar defects, tooth loss, and upper and lower malocclusion.

Medial maxillary alloplasia is known as a Tessier type 4 cleft [41, 47], which is caused by failed fusion of the medial third of the maxillary bone. The split begins at the orbital rim and extends to the alveolar bone *via* the infraorbital foramen. Symptoms of medial maxillary dysplasia include the orbital rim resembling the piriform aperture, the orbital floor resembling a funnel-shaped concavity, a hypodeveloped nose, and a complete alveolar cleft. The most common type of alvcolar cleft occurs between the lateral incisor and canine; an alveolar cleft between two incisors is rare.

Lateral maxillary dysplasia is known as a Tessier type 5 cleft [41, 47]. Symptoms of lateral maxillary dysplasia include orbital floor defects located lateral to the infraorbital foramen. The alveolar cleft occurs between the canine and the first molar. Lateral maxillary dysplasia can occur unilaterally or bilaterally, but lateral maxillary dysplasia is very rare.

4.2.4.2.3. Zygomatic Dysplasia

Zygomatic dysplasia, known as Tessier type 6, 7, and 8 clefts [41], usually occurs bilaterally by inborn autosomal dominant inheritance. Symptoms of zygomatic dysplasia include zygomatic bone defects, flattened cheeks, and exophthalmos.

4.2.5. Discussion

Classification of craniofacial deformities is mainly based on their localization. Some authors have classified such deformities according to the pathogeny, such as congenital factors, acquired factors, trauma, or disease. Other authors have classified craniofacial deformities according to the defect site, such as the soft *versus* hard tissue. Still others have classified these deformities according to a simple description of their physical characteristics. This chapter discusses craniofacial defects with respect to four aspects: periodontium defects, craniofacial tumors, trauma, and congenital defects. This discussion includes the etiology, clinical manifestations, disease classification, and treatment of craniofacial defects.

Diseases of the periodontium, which comprises the gingiva, periodontal ligament, and alveolar bone, are classified according to their pathogeny. Factors include malnutrition, disease, and injury caused by periodontal deformation, trauma, and gingival recession. Gingival diseases and destructive periodontal disease are the most common types of periodontium defects.

Craniofacial tumors, which are common in the clinical setting, can be categorized as benign or malignant. They can also be classified into two types according to the type of operation and range of tissue defects: soft tissue-only defects and maxillofacial combined resection (soft tissue and jaw bone defects).

Craniofacial trauma, a major threat to public health, can be divided into maxillofacial soft tissue injury and maxillofacial fractures. Maxillofacial fractures are in turn classified into two types: fractures of the jaws (mandible and maxilla) and fractures of other maxillofacial structures, including the zygoma, orbital floor, and nasal bone.

A cleft refers to failed fusion of two adjacent embryonic facial processes. Cleft lip and palate is the most common congenital deformity in the oral and maxillofacial region. Craniofacial dysplasia can be classified as internasal or nasal dysplasia, maxillary or midface dysplasia, zygomatic dysplasia, and other types.

Generally speaking, the classification systems for craniofacial defects are varied. However, establishment of an effective classification system will facilitate a greater and more accurate understanding of craniofacial defects. The previous discussion has presented various classifications of craniofacial defects used by foreign experts and scholars with the goal of providing useful references or oral health care professionals.

4.2.6. Conclusions

In this section, classifications of craniofacial defects accord to periodontium defects, craniofacial tumors, trauma, and congenital defects were discussed in both text and illustrative form with reference to the clinical aspects of common and rare craniofacial defects. We believe that this classification of methods is simple and clear, easy to understand and remember, and will have guiding significance for future clinical work.

4.3. CURRENT TECHNIQUES AND LIMITATIONS OF THERAPY FOR CRANIOFACIAL DEFECTS: COMPOSITE MATERIALS AND SELF-TRANSPLANTATION

4.3.1. Autologous Tissue Transplants

The term "craniofacial" includes the oral and facial soft tissue, maxillofacial bone (upper jaw, lower jaw, cheekbones, *etc.*), TMJ, salivary glands, neck, and other structures. Because the color, texture, flexibility, thickness, and contour of local flaps are consistent with the characteristics of craniofacial tissue, regional flaps can be used to repair many small and relatively uncomplicated defects in the craniofacial region [51]. Any flap can be divided into two categories according to the blood supply pattern: flaps with a random blood supply and flaps with an axial blood supply [52]. Flaps with a random blood supply mainly include Z-plasty flaps, W-plasty flaps, V-Y advancement flaps, and bilobed flaps. Such flaps have random blood circulation patterns and are often used for restoration of small, full-thickness maxillofacial defects. Axial pattern flaps are anatomically defined; their blood is supplied by "pedicle vessels," which include at least one arterial supply and a venous drainage system. Many types of flaps exist, such as simple skin or subcutaneous tissue flaps, and an axial composite flap comprises at least two types of tissue [51]. Different means of repair are required for injuries and defects of a variety of causes and degrees. For example, skin and soft tissue defects may require repair *via* neighboring flaps or free flaps, while soft and hard composite defects often require more complex tissue flap repair techniques.

4.3.1.1. Characteristics of Autogenous Bone Grafts

Reconstruction of craniofacial defects, which are mainly caused by cancer excision, trauma, and congenital deformities, poses a huge challenge for oral and maxillofacial surgeons. Craniofacial defects have various causes and usually have a negative effect on the appearance and function of the face. For example, they may result in facial asymmetry, maxillofacial deformities, chewing force reduction, and voice abnormalities, all of which may seriously affect the psychological health of the patient.

Restoration of both function and aesthetics when performing craniofacial defect reconstruction after continuity resection is very challenging. Current techniques always fall short of fulfilling both demands. Craniofacial defect reconstruction is indicated for stabilizing the residual segments, restoring masticatory function, and maintaining facial contour. These aims can be achieved by using reconstruction plates and bone grafts for the replacement of a missing craniofacial segment.

Because of several factors, excessively large maxillofacial defects far exceed the repair capacity of autogenous tissue transplantation, seriously affect the natural contours of the face, interrupt normal blood circulation, and increase the wound infection rate. Thus, reconstruction of large maxillofacial defects is a substantial challenge for reconstructive surgeons [53, 54]. Furthermore, defect reconstruction will be more difficult in patients in poor physical condition. For example, skin damage and osteoporosis can be found in oncologic patients undergoing radiotherapy, and multiple tissue defects such as soft tissue, bone, and nerve defects can be found in military populations [53, 55, 56]. Autologous tissue transplants for defect reconstruction are commonly divided into two types: flaps with vasculature and grafts without vasculature [57]. The use of flaps for reconstruction of the craniofacial region is preferred, although the technical requirements of harvesting or transferring flaps are higher and fewer donor sites are available than with the use of grafts [57 - 59]. In any case, the geometry of the donated tissues does not match the shape of the defect; thus, surgeons must shape the graft to restore the inherent morphology of the defects to the greatest extent possible. Because fewer potential donor sites are available and the morphology of donor tissues is very important, autologous flaps generated by bioreactors have become an alternative choice for restoration of large craniofacial defects [60]. *In vivo* bioreactors include scaffolds or chambers that are placed within the patient's body where tissue can grow and be harvested for future transfer [58]. The application of bioreactors in reconstruction of craniofacial defects is usually a two-stage strategy: first, the bioreactor maintains the anatomical space of the defect, and second, new tissue can be grown in the device [59]. The use of bioreactors involves a combination of growth factors and scaffold materials, which have been used in preliminary investigations of human craniofacial defect reconstruction [61 - 63]. However, there are also some negative effects of the application of growth factors to maxillofacial defect reconstruction. For example, growth factors may promote tissue overgrowth, cause nerve impaction, and promote regulatory challenges [64]. For these reasons, some researchers have evaluated the use of bioreactors without growth factors to reconstruct craniofacial defects. Specifically, different scaffolds filled with polymer-based chambers have been transplanted onto the periosteum of goat ribs for tissue regeneration [65 - 69]. These chambers may contain the accompanying intercostal artery and vein after tissue ingrowth, thus matching the dimension of the flap with that of the implanted tissue chamber [65, 69]. Many types of scaffolds, such as devitalized autografts, morselized autografts, and polymer scaffolds, have been researched as potential scaffold materials. The results of these studies have confirmed that autografts can regenerate the greatest mass of bone tissue [66, 69]. In addition, these tissue chambers were harvested from the rib at different time points, and the results confirmed that the optimal duration for growth of ossific tissue is 9 weeks [66, 69]. Regenerated vascularized bone flaps have been found in sheep models and clinical cases after using this method [65 - 70].

Although this system can be successfully used to regenerate ossification tissue with a specific dimension, it has not been used in the reconstruction of maxillofacial bone defects in animal models. Additionally, autogenous bone may be beneficial to bone regeneration because of the presence of endogenous growth factors in the extracellular matrix, but its harvest can cause complications at the donor site. Combined application of artificial synthetic bone and autogenous bone can reduce the amount and incidence of donor tissue complications [71, 72]. For example, different proportions of HA and beta-tricalcium phosphate (β-TCP) can be synthesized as biphasic ceramic composite graft materials, which have shown to be effective in the restoration of craniofacial bone defects [73, 74]. In one animal experiment, tissue chambers filled with different proportions of synthetic particles were implanted into goat rib periosteum to repair the tissue defect. These tissue chambers were harvested after 9 weeks of growth, and microcomputed tomography and tissue section examination were used to evaluate the influence of initial synthetic graft union on the quantity and quality of tissue growth. In this *in vivo* study of three animals with mandibular defects, the internal structures of the bioreactor are shifted as vascularized free flaps. At 12 weeks postoperatively, the mandible was evaluated with respect to viability of the regenerated bone and integration of the regenerated bone and natural bone. The aims of this study were to determine the feasibility of shifting a bioreactor to a vascularized free flap *in vivo* to repair cranial and maxillofacial defects and to evaluate the effect of synthetic bone grafts on tissue growth.

Two major clinical problems faced by reconstructive surgeons are too many bones requiring repair and insufficient bones available for repair. Cases involving too many bones requiring repair, such as craniosynostosis and ankylosis of the TMJ, are rare. Although conditions involving excess bone are generally not rectified by applying skeletal tissue engineering strategies, the consequences of surgical procedures to correct bony excess may in turn lead to skeletal deficiencies. In either event, the clinical problems associated with inadequate craniofacial bones should be the subject of many reviews. Craniofacial bone defects and inadequate bone caused by congenital or acquired factors are very common. Acquired factors include tumor resection, trauma, and infection, while the most common congenital factor is cleft lip and palate. The requirements of craniofacial defect reconstruction vary with the defect site. The requirements of reconstructing the mandible, which is a structure that bears force and moves, are substantially different from those of other bones in the craniofacial region. Despite these different requirements, the gold standard for craniofacial defect reconstruction is autografting, and such grafts are usually the ribs or tibia. Despite the good results that can be obtained from defect reconstruction using autografts, this technique also has some disadvantages, such as limited donor sites, donor site complications, an unpredictable absorption rate, and infection. Because of these potential problems associated with autografts, synthetic materials have become an alternative for defect regeneration and have been used in both pilot and clinical studies. Other restoration strategies for bone defects are also associated with problems, including potential disease transmission and infection. Because of limited donor sites for autografts and the complications associated with nonautologous tissue materials, other alternative bone induction strategies have been recommended for use in the reconstruction of craniofacial defects. Current studies of craniofacial defect reconstruction should be reviewed. Bone grafting enhances bone regeneration in many conditions in both orthopedic and maxillofacial/oral surgery. Although autologous bone grafts are considered the gold standard bone grafting material with which to enhance bone regeneration and repair bone defects, harvesting of such grafts is associated with donor-site morbidity and restricted availability. Furthermore, the alternative use of allografts has certain drawbacks, including high cost, the need for sterilization, and issues of processing and storage; the most important is the potential immunogenic host response to disease transmission and foreign tissue.

Many studies have been carried out to overcome the limitations of autologous grafts and the disadvantages and allogeneic bone grafts. These studies have led to the development of new orthobiologic materials for use as either alternatives or adjuncts to traditional grafting materials for the management of clinical conditions requiring enhancement of bone regeneration. Several bone grafting materials other than autografts and allografts are now commercially available for bone regeneration procedures. These include demineralized bone matrix, synthetic scaffolds such as calcium phosphate and HA, autologous bone marrow aspirate, bone morphogenetic proteins, and other factors. On one hand, rapid vascularization is an important process for prompt and long-term survival of bone graft materials. Moreover, the number of regenerated vessels influences the stimulation from surrounding tissues, which allows pre-existing vessels to commence infiltration into the graft material itself. Angiogenesis is regulated by a variety of growth factors, particularly vascular endothelial growth factor (VEGF), which is secreted by stroma cells and inflammatory cells to induce blood vessel ingrowth. A cellular substitute material usually requires more time for cell proliferation or vascularization and matrix formation than do autogenous cancellous bone grafts. Osteoconduction appears to be optimized in devices that mimic both bone structure and bone chemistry.

Tissue engineering can be used in place of autologous or allogeneic bone grafting and has received a great deal of attention as an alternative technique for tissue defect reconstruction. Mesenchymal stem cells (MSCs) provide an ideal cell source for the regeneration of craniofacial tissues for the following reasons: they exist *in vivo* and are easy to harvest, they have the capacity for proliferation and differentiation, they have low immunogenicity and tumorigenesis, and no ethical or legal restrictions exist. Ideal cells for cell-based tissue engineering should possess all of these features.

BMSCs can differentiate into various cell types under different conditions *in vitro*, such as osteoblasts, chondrocytes, adipocytes, nerve cells, muscle cells, and others. A number of studies have involved transplantation of BMSCs into muscle tissue defects in animal models, and the results showed muscle tissue regeneration after a period of time. These findings have great clinical significance for myocardial regeneration. To obtain reliable clinical data on bone substitute materials, a suitable experimental model is needed to determine the comparability of materials and the repeatability of results. Importantly, complete healing of some artificial defects will not occur naturally during the life of animal models because of their size; such defects are called critical-size defect (CSD) and were initially described by Hollinger [75].

The biology of rabbits is reportedly similar to that of human beings, and rabbits have been recognized as a good model for the study of craniofacial regeneration. A review of the literature revealed multiple widely accepted experimental CSD designs in rabbits [76 - 78]. Lu and Rabie [76] conducted multiple studies involving craniofacial bone defect repair in rabbits with a 10 mm×15 mm defect without loss of craniofacial continuity. The use of MSCs in regenerative medicine is still in its infancy, and effort should therefore be exerted toward establishing novel cell-based and composite bone graft substitutes that achieve predictable results. Various studies have been conducted to evaluate the role of autogenous, undifferentiated BMSC transplantation for craniofacial bone defect repair in rabbit models.

On the other hand, autografts have been considered the gold standard for the restoration of congenital or acquired bone defects in humans [79]. However, these surgical procedures often lead to complications. Tissue engineering induces bone formation with fewer complications and similar efficiency compared with surgery. This has led to the development of BTE as a new treatment strategy [80, 81]. The concept of tissue engineering was introduced by Langer and Vacanti [21] as an "interdisciplinary field that applies the principles of engineering and life sciences toward the development of biologic tissues that restore, maintain, or improve tissue function." Conceptually, tissue engineering can be divided into three interconnecting components: (1) the carrier/scaffold, (2) the cells, and (3) an environment or recipient site in which the composite is placed [82].

Researchers have made unremitting efforts to maximize the effects of these elements and improve the regenerative capacity of tissue engineering. Nonimmunogenic features of MSCs and the advanced nature of biomaterial technology have made remarkable contributions to BTE [83]. Based on its solid scientific foundation, BTE may represent a bridge between reconstructive surgery and orthopedic surgery and may allow for defect reconstruction with few complications and high efficiency. The following six stages are necessary to realize effective BTE: (1) selection of the extracellular matrix scaffold, (2) isolation and purification of stem cells, (3) expansion of the cells, (4) seeding of the cells on the scaffold, (5) implantation of the composite into animal bone defects, and (6) evaluation of the results of BTE [82, 84 - 86].

Calcium phosphate matrices have been used to repair CSDs in animal models of bone defects [87, 88]. Alphat α-TCP and β-TCP are the main types of calcium phosphate used [89]. The similarity of these two materials lies in their chemical structure. However, they differ in their crystal structure; the solubility of α-TCP is higher than that of β-TCP [90]. Alpha-TCP may exhibit higher dissolution and porosity, leading to better osteoconductivity and easier cell adhesion when added to calcium sulfate hemihydrates. The bovine bone mineral Bio-Oss collagen exhibits both biocompatibility and osteoconductivity [91]. The sponge-like structure and mutual connection of the pores are conducive to cell adhesion [92 - 94]. The high performance of these biomaterials has promoted the development of BTE.

According to the Mesenchymal and Tissue Stem Cell Committee of the International Society for Cellular Therapy, the criteria for judging MSCs are as follows [92 - 96]. They must have (1) plastic adherent features under standard culture conditions; (2) high expression of surface markers cluster of differentiation CD73/CD90/CD105 and no expression of CD11b/CD79, CD14/CD34/CD45, or CD19/HLA-DR; (3) the capacity to differentiate into chondrocytes, osteoblasts, and adipocytes *in vitro*; and (4) a corresponding antibody that will positively react on surface markers such as CD73, CD90, and CD105. These criteria are crucial for determining MSCs. The immunophenotype of muscle MSCs, which are derived from the orbicularis oris of patients with cleft lip and palate, has been characterized by some scholars [97]. BTE strategies based on these cells may be a promising alternative to autogenous bone grafts.

To compare the efficacy of autologous bone grafts and BTE, some scholars transplanted autologous bone and a composite comprising α-TCP and bovine bone mineral matrices associated with muscle-derived MSCs into alveolar bone CSDs in rat models. Better outcomes were obtained in the BTE group, providing a basis for the application BTE strategies in reconstruction of alveolar clefts in human patients [98, 99].

4.3.1.2. Characteristics of Free Flaps

The free flap concept was first proposed in 1983 by the Finnish surgeon Asko-Seljavaara [100]. Free flaps have many perforator vessels that can be safely used for tissue restoration and reconstruction within the range of the perforasome [101]. The perforator located nearest to the defect allows for the simplest flap design and the shortest angle of rotation. The use of pedicle perforator flaps to reconstruct medium-sized craniofacial defects is a good choice because it is most likely within the secure borders of the applied perforasome and the donor site can be closed directly. The perforators were readily localized with color Doppler ultrasonography and consistently found where expected according to the anatomical description [101,102]. When the perforator was localized, the flap was released, allowing for the necessary rotation in the absence of tension. In contrast to most previous studies, the perforator was not dissected down to the axial vessel [103 - 105]. Different types of flaps were used, ranging from simple elliptical flaps to complicated multilobular flaps, also known as cogwheel flaps. Many reports to date have shown that this technique is simple and reliable, although sample sizes have been limited. Common complications of free flaps include epidermolysis, venous stasis, and partial tissue necrosis that heals by secondary intention without further management. The degree of flap congestion in the lower limb or distal forearm is greater than with other types of flaps [105 - 109]. When a flap with a short pedicle is rotated 180° on the limb, potential rotation of the pedicle along with the rotation of the flap could cause vascular congestion. However, other factors such as insufficient venous drainage and the effect of gravity also contribute to vascular congestion [110]. The incidence of complications in the limbs is largely unknown and varies among reports. The incidence of venous congestion ranges from 17.2% to 25.8%, and that of total flap loss ranges from 1.1% to 5.6%; partial flap necrosis is the most common complication [105 - 107]. The incidence of complications in the upper limb and in the rest of the body is not as well documented, ranging from 19.3% to 23.0% [108,109]. Complications will always occur. Although the incidence of complications is high, most complications are relatively slight. A simple solution is intensive flap care. The freestyle pedicle flap reduces the use of skin grafts and has become the preferred flap type for the reconstruction of medium-sized defects.

Reconstruction after resection of craniofacial cancer is usually challenging, especially when the tumor extends to the base of the skull. Vascularized flaps, whether free or pedicled, are one option. Free flaps have the added benefit of providing access to more cephalad anterior cranial base defects without the constraints of the rotation arc of pedicled flaps.

Use of local flaps for facial reconstruction and use of prosthesis for repair of large-scale cranial defects often resulted in aesthetically inadequate outcomes only a few decades ago. In addition to the stigma caused by visible disfigurement, the patients were impaired by limited digestive tract and upper airway function because chewing, breathing, speech, and swallowing were affected in accordance with the extent of the defect. Such side effects resulted in psychosocial disabilities and were often poorly accepted by the patients [111].

In the 1980s, improvements in microsurgical techniques established free tissue transfer as a standard procedure in primary and secondary reconstruction of the craniomaxillofacial region. When neither the orbit nor facial skin is affected, microvascular free flap transfer allows for early rehabilitation of swallowing, speech, and, with the additional use of an osseous component, chewing. This technique also helps to support and restore the facial contour. However, the aesthetic outcome is not always pleasing, although defect closure with microvascular flaps results in functional effectiveness [112]. In particular, a better aesthetic result can be achieved after eye and eyelid enucleation by adding episthetic treatment to microvascular free flap reconstruction [113]. Craniofacial defects are usually attached to extraoral implants inserted into the periorbital area [114]. Some present studies show our experience in applying these complementary methods for the treatment of extensive midface post-tumor resection defects and orbital defects.

The construction of facial normality or surgical restoration should follow Gillies' classic principle of using "similar" organizations. This applies to neuromuscular dynamic functions, the facial skin envelope, the soft tissue volume, and the bone and cartilage supportive framework. An aesthetically pleasing face created by a combination of proportion and balance can be divided into several facial aesthetic subunits [115, 116], such as the nose, periorbital region, auricles, forehead, and commissural structures. Because regional and local flaps match facial tissue features such as texture, color, pliability, contour, thickness, and the absence or presence of hair, they can be used to repair many small, relatively less complex craniofacial defects [51]. Flaps can be roughly divided into two categories: those with a random or axial vascular supply [117].

Many flaps that are used for restoration of smaller, full-thickness defects on the face, such as W-plasty, Z-plasty, V-Y advancement, bilobed, and rhomboid flaps, have a random vascular supply. The definition of an axial flap is based on the anatomical structure of the blood vessels, called the "pedicle", which includes at least one vein and one artery. There are many types of flaps ranging from simple skin or subcutaneous tissue flaps to composite flaps, which comprise various types of flaps [51].

Finally, the choice of skin flap depends on the features of the defect that needs to be repaired. Defects can be evaluated according to the affected anatomical structure, such as the inner nasal lining, cartilaginous support structure, or architectural bony structure as well as according to the available external soft tissue coverage [118]. The purposes of nasal reconstruction are to achieve an optimal aesthetic appearance and maintain a patent airway [118]. Secondary repair of small defects in certain locations can achieve satisfactory results, but adverse tissue notches or distortion may occur in other areas such as the nasal tip or ala [119]. Full-thickness flaps harvested from the supraclavicular region and pre- or post-auricular area may provide a suitable organization match for skin defects with a healthy graft bed. Small chondrocutaneous composite flaps obtained from the ear have been used to reconstruct three-layer defects, but their survival depends on a well-vascularized wound edge [120].

4.3.1.3. Flap Classification (Local Flaps, Pedicled Flaps, Vascularized Flaps, and Microvascular Free Tissue Transfer)

Local Flaps

Local flaps are currently considered the best option for reconstruction of moderately sized defects of the face [121]. Axial-pattern forehead and nasolabial flaps are workhorse flaps used for coverage of the perinasal and nasal areas based on their rich vascularity and reliable and consistent anatomy. These characteristics enable a flexible, long flap on a narrow pedicle [122, 123]. However, the primary limitations of local flaps are their bulkiness and limited range of motion at the pedicle site, necessitating a secondary delayed procedure. The importance of craniofacial defects in successful reconstruction is obvious, and a simple single-stage approach is therefore the first choice if possible.

Loose skin is present near the nasal area and can be used for flap reconstruction of nasal skin defects whose diameter is less than 2 cm. Such flaps include bilobed, banner, glabellar, nasolabial, dorsonasal, and cheek flaps [124]. The use of forehead flaps based on supratrochlear vessels from one or both sides is an excellent way to reconstruct nasal defects [124]. The temporomastoid flap can be used to transfer mastoid skin and/or ample posterior auricular skin based on the fact that it contains the superficial temporal artery, can cover the total nasal region, and can comprise some auricular cartilage when needed [125].

Restoration and reconstruction of nasal defects is very important for airway patency and aesthetic nasal projection. The midline can be supported through the nasal bones, the front edge of the medial crus, and the nasal septum. Additionally, lateral support on both sides can be reconstructed using bone grafts and/or cartilage. Like skin grafts, bone and cartilage grafts should be used when reconstruction of the skin and lining. They require a well-vascularized bed to survive. Local septal flaps may be helpful for midline support. The nasal lining needs to be reconstructed to prevent contraction through scarring and to protect the structural support [126]. Various folded flaps described above can be used to simultaneously repair both the external skin and internal lining. Alternatively, local external skin flaps can be resurfaced on their underside with full-thickness skin grafts. Local mucosa can also be used to reconstruct defects, or mucosa from other sites such as the oral cavity can be used as grafts. Reconstruction of any other facial defects can be performed using similar principles. Defects must be assessed in terms of their location, anatomical structure, and aesthetic subunits when designing local flaps. The reconstruction principle of each structure within the face is unique and should be followed [127, 128].

Perforators in the craniofacial region usually have a stable anatomical structure. In theory, all well-known flaps depend on the first perforator in the rotation axis. Random flaps seem reliable and are good choices for restoration of the facial subunits, especially the periorbital, periocular, and nasal areas, with good results using a one-stage strategy.

Pedicled Flaps

The development of modern flap surgery and microsurgery has relied on clinical observations and anatomic structures of the flaps. The axial pattern flap was first proposed by Jackson and McGregor in the 1970s [129, 130]. This marked the start of a decade of use of myocutaneous and fasciocutaneous flaps in 1980s, followed by the more widespread use of perforator flaps in the 1990s [131, 132]. Ian Taylor showed that the skin is supplied by perforators that can be directly applied during flap surgery [133]. The original description of the angiosome was provided by Taylor and Palmer [134] in 1987; this definition made the available skin area narrower than those supplied by a single perforator [135]. An important principle of perforator flap application is clarification of the range of safe flap application. Knowledge of the safe boundary of the perforasome will promote its more widespread use. Reconstruction of medium-sized facial defects using freestyle perforator flaps is safe [136]. Thus, many studies have involved reconstruction of moderately sized defects of the extremities and torso using freestyle pedicled perforator flaps.

Vascularized Flaps

Vascularized flaps provide durable and reliable reconstructive options for anterior and middle skull base defects after resection of large, complex malignant neoplasms of the craniofacial region. Free flaps, such as anterolateral thigh flaps, radial forearm free flaps, and thoracodorsal artery perforator free flaps, are well suited for these defects because they have a low incidence of complications, low donor-site morbidity, and good aesthetic and functional results. The quality of life of patients with serious facial injuries is severely impacted. Over the past few decades, the use of facial vascularized allografts has shifted from animal experiments to clinical application. Such grafts are mainly used for reconstruction of defects caused by trauma, tumors, congenital deformities, and infection. Their use may now be regarded as a feasible alternative for restoration of complicated craniofacial defects because reconstruction using autografts still provides unsatisfactory results. The effect of reconstruction using allografts is better than that of autografts in terms of satisfactory anatomical reconstruction, desirable appearance, and functional recovery. With an improved functional status, patients can recover their ability to perform facial expressions and functional movements such as smiling, talking, and chewing. The ideas raised in the 1997 movie *Face/Off* have now been realized in the clinical field.

Microvascular Free Tissue Transfer

In the past, the only means to reconstruct large, complex facial defects was the use of combined regional pedicle flaps with local flaps and nonvascularized grafts. Unfortunately, this resulted in significant donor site complications such as widespread scarring and functional impairment caused by tissue contraction [137]. Microvascular free tissue transfer (MFTT) was proposed in the 1960s and 1970s and has since developed into a reliable reconstructive technique. MFTT has completely changed surgeons' ability to repair more complex defects by improving surgical outcomes.

An MFTT, called a "free flap", is an autologous vascularized graft. Axial pattern flaps with vascular pedicles of appropriate diameter can be converted to free flaps. This involves three major steps: (1) complete dissociation of the flap from the donor site, (2) microvascular anastomosis of the recipient vessels and vascular pedicle of the flap to reconstruct the flap blood supply, and (3) the intervention phase of flap ischemia. Numerous studies have shown that the survival rate of free flaps is >95%.

Free flaps have many generic advantages. They can be designed to contain many types of organization and oriented in many configurations to improve inset [138]. They can also be obtained from almost any part of the body to maximize suitability for the defect and tissue availability. Moreover, they usually require less time and fewer surgical stages to complete restoration than do other reconstruction techniques. Improvements in the harvesting and use of free flaps have reduced the incidence of donor site complications [139, 140]. MFTT has two main advantages in the reconstruction of facial defects. First, it can be used to repair complex or larger defects, the size of which exceeds the reparative ability of conventional methods. Second, it has a decisive advantage over other technologies such as regional flaps, local flaps, and/or various grafts is that it has become more suitable for the rehabilitation of the facial soft and skeleton tissues defects. Modified MFTT is now more commonly used in facial reconstruction.

4.3.2. Use of Composite Materials for Craniofacial Defects

Traditional facial and cranial bone reconstruction techniques have a long history and have achieved good clinical results [141 - 143]. However, they also have some disadvantages, such as donor site complications [144]. Composites avoid scar formation, reduce the incidence of donor site complications, reduce costs, shorten the treatment time, and utilize known components [145]. Metallic implants have also been used [146] or combined with HA cement or bone in craniofacial bone reconstruction [147]. Although these materials are readily available for immediate use, they require pretreatment and remodeling to match the defect. Additionally, their resistance to traumatic injury is uncertain. Additive manufacturing technology has been used in many medical fields. Personalized skull implants can be produced based on preoperative 3D additive manufacturing models, which improve the accurate of defect reconstruction [148]. Cranial planting materials are more widely applied, and dental implants have become one of the most effective means to repair defective or missing teeth.

4.3.2.1. Features of HA

Bone tissue is a biocomposite that consists of a collagen matrix containing dispersed HA nanoplatelets. HA has excellent biocompatibility and high hardness, bioactivity, and osteoconductivity. Biocomposites used for load bearing should have the advantages of HA [149]. However, synthetic composites composed of HA and collagen are generally used as biodegradable scaffolds because of their poor mechanical properties [150, 151]. To enhance the mechanical properties of polymers, nondegradable biocomposites such as high-density polyethylene (HDPE) have been considered to serve as the matrix of HA-reinforced polyethylene composite (HAPEX™) [152 - 154]. A 4:6 ratio of HA and HDPE will significantly improve the elastic modulus of the composite. However, HAPEX has low tensile strength and is mainly used for middle ear implants, orbital floor reconstruction, and maxillofacial surgery. In addition, large HA particles are generally a few micrometers in diameter and easily debond from the cracks and the polymer matrix into small pieces when conducting mechanical tests [155].

Recent advances in nanotechnology have provided nanocomposites with superior mechanical and physical properties [156] and biocompatibility [157, 158]. Synthetic nanoscale HA has greatly contributed to the proliferation and adhesion of osteoblasts. Carbon nanotubes have an exceptionally high Young's modulus and electrical conductivity and show good biocompatibility toward human cells such as myoblasts, osteoblasts, and neurons [159 - 162]. In some biomedical applications, carbon nanotubes and nano-HA have been used as reinforcing fillers for polymers in forming functional composites. However, dispersion of carbon nanotubes throughout the polymeric matrix is difficult because the nanotubes are curled in shape. This leads to inferior mechanical performance of the resulting composites. In contrast, carbon nanofibers (CNFs) are straight and thus more effectively disperse throughout the polymeric matrix of biocomposites. Additional advantages of CNFs are their excellent biocompatibility and lower production cost. In this regard, CNFs have been incorporated as polymers into functional composites for biomedical applications [163, 164]. CNFs are usually synthesized through high-temperature catalytic decomposition of hydrocarbons on transition metal catalysts such as cobalt, iron, and copper nanoparticles [165 - 167]. Helical and straight CNFs can be selectively prepared through appropriate regulation of the raw gas composition, reaction temperature, and shape of the metal nanoparticles. In the latter case, Jian *et al.* [164] reported that spherical copper nanoparticles were conducive to the formation of straight CNFs, while polyhedral or facetted copper catalysts produced helical CNFs.

The addition of HA nanorods (20% by weight) to polypropylene (PP) can enhance its biocompatibility and mechanical properties. PP exhibits excellent dimensional stability, flexibility, chemical resistance, and durability. Ductile HDPE can be substituted by PP as biomedical implants undergo cyclic stress because PP has higher fatigue resistance than does HDPE. One study [168] evaluated the effect of the addition of CNF on the thermal, structural, and mechanical behaviors of PP and PP/20 wt% HA nanorod composites. In the latter case, low loadings of CNF (0.5–2.0 wt%) were added to the PP/20 wt% HA nanorod composite to generate heterofunctional conjugates with higher mechanical performance and biocompatibility. The authors' purpose was to develop polymer composite grafts with good biocompatibility and suitability for bone defect restoration and reconstruction, especially in craniofacial applications.

4.3.2.2. Types and Characteristics of Different Implants

Reconstruction of large craniofacial skeletal defects is one of the greatest challenges for maxillofacial surgeons. Long-term clinical experience has shown that the gold standard for bone reconstruction is autologous bone transplantation. However, there are some drawbacks to this approach, such as donor site complications and the limited amount of autologous tissue for large defect reconstruction. Therefore, implants are needed to reconstruct craniofacial defects. The most important factor in the use of biomaterials for bone reconstruction is a sufficient understanding of these biological materials, such as the exact components of the composite material and the advantages and disadvantages of each material. The existence of free radicals and residual polymethyl methacrylate (PMMA) homopolymer before or after polymerization and the presence of free and residual PMMA monomers before and after polymerization can cause a stimulus response. This response can be minimized [165] when the graft is fabricated in controlled conditions during the preoperative processes. Compared with traditional methods, fabrication in controlled conditions can shorten the operation time and reduce the total cost of defect reconstruction despite the fact that the cost of the implant is high. The price of implants may decrease in the future as new industrial manufacturing processes emerge. However, long-term clinical follow-up of more patients is needed before drawing conclusions regarding current composite transplantation. State-of-the-art techniques in large cranial repair involve the use of alloplastic implants, autologous bone transplantation, or combinations thereof [166]. These techniques are associated with bone resorption and skin protrusion/infection, respectively, which can limit their clinical use in difficult cases. Current synthetic implant materials for cranioplasty are produced using different metals, polymers, or sintered HA.

Metallic and polymer materials do not possess bioactive characteristics and are inert. A great number of bioactive compounds, including bioceramics and osteoinductive cytokines, have stimulatory effects on bone growth. However, they are limited to the repair of smaller defects at orthotopic sites, bone fusion procedures, and healing fractures [167, 169]. Regeneration of larger segmental bone defects in humans is more challenging. Many operative procedures are generally needed for such cases, and reconstruction typically necessitates the use of scaffolds supplemented with autologous bone or cell transplantation to possibly gain a healing response [170, 171].

Treatment of large or complex bone defects necessitates a scaffold that provides osteoconduction and osteoinduction, adjustable geometry in the operating room, and sufficient mechanical strength. One method not previously explored is the use of injectable molds and bioceramics that are shaped outside the body. After hardening, the implants can be implanted and sterilized. By molding bioceramics on a titanium mesh in a mosaic design, good handling properties can be combined with mechanical stability. Injectable calcium phosphate bioceramics can be found in both brushite- and HA-forming cements in the field of therapeutic chemistry [172]. A third category of therapeutic chemistry involves the use of monetite-forming cements, although this has not been as extensively explored as the first two categories [173]. Brushite cements resorb relatively quickly *in vivo*, whereas apatite cements are considered to be almost insoluble [174]. Monetite has a solubility and resorption rate between those of apatite and brushite [175]. Other noncementitious phases in addition to the three above-described cement phases may be added to obtain specific properties such as β-inorganic pyrophosphate or β-TCP [175,176]. The use of an adjustable scaffold with a specific calcium phosphate composition, interconnected by a titanium mesh and molded in a mosaic design, has been described in the repair of a cranial defect in a severely injured patient.

The ideal implants should be lightweight, strong, nonmagnetic, and easy to extend and fix; they should also exhibit high biocompatibility, long-term stability, and low thermal conductivity. A number of these requirements can be met through the use of fiber-reinforced composites (FRCs).

FRCs composed of reinforcing synthetic fibers are adhered to the surrounding thermosetting or thermoplastic polymer scaffolds by a coupling agent. Glass fibers are the most frequently used. Dimethacrylate-based polymers are a commonly used thermosetting material, while polyetherketone is a commonly used thermoplastic material. Dimethacrylate-based glass FRC has acceptable osteoconductivity and biocompatibility.

PMMA and bioactive glass (BAG) have been used in clinical applications. PMMA has advantages in terms of intraoperative remodeling, but it is an exothermic material and does not attach chemically to the bone. BAG is a rigid and brittle material. It cannot be bent or shaped in the reconstruction process. Mechanical preparation of brittle BAG is difficult because it may cause the implant to develop microfractures, leading to unsafe individual implant formation. Antimicrobial BAG has the ability to chemically attach to the surrounding bone, but it still cannot be bent or shaped. The application of a new composite graft comprising PMMA and BAG in craniofacial reconstruction increased the application efficiency of PMMA and BAG.

BAG has the ability to combine bone and multiple composite materials. Of these applications, clinical studies involving the use of BAG S53P4 to fill the frontal sinus on mastoid cell area, and reconstruct nasal perforations have been reported. There is an increasing interest in composite materials such as PMMA and FRC implants, which combine the advantages of their component materials.

BAG has been used in the treatment of various clinical diseases [177 - 179]. The influences of BAG on the clinically important bacteria have been experimentally investigated. Microbes die when placed in contact with BAG [180, 181].

The purpose of many scholars is to assess the validity and compatibility of personalized bioactive FRC grafts in upper craniofacial reconstruction.

4.4. *IN VITRO* STUDIES OF CRANIOFACIAL REGENERATION

The craniofacial area comprises bone and cartilage as a scaffold that supports secretory organs, muscles, subcutaneous structures, and delicate skin. These components are connected by a large number of ligaments and dynamically interact through a large vascular network [182]. Craniofacial defects secondary to trauma, cancer, infection, and congenital malformation are common in the clinical setting. A major challenge in the repair of such defects is the limited number of donor sites, such as bone, muscle, nerves, skin, and mucosa, that are suitable for the restoration of local defects [183]. Even the most advanced autologous reconstructive techniques, such as composite free tissue flaps, are usually insufficient to repair a wide range of maxillofacial defects [184]. Therefore, it is necessary to develop new approaches to overcome these difficulties. Tissue engineering technology has provided new ideas and opportunities for craniofacial regeneration in recent years. Compared with transplantation, tissue engineering provides new hope for defect regeneration and avoids the use of immunosuppressants. Tissue engineering is defined as "an interdisciplinary field that applies the principles of engineering and the life sciences toward the development of biologic substitutes that restore, maintain, or improve tissue function" [185]. The modern concept of tissue engineering technology was presented in the late 1980s, when stem cell technology was developed. Tissue engineering is based on three elements: The seed cells or stem cells, (embryonic stem cells [ESCs], multi-induced stem cells, and adult stem cells), biological scaffolds (such as collagen, bone, or synthetic matrix), and physiologic signaling molecules (morphogenes, growth factors, and adhesins) [186]. Key problems of tissue engineering technology that need to be preferentially solved include the directional induction of seed cells, types of tissue regeneration, and how biological factors improve regeneration material compatibility. This section of Chapter 4 mainly discusses the basic experiment, which is divided into *in vitro* and *in vivo* parts. Ideal results of *in vitro* experiments are a prerequisite for the success of *in vivo* experiments.

4.4.1. Cell Source

Regenerative medicine and tissue engineering have rapidly developed in the last 20 years and have become increasingly sophisticated with the development of stem cell technology [187]. The most important factor in the success of tissue engineering is the ability of stem cells to proliferate to adequate numbers of cells that maintain the desired phenotype and execute appropriate biological functions [188]. Stem cells are undifferentiated cells with a self-renewal capacity that both produce more stem cells and differentiate into different cell lineages under appropriate conditions [189, 190]. Based on their cellular origin and plasticity, seed cells can be divided into three categories: ESCs, induced pluripotent stem cells (iPSCs), and adult stem cells [191]. Because of these advantages, seed cells are currently being widely studied regarding their potential for application to craniofacial tissue engineering. The use of seed cells would completely rebuild of craniofacial defects, which are difficult to restore using traditional methods [192]. This section will elaborate on the fractional cultivation, identification, and biological characteristics of ESCs, iPSCs, and adult stem cells.

4.4.1.1. ESCs

ESCs, the only true totipotent stem cells, are isolated from the inner cell mass of 5- to 7-day-old blastocysts. ESCs are pluripotent during embryonic development. Early embryonic blastocysts comprise about 50 to 150 cells, which can differentiate into cells derived from all three primary germ layers (ectoderm, mesoderm, and endoderm). With sufficient stimulation, ESCs can differentiate into more than 200 types of adult cells, but they do not generate any unnecessary placenta or embryonic membrane. ESCs are able to proliferate indefinitely *in vitro* and can theoretically differentiate into any type of cell or form an entire organ [188, 193 - 195]. The most significant potential clinical application of ESCs is in regenerative medicine, where they may be used in the development of cell replacement therapies [196 - 199]. Many attempts were recently made to guide the differentiation of ESCs into osteoblasts and explore the effect of ESCs in bone regeneration [200, 201]. The Federal Drug Administration has approved the use of ESCs to cure paralysis caused by spinal cord injury [202] and to treat eye diseases in human studies [203]. ESCs have also been used for the regeneration of craniofacial defects, including but not limited to the defects of the mucosa, periodontal tissue, and alveolar bone [204 - 207]. Ohazama [208] reported that ESCs that had reorganized with the oral epithelium on the 10th day of embryonic development expressed unique gene loci for odontogenic mesenchymal cells, such as Pax9, Msx1, and Lhx7. This suggests that ESCs can respond to the induced signals sent by embryonic dental epithelium. Although such studies are beneficial to understanding the basic process of tooth regeneration and development, it is still necessary to solve several major problems before implementation of ESCs into clinical practice: tumorigenicity, immunogenicity, and ethical issues.

The most common way to induce initial differentiation in culture is to form 3D spherical structures, also known as embryoid bodies, in suspension culture because after *in vivo* implantation, they have similarities with embryonic tissue. These polymeric structures comprise derivatives of all three embryonic germ layers [209 - 211], and their formation is a prerequisite to production of the most mature somatic cell lineages when ESCs differentiate into a large number of endoderm-like cells in monolayer culture (which can be identified when the ESC colony flattens and develops a "rough" appearance).

Human ESCs have the ability to form any cell, tissue, or organ and are available "on demand" for tissue restoration, replacement, or regeneration. This is expected to completely change human disease treatment modalities. In the shorter term, 3D ESC-derived "organ" cultures could be applied in pharmacological experiments to simplify the process of drug testing, reduce the quantity of experimental animals used, and show the mechanical details of differentiation itself [212].

4.4.1.2. iPSCs

A large number of recent studies on nuclear reprogramming have shown that one cell can be transformed into other irrelevant cells by changing genetic expression patterns, which leads to the generation of iPSCs [213 - 215]. There are broad prospects for the application of iPSCs in regenerative medicine. Recent reports have shown that human and mouse iPSCs have the capacity to induce bone based on differentiation protocols for ESCs [216]. Other studies have found that iPSCs can induce periodontal tissue regeneration in nude mice. Similar results were obtained using silk scaffolds and SATB2-modified iPSCs to restore calvarial bone CSDs in mice [217, 218]. However, establishment of an optimized induction strategy and related mechanisms requires further study and improvement.

The therapeutic potential of iPSCs is quite large because they are particular stem cells, and neither immunological barriers nor racial or political issues need to be considered. In addition, they can be harvested from easily accessible and abundant sources of tissue such as the donor's hair, adipose tissue, or skin. These advantages make iPSCs superior to other stem cells. Furthermore, these cells can be harvested from individual patients; therefore, they are more widely used than ESCs for the treatment and study of human diseases by developing disease models and transferring gene-corrected autologous stem cells on a patient-by-patient basis. This is especially important for diseases lacking abundant animal models or *in vitro* experiments. These mutation- and disease-specific cell lines provide an opportunity to display the development path of complex medical conditions.

In 2006, Yamanaka and Takahashi [219] reported that iPSCs were successfully acquired from embryo and fibroblast cells of adult mice by the ectopic coexpression of four genes: Klf4, Sox2, Oct4, and c-Myc. The expression of these genes was enough to reprogram somatic cells to a state of multiple potential. Successful reprogramming also rapidly converted iPSCs into a wide variety of other cell types, including melanocytes, neural stem cells, stomach and liver cells, mature B cells, pancreatic β-cells, adipose stem cells, and keratinocytes, Reprogramming skill displayed a universal ability to change the identity of the cells. In dentistry, iPSCs can be harvested from various types of dental tissues and cells, including stem cells from the apical papilla, dental pulp stem cells (DPSCs), stem cells derived from human exfoliated deciduous teeth, buccal mucosa fibroblasts, tooth germ progenitor cells, gingival fibroblasts, and periodontal ligament fibroblasts [220 - 223]. DPSCs are more efficiently reprogrammed than dermal fibroblasts and express high levels of the marker gene of endogenous reprogramming factors such as KLF4 and c-Myc and/or ESCs [224]. Because it is convenient for dentists to obtain these cells, iPSCs from dental tissues are expected to become a promising cell source for tissue regeneration [225].

Some challenges must be overcome before this technology is applied within the context of a clinical trial. However, gene manipulation coupled with cell therapy strengthens both regulatory and safety issues. For example, integration of lentiviral or retroviral DNA into the chromosome increases the risk of silencing essential genes or evoking oncogenes [226]. These problems can be partially resolved using adenoviruses, piggybacking, Cre/LoxP recombination, plasmid constructs, and some nonviral methodologies such as microRNA and small molecules and/or proteins [227 - 230]. Importantly, although increasingly more research is being performed to develop more efficient methods to generate iPSCs, transplantation *in vivo* could lead to teratoma formation. This is a great obstacle that must be overcome before clinical application. Another important additional problem is that the generation of these cells remains a long process with low reprogramming efficiency (generally reported to be <0.01%). To overcome this disadvantage, researchers have widened the application of ES cell-specific micro-RNA and some small molecules such as signaling molecules and direct epigenetic modifiers [198, 231, 232].

4.4.1.3. Adult Stem Cells

Adult stem cells are pluripotent stem cells because their potential is generally limited to one or more lineages of specialized cells [196]. Their use avoids the ethical controversy associated with ESCs [233].

It is possible to acquire adult stem cells through the following:

1. BMSCs: The first successful stem cell therapy is likely to be bone marrow transplantation. Actually, peripheral blood stem cell acquisition is currently being used to substitute bone marrow puncture [233, 234].
2. Adipose-derived adult stem cells (ASCs): These cells may be isolated from human adipose tissue, which is normally carried out by liposuction [235].
3. Umbilical cord stem cells: These cells can be harvested from the umbilical cord blood [236, 237].
4. Amniotic fluid-derived stem cells: These cells can be obtained at the time of delivery or by aspiration of the amniotic membrane during the course of genetic screening [238].
5. Dental stem cells (DSCs): These are the most readily available stem cells. They can be isolated from the healthy dental tissue of either deciduous or permanent teeth, including the periodontal ligament, the apical area of developing teeth, and other structures of the tooth. Craniofacial stem cells, including DSCs, originate from mesenchymal cells and neural crest cells during development [239]. Two main types of cells are involved in the formation of dental hard tissue: epithelium-originating ameloblasts, which induce the formation of enamel, and mesenchymal-derived odontoblasts, which are responsible for dentin generation [240].

4.4.1.3.1. MSCs

MSCs were first successfully isolated by Friedenstein *et al.* [241] in the 1960s. These cells are called bone marrow stromal cells. When transplanted, MSCs have the capacity to generate adipose cells, cartilage cells, osteocytes, and fibroblasts [241 - 243].

MSCs are the most thoroughly studied stem cells in the field of craniofacial bone defect regeneration. The use of MSCs as seeds on poly (L-lactic-co-glycolic acid) scaffolds is reportedly effective in the reconstruction of pig mandible defects. On bone X-rays, tissue-engineered bone exhibits uniform radiation density and a uniform, unclear interface between the natural and rebuilt bone [244]. Mendonça *et al.* [245] demonstrated that bone marrow cells cultured in a particular osteoinductive environment (created through clots rich in autogenous platelets and fibrin as well as membranes containing a mineral matrix of HA and β-TCP) had the capacity to promote callus formation and resultant bone regeneration when implanted into huge maxillary defects. This strategy was also applicable to reconstruction of skin, nerve, and vessel defects within the region of implantation [245]. MSCs also have positive impacts on cranial suture closure. When MSCs were transplanted into bone defects around the sagittal suture of adolescent rats 16 to 24 weeks postoperatively, many newly formed bones were observed, and the suture-like gap was also filled with new bone. These capacities of MSCs are clinically significant because cranial sutures are important to skull growth, and MSC transplantation is expected to be an effective technique with which to repair skull defects in patients with incomplete skull development [246]. Other researchers have evaluated the use of stem cells and appropriate scaffolds in the regeneration of complex tissues. Chen *et al.* [247] reported that MSCs can differentiate into osteoblasts and chondrocytes through the use of double gene-activated scaffolds for bone and cartilage regeneration. One layer is a chitosan gelatin scaffold activated by plasmid transforming growth factor β (TGF-β) for cartilage formation, and the second layer is a plasmid bone morphogenetic protein 2-activated HA/chitosan-gelatin scaffold for cartilage formation. The double scaffold was seeded with MSCs and implanted into osteochondral defects of the rabbit knee joint. The significance of this approach is that both the scaffold design and delivery of local genes can induce the differentiation of single stem cells into various cell lineages. This is extremely important in the reconstruction of tissues comprising different cell types [247]. According to recent reports, bone marrow stromal cells are able to express amelogenin.

Origin of MSCs

The individual development of MSCs is complex and the exact source of MSCs is unclear, mainly because of the lack of a unique MSC marker. There are three possible origins of MSCs, as described below.

Epithelial–mesenchymal Transition

Epithelial–mesenchymal transition is a cellular process in which epithelial cells lose their epithelial characteristics and obtain mesenchymal cell characteristics. During this process, the epithelial cells lose their junctional structure and polarization, undergo cytoskeleton reorganization, and are transformed into spindle-shaped mesenchymal-like cells [248]. In association with these morphological changes, the expression of epithelial genes is down regulated as the expression of MSCs gene is upregulated. Upon the completion of epithelial–mesenchymal transition and expression of MSC markers, these cells have the capability for multidirectional differentiation [249 - 251].

Pericyte Origin

Pericytes, which surround capillary endothelial cells, have the ability to contract. Pericytes express some MSC markers such as platelet-derived growth factor receptor β and CD146 [252, 253], and MSCs also express some pericyte markers such as pericyte-associated antigen 3G5 [254]; this indicates a potential link between pericytes and MSCs. Crisan *et al.* [255, 256] showed that MSCs have a perivascular origin in multiple human organs. Later studies showed that smooth muscle cells and pericytes have the capacity to differentiate into adipocytes, osteocytes, chondrocytes, osteoblasts, and nerve cells [257 - 260], indicating a potential perivascular origin of MSCs. Notably, MSC-like cells can be harvested from avascular tissues such as the intervertebral discs, and these cells can express CD166 and CD146 as MSC markers and have the potential for multilineage differentiation. These findings indicate that pericytes may not be the only source of MSCs [261 - 263].

Adventitial Cells

According to previous reports, human vascular adventitial fibroblasts in pulmonary arteries contain MSC-like cells. These cells express MSC marker genes and undergo adipogenesis and osteogenesis in appropriate culture conditions [264]. Corselli *et al.* [265] recently isolated a CD146−/CD34+/CD45−/CD31− cell population from the tunica adventitia vasorum. This cell population expressed MSC markers and exhibited multipotency differentiation similar to that of BMSCs [265]. Thus, two types of perivascular MSC progenitors may coexist: pericytes in capillaries and adventitial cells around larger vessels.

Isolation of MSCs

Stem cells are believed to exist in specialized "stem cell nests" in perivascular areas. To isolate the same type of MSCs, cells are first separated from their tissue of origin by either explant culture or enzyme digestion [266]. The MSC population can then be isolated and identified from the various kinds of cells by the following four methods: (a) selection based on physiological criteria, including proliferation, morphology, and differentiation properties; (b) immunohistochemical staining; (c) immunomagnetic bead selection; and (d) fluorescence-activated cell sorting.

Functions of MSCs

Although the precise mechanisms in which MSCs are involved remain incompletely understood, there are four main functions of MSCs, as described below.

Homing Effect

When MSCs are transplanted into the host body, they migrate to tissue-damaged or injured sites. This specific capability is called homing efficiency [267, 268]. A group of molecules called homing-related molecules induce MSC migration. There are three main categories of homing-related molecules: (a) adhesion molecules, such as vascular cell adhesion molecule 1 and intercellular adhesion molecule 1 [269]; (b) chemokines, such as chemokine ligand 2–chemokine receptor 2 and chemokine ligand 12–chemokine receptor 4 [270, 271]; and (c) matrix metalloproteinases. Additionally, inflammatory cytokines can regulate homing-related molecules; for example, vascular cell adhesion molecule-1 and interleukin 1 upregulate intercellular adhesion molecule-1 and tumor necrosis factor α, leading to the chemostasis of MSCs [272, 273].

Plasticity and Multilineage Differentiation

Plasticity refers to the capability of MSCs to differentiate into functional and mature cells of the same or a different germ layer of origin [274]. Numerous studies have proven that BMSCs have the ability to differentiate into multiple types of cells in tissues including cartilage, bone, skeletal muscles, and adipose tissue [275 - 279]. BMSCs can also differentiate into endothelial cells [280], epithelial cells [281, 282], neural cells [283], cardiomyocytes [284], and hepatocytes [285]. The plasticity of MSCs represents the cytological foundation of regenerative medicine. There are four mechanisms of plasticity: (a) stem cells derived from the embryonic stage are stable during development and remain unchanged in postnatal tissues; (b) "true" postnatal stem cells are located in tissues such as those comprising multiple adult progenitor cells [286]; (c) MSCs dedifferentiate, transdifferentiate, or redifferentiate [287]; and (d) the fusion of donor cells and receptors triggers the exchange of genetic information, which leads to the transdifferentiation of MSCs [288, 289]. The plasticity of MSCs can be enhanced by changes in the environment, stimulation of exogenous growth factors and regulated by genetic manipulation. For example, previous studies have revealed that hypoxic conditions enhance differentiation of BMSCs into osteogenic cells [243] and that MSCs that have been genetically engineered with the osteogenic factor bone morphogenetic protein 2 (BMP-2) promote osteogenesis [290, 291].

Immunomodulatory Effect

MSCs also reportedly have an immunosuppressive effect [292, 293]. Inflammatory stimuli such as prostaglandin and nitric oxide, as well as the cytokines interleukin 1, tumor necrosis factor α, and interferon γ, can stimulate MSCs to produce immunosuppressive factor E2 [294, 295]. MSCs can regulate both the adaptive and innate immune systems. They can also inhibit T-cell and B-cell proliferation [296] and inhibit the maturation of monocytes [297]. Based on their immunosuppressive effect, MSCs have been used to treat patients with immune diseases such as multiple sclerosis [298], rheumatoid arthritis [299], and systemic lupus erythematosus [300]. However, the clinical use of MSCs as an immuno-suppressive agent requires greater evidence. In particular, well-controlled, randomized studies will further our understanding of their immunomodulatory effect.

Trophic Effects

Once MSCs have been transplanted to the injured site, they produce trophic factors including chemokine ligand 2, insulin-like growth factor, interleukin 6, VEGF, fibroblast growth factor (FGF), and hepatocyte growth factor under the local stimuli of the microenvironment. These trophic factors can promote cell proliferation and enhance angiogenesis and antiapoptosis [301]. A recent study showed that conditioned medium from MSCs can promote bone regeneration, suggesting that paracrine factors secreted by stem cells can promote osteogenic differentiation [302].

4.4.1.3.2. ASCs

With the increased worldwide incidence of obesity, the therapeutic use of subcutaneous adipose tissue is also increasing [303]. Approximately 400,000 liposuction procedures are performed in the United States each year, and the amount of lipoaspirated fat tissue ranges from 100 to 3000 ml [304]. As discussed below, ASCs are pluripotent stem cells that have a wide range of therapeutic applications.

Stromal cells have features of preadipocytes. They can be isolated from adipose tissues of adult patients, proliferated *in vitro*, and induced to differentiate into adipocytes [305 - 308]. Adipocyte differentiation is a complicated process accompanied by a series of changes in gene expression, cell morphology, and hormone sensitivity and was conducted in a preliminary study in mice preadipocyte cell lines instead of human preadipocytes. This section focuses on *in vitro* culture of stromal cells isolated from human adipose tissue. A growing number of studies are showing that these cells are pluripotent and have the ability to regenerate bone, nerves, and cartilage [309 - 311].

Adipose Tissue-Derived Stromal/Stem Cell Isolation

Rodbell *et al.* [312, 313] were the first to isolate adipose stem cells from adipose tissue in the 1960s. The authors finely chopped rat fat pads and extensively washed them with phosphate-buffered saline (PBS) to eliminate hematopoietic cells. The obtained tissue fragments were incubated, digested, and centrifuged with collagenase to separate floating mature adipocytes from the pelleted stromal vascular fraction (SVF) (Fig. **4**).

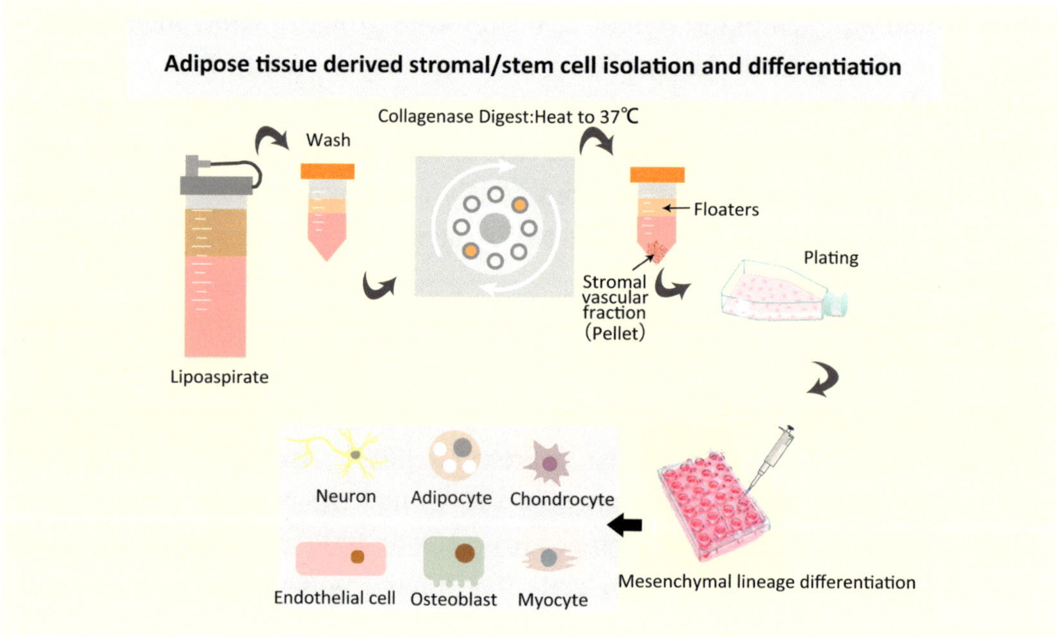

Fig. (4). Isolation and differentiation of ASCs from lipoaspirated tissue.

The SVF constituted a heterogeneous cell population comprising fibroblasts, pericytes, circulating blood cells, endothelial cells, and preadipocytes (also known as adipocyte progenitors) [312, 313]. The plastic adherent population which enriched for the "pre-adipocytes is selected from the SVF by the final isolation step. This method was subsequently modified to isolate cells from human adipose tissue samples [314 - 318]. Human adipose tissue fragments were originally chopped by hand; however, this technique has been simplified with the development of liposuction surgery. During liposuction surgery, orthopedic surgeons inject physiological saline solution containing anesthetic and/or epinephrine into the subcutaneous tissue through a catheter and subsequently harvest tissue and liquid under suction [319]. The size of the finely minced tissue segmentsis determined by the size of the cannula applied. Researchers have determined that liposuction does not significantly change the viability of isolated SVF cells [320, 321]. In fact, the features of adipocyte progenitors can be identified directly within the pumped liposuction fluid and SVF derived from the tissue fragment digests [322]. However, ultrasound-assisted liposuction can reduce the quantity of cells recovered from organization digests, as is their proliferative capacity [323]. Restoration of ASCs can be further improved by regulating the centrifugal speed [322]. Studies have concluded that optimal cell recovery can be obtained using a centrifugal speed of 1200 × *g* [322]. Adipose tissue is collected by liposuction aspiration or aspiration biopsy. The duration of time for which adipose samples are stored at room temperature before use should preferably not exceed 24 hours. ASCs can be isolated from adipose tissue by extensively washing the tissue with PBS containing 5% penicillin/streptomycin (P/S). After removal of the tissue fragments, the sample is placed in a sterile tissue culture plate digested by PBS containing 0.075% collagenase type I and 2% P/S. The adipose tissue can be chopped using ophthalmic scissors, and a 25- or 50-ml pipette is then used to stir the sample up and down several times to promote digestion. The sample is placed in a cell incubator with a stable environment of 37°C and 5% CO_2 for 30 min, and 5 ml of α-MEM containing 20% heat-inactivated fetal bovine serum (FBS) is then added to neutralize the activity of collagenase type I. After disintegration of the sample, it is transferred to a 50-ml centrifugal tube, and the formation of solid aggregates is avoided. ASCs can be obtained from the SVF by centrifuging the sample at 2000 rpm for 5 min. Finally, to complete the isolation of the stromal cells from the primary adipocytes, the samples are removed from the centrifuge and violently shaken to completely destroy the precipitate and mix the cells. The centrifugation step is then repeated. After spinning, all of the collagenase solution above the culture pellet is carefully aspirated to avoid cell interference. The pellet is then resuspended in 1 ml of lysis buffer, incubated for 10 min on ice, washed with 20 ml of PBS containing 2% P/S, and centrifuged for 5 min at a rotation speed of 2000 rpm. The supernatant is then aspirated; the cell pellet is resuspended in a maximum of 3 ml of α-MEM stromal medium containing 20% FBS, 1% L-glutamine, and 1% P/S; and the cell suspension is filtered through a 70-μm cell strainer. Additional cells can be obtained by using an additional 2 or 3 ml of stromal medium as a wash through the cell strainer. The cells are grown in a lysine-coated culture plate and incubated at 37°C in 5% CO_2. The amount of adipose tissue seeded in each well of a 12- or 24-well plate is about 500 or 150mg, respectively.

Culture and Expansion of ASCs

All medium must be removed from the wells after 72 hours of incubation. Notably, the percentage of preadipocytes derived from the SVF after digestion depends on the patients. If the proliferation of SVF is ineffective, the concentration of FBS in the culture medium can be increased to 25%, which would lead to more premature adipogenesis. Signs of deterioration such as cytoplasmic vacuolization, particle aggregates around the nucleus, and/or peeling of the cells off the plastic surface may indicate a toxic or insufficient medium, aging of the primary cells, or microbial contamination.

The cells are washed with pre warmed PBS (1% antibiotic can be added). The purpose of using the pipetting solution in the cell layer several times is to completely separate the cells of interest from blood cells and/or tissue fragments. An appropriate amount of fresh medium is added to each well according to the volume of the culture plate. The cells are incubated in a humidified tissue culture incubator that maintains them in a stable environment of 37°C and 5% CO_2. Until the cells reach 80% confluence, they can be harvested or directly induced to undergo multidirectional differentiation. To harvest viable ASCs, approximately 500 μl of warm sterile PBS must be added to the wells and maintained for 2 min. A 500-μl volume of trypsin/0.5% EDTA solution then replaces the PBS. The cells are then incubated in the incubator for an appropriate duration of time. They are observed under a microscope until more than 90% of the cells are confirmed to be detached, and 500 μl of stromal medium containing FBS is then added to neutralize the trypsin reaction. The stromal medium containing the suspended cells is obtained, and the cell suspensions are transferred from the well to a sterile 2-ml cryopreservation tube. The tube is centrifuged at 1200 rpm for 5 min. The supernatant is removed, and 250 μl of culture medium is used to resuspend the cells. Cell counts are then performed using trypan blue: the volume ratio of trypan blue to cells is 8:1; that is, 87.5 μl of trypan blue is added to 12.5 μl of suspended cells. A hemocytometer is used to count the cells, which are then replated in accordance with the capacity of the cell culture plate. When amplified from a frozen vial, the cells are rapidly (1–2 min) thawed at 37°C and immediately seeded in the plate with complete stromal medium. The medium is changed on the second day and every 3 days thereafter.

Characterization of ASCs

Although MSC surface markers were identified and defined as early as 2006 [323], the immunophenotype characterization of ASCs has been fiercely debated for a long time. Multiparameter flow cytometric analytic and sorting strategies have been applied because of the inherent heterogeneity of SVF. Based on the perivascular marker CD146, the endothelial marker CD31, the hematopoietic marker CD45, and the stem-stromal markers CD117 (c-kit), CD105, CD90, and CD34, four different cell populations from SVF have been defined in uncultured conditions: putative ASCs (CD146−, CD117−, CD105−, CD90+, CD45−, CD34+/−, and CD31−), vascular smooth muscle cells or pericytes (CD146+, CD117+, CD105−, CD90+, CD45−, CD34+/−, and CD31−), endothelial progenitor cells (CD146+, CD117+, CD105−, CD90+, CD45−, CD34+,and CD31+), and hematopoietic cells (CD45+) [324]. Studies on whole adipose tissue have demonstrated that the progenitor components/stem components, organized around small vessels in a ring, are dominated by a prevalent supra-adventitial layer of CD34+ cells displaying MSC-like multipotentiality [324, 325]. The surfaces of the supra-adventitial ASCs are colonized with CD146+ perivascular cells or pericytes, and these cells surround arterioles and venules [324,326]. A component of proliferative CD31+ and CD34+ endothelial progenitor cells is related to the luminal layer [327].

Compared with extensively cultivated ASCs, early-passage ASCs and freshly isolated SVF cells express higher levels of human leukocyte antigen-DR, CD117 (c-kit), and stem cell surface markers such as CD34, along with lower expression of stromal cell markers such as CD105, CD90, CD73, CD54, CD34, CD29, CD13, and MHC I [328, 329]. It seems that CD34− ASCs exert higher plasticity, while CD34+ ASCs have a greater proliferative potential [330, 331].

The International Society for Cellular Therapy and the International Federation for Adipose Therapeutics and Science recently defined the minimal criteria for the identification of ASCs and provided preliminary guidance for the scientific community working with ASCs [332]. Cells in the SVF are identified by the combination of the following surface markers: CD45−, CD34+, and CD31−. CD105, CD90, CD73, and CD13 should also be evaluated in the analysis of stromal/stem cell markers. In culture, like BMSCs, ASCs are positive for CD105, CD90, CD73, and CD44 and negative for CD31 and CD45. Unlike BMSCs, ASCs are negative for CD106 and positive for CD36. Finally, a multilineage differentiation assay should be performed to identify ASCs.

Immunophenotypic Characterization of ASCs

ASCs from passage II were used for immunophenotypic characterization. The cellular suspension was centrifuged (300 × *g*, 4 min) and the cells were trypsinized. ASCs were stained with antibodies conjugated with phycoerythrin (PE), allophycocyanin (APC), or fluorescein isothiocyanate (FITC): CD90-APC, CD105-PE, CD45-FITC, CD73-PE, CD16-PE, and CD34-APC (BD Biosciences, San Jose, CA, USA). In total, 5×10^5 cells were resuspended in 200 μl of PBS and incubated with PE-conjugated, APC-conjugated, or FITC-conjugated antibodies for 20 min in a dark environment at room temperature. Flow cytometry was used to analyze the antibody-specific fluorescence channels of the samples. The experimental results of flow cytometry indicated that these cells expressed CD105, CD90, and CD73 and were accepted as MSC surface markers, while they exhibited negative expression of hematopoietic cell markers CD45, CD34, and CD16.

In conclusion, ASCs provide a new method for studying novel treatments for a large number of acquired and inherited diseases. ASCs may also be beneficial for identifying novel molecular targets for drug development. Many important scientific and medical problems need to be resolved before ASCs are applied in human beings. These problems include how to improve massive manufacturing methods with corresponding quality guarantee and quality control to produce cells in compliance with the current Good Manufacturing Practice regulations.

4.4.2. Tissue Regeneration

4.4.2.1. Bone Regeneration

Bones are dynamic organs that undergo continuous remodeling to maintain tissue homeostasis, modify morphology and shape, and repair fractures [333]. Currently available clinical treatments are limited to allografts, osteomyocutaneous flaps acquired from an autologous donor site, microvascular bone, and distraction osteogenesis [334 - 336]. Bone "free flaps" harvested from the iliac crest, scapula, rib, or fibula represent the gold standard treatment because they are equipped with all essential elements for regeneration, including cellular precursors, differentiated bone cells, and proper differentiation/growth factors. The main disadvantage of free flaps is the skeletal defect induced at the donor site. Other technical problems include maintenance of the arterial and venous blood flow of the flap to prevent an insufficient blood supply to the recipient site, especially in patients who have received radiation therapy; peripheral vascular disease; a poor general clinical condition; and a wide range of bone defects requiring repair [337]. Therefore, the purpose of scientific research is to bypass the need for allografts or autologous transplantation in the restoration of large bone defects (either post-traumatic or created by a surgical operation or trauma) in which spontaneous recovery cannot be expected. *In vitro* and *in vivo* studies have suggested the potential of ASCs to differentiate into mature osteoblasts that improve bone healing and produce the paracrine effects that facilitate migration and differentiation of resident precursors.

Some studies have found that demineralized bony matrix is capable of ectopic ossification, and this led to the discovery of a number of signaling molecules for osteogenic protein [338]. These signaling molecules are termed BMPs. They are capable of osteogenesis and osteoinduction and are widely used in BTE [339]. Currently investigated bone signaling molecules include platelet-rich plasma (PRP) (containing platelet-derived growth factor), BMP-2, and BMP-7 (osteogenic protein 1). Scaffold biomaterials for bone engineering include ceramics (TCP, HA), polymers (polyacids), natural materials (bovine collagen, demineralized bony matrix), and composites of these biomaterials [339]. Both MSCs and ASCs are ideal choices for bone engineering because they have the capacity to differentiate into chondrocytes and osteoblasts [340]. Bone marrow stromal cells are the current choice for *in vitro* cultivation. However, MSCs (*e.g.*, ASCs) are presented in some organizations and can be used for implant colonization *in vivo* [341]. Several strategies have been successfully used in clinical reconstruction (Fig. **5**); each one involves various components of the afore mentioned triad [342].

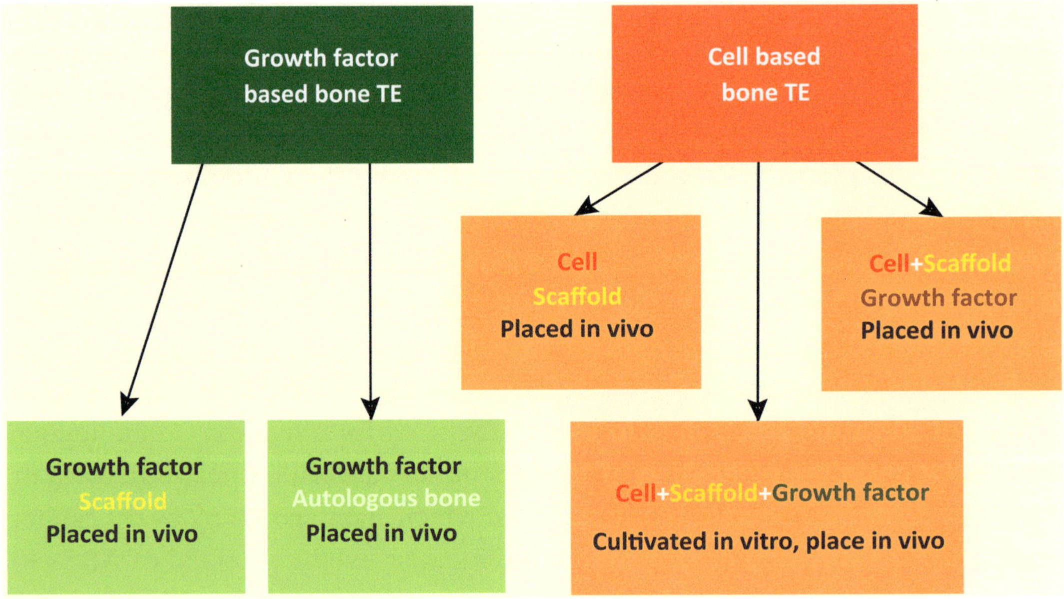

Fig. (5). Translational strategies in the tissue engineering of bone. TE, tissue engineering.

4.4.2.1.1. Osteogenic Potential of ASCs and their Role in Bone Regeneration

Cell-based approaches are now widely considered the most effective for bone formation and regeneration because they are capable of effectively maintaining the physiologic osteogenic process *in vivo*. In fact, the greatest value of ASC application is bone reconstruction/regeneration [343, 344]. The secretome of ASCs [345, 346] and the SVF [347, 348] contains different endocrine factors (adipokines) that activate bone remodeling [349, 350]. Specifically, VEGF, present in the secretome of both ASCs and the SVF, plays an important role in the restoration of bone defects and fractures. VEGF is capable of activating the formation of a new network of blood capillaries, which is required during bone regeneration [351]. Furthermore, VEGF plays a major role in the collection of hematopoietic stem cells that contribute to the formation of new bone [352].

4.4.2.1.2. Osteogenic Differentiation

When cells reach 80% confluence, preadipocytes can be differentiated to undergo osteogenesis. Osteogenesis can be induced using culture medium containing 50 μM ascorbate-2-phosphate, 2 mM β-glycerol phosphate, and 1 nM dexamethasone. The preadipocytes are induced in this osteogenic medium for approximately 14 days, and the medium is exchanged every third day. After fixation in 10% formalin, the preadipocytes are stained with 40 mM alizarin red (pH 4.1) to assess their mineralization.

4.4.2.1.3. Scaffolds for ASCs in Bone Repair

Scaffolds for osteogenesis should mimic bone structure and morphology, optimize integration into the surrounding tissue, and provide a good microenvironment conducive to ASC adhesion, proliferation, and differentiation. The macro-architecture and micro-architecture of the scaffold are highly dependent on the production process [353, 354]. HA, specifically $Ca_{10}(PO_4)_6(OH)_2$, is a well characterized biomaterial that is currently used in clinical applications in different forms. HA is suitable for integration into or substitution of damaged or diseased bone tissues because it is similar to the mineralized bone phase and provides the basic ion in the process of bone resorption [355, 356]. Additionally, β-TCP, specifically $Ca_3(PO4)_2$, has been shown to be suitable for clinical use as a carrier for MSCs because of its crystallographic and chemical similarities to the inorganic phase of native bone [357, 358]. Biphasic calcium phosphate (BCP) is a homogenous composite of β-TCP and HA [359]. Different composition ratios of β-TCP and HA will lead to great differences in the dissolution and absorption properties of BCP. Unfortunately, the poor mechanical properties of calcium phosphate ceramics lead to brittleness and fractures [360]. Several *in vitro* and *in vivo* studies have identified an osteoinductive role of biomimetic scaffolds on ASCs [361]. In particular, Marino *et al.* [362] revealed that β-TCP matrix alone is sufficient to activate the differentiation of ASCs into an osteoblastic phenotype regardless of whether the cells are grown in a differentiative or proliferative medium. Additionally, Liao *et al.* [360] compared the osteogenic potential of porcine ASCs among three scaffolds (collagen I-coated polycaprolactone/TCP [PCL - COL], PCL/β-TCP [PCL - TCP], and PCL) to identify the optimal scaffold for bone engineering. Their *in vitro* study showed that porcine ASCs had the best osteogenic differentiation potential for PCL-TCP-COL scaffolds as demonstrated by the highest alkaline phosphatase activity and osteocalcin mineralization and expression [361]. An experiment in nude mice showed better vascular tissue and woven bone formation in the PCL-TCP-COL group than in the PCL group. Moreover, both *in vitro* and *in vivo* studies found that the osteogenic ability of porcine ASCs was improved by coating COL onto the PCL-TCP scaffolds. In addition, Arrigoni *et al.* [363] compared neoformed bone tissues obtained by treating tibial CSDs with either HA–autologous ASC constructs or HA alone with respect to their biomechanical, immunohistochemical, and histomorphometric properties. The study demonstrated that tibial defects treated with rabbit HA–autologous ASC constructs healed more effectively than did those treated with HA alone [272].

Silicone-, magnesium-, and calcium-containing ceramics, such as akermanite ($Ca_2MgSi_2O_7$), display better degradation rates and mechanical properties than do other bioceramics and reportedly improve the osteogenic commitment of MSCs [364, 365]. As shown by Liu *et al.* [357], the attachment and proliferation of human ASCs on β-TCP and akermanite were similar *in vitro*, and osteogenic differentiation of ASCs on the akermanite was significantly higher than that on β-TCP after 10 days of culture. Zanetti *et al.* [364] recently observed that ASCs cultured in osteogenic medium for 21 days prior to seeding onto akermanite-based scaffolds exhibited greater osteocalcin expression and calcium deposition than did cells seeded onto β-TCP and PCL.

About 3000 published surveys in the scientific databases have defined ASCs as the safest and most effective cell type for regenerative medicine approaches. Bone regeneration is currently the most promising field for translation of experimental ASCs protocols to the clinical setting. Nonetheless, research on bioscaffolds is rapidly developing in the field of ASC applications in craniofacial regeneration, allowing cells to proliferate, differentiate, and be exposed to growth factors and cytokines. In conclusion, these studies highlight the advantage of using ASCs in combination with biomimetic scaffolds to provide a very effective strategy for bone regeneration.

4.4.2.2. Cartilage Tissue Regeneration

Articular cartilage defects, especially partial thickness defects, do not readily heal because of the low mitotic activity of the residing chondrocytes and the avascular nature of cartilage. Some studies have indicated that autologous chondrocyte implantation is an effective method for treating cartilage defects [366]. However, some disadvantages of this technique include the requirement for two separate surgical procedures, the difficulty in obtaining enough chondrocytes, donor site morbidity, and the slow rate of chondrocyte proliferation.

Because of their multidifferentiation potential, the ability of iPSCs to differentiate into chondrocytes for cartilage repair has gradually gained researchers' interest. Recent research has indicated that iPSCs have proliferation and differentiation capabilities similar to those of human ESCs and display great potential for regenerative medicine [367]. Human iPSCs can be harvested from fetal neural stem cells transfected with a polycistronic plasmid vector carrying a plasmid expressing the human OCT4 gene or the mouse Klf4, Sox2, Oct4, and c-Myc genes to form cartilaginous tissue *in vitro* [368]. Additionally, some studies have reported that human osteoarthritic chondrocytes (OCs) respond similarly to non-diseased passaged dedifferentiated chondrocytes by the strategy of co-culture [369]. Wei *et al.* [370] investigated whether these OC-derived iPSCs can differentiate into chondrocytes and whether iPSCs can be derived from Ocs by the expression of transcription factors. To enhance the differentiation of iPSCs into chondrogenic cells, the authors used lentivirus to infect iPSCs seeded in an alginate matrix and then co-cultured these iPSCs with chondrocytes and TGF-β1 *in vitro*. A large number of data indicated that iPSCs can be obtained from OCs through defined factors. This combination strategy contributes significantly to chondrogenesis of OC-derived iPSCs.

Some studies have been performed to minimize the number of factors necessary for reprogramming, and the results indicated that both Klf4 and Oct4 are adequate to reprogram neural stem cells that have differentiated into pluripotent stem cells, (two-factor Oct4 and Klf4, OK iPSCs) [371]. Kuboth *et al.* [372] then assumed that two-factor iPSCs can differentiate into chondrocytes *in vitro*, as evidenced by the expression of cartilage-associated genes and the appearance of cartilage nodules and proteins with efficiency comparable to that of murine ESCs. The authors found no statistically significant difference in the efficiency of chondrogenic nodule differentiation between iPSCs and ESCs. Additionally, they demonstrated the induction of iPSC chondrogenic differentiation by certain members of the TGF-β1 family.

MSCs were derived from a human dermal fibroblast iPSC line to formulate essential criteria for the production of articular-like chondrocytes from human iPSCs [373]. Using a standard high-density culture medium typically reserved for the chondrogenic induction of PSCs and human MSCs, the researchers found that an abundant mesenchymal progenitor cell population from differentiated human iPSCs and an appropriate high-density culture environment are essential factors for improving the potential of iPSCs to differentiate into chondrogenic cells and restricting the heterogeneity of the differentiated progeny. In addition, recent statistics have indicated that compared with pellet culture differentiation, the micromass cultures formed by BMP-2 treatment of iPSC-derived MSC-like cells lead to a phenotype variety of typical of articular chondrocytes.

One of the major challenges in using iPSCs for either *in vitro* modeling or therapeutic applications is the difficulty in acquiring uniform differentiation to the specific cell type of interest [374]. Diekman *et al.* [375] used murine iPSCs to create tissue-engineered cartilage; an initiation population of cells was successfully pre differentiated toward the chondrogenic lineage. The authors noticed that the most forceful chondrogenic differentiation occurred in a new system with BMP-4 treatment in micromass culture using Col2 expression to isolate differentiated cells. They then cultured these cells for several passages in culture medium with basic FGF (bFGF) and maintained the cells for 21 days in the presence of TGF-β3 micropellets.

All of these studies indicated that it is beneficial to induce differentiation of iPSCs into mesenchymal cells before the beginning of chondrogenic differentiation.

4.4.2.2.1. Chondrogenic Differentiation

Cell pellets are cultured in chondrogenic medium to induce chondrogenic differentiation. Such medium is supplemented with high-glucose Dulbecco's Modified Eagle's Medium containing 10^{-7} M dexamethasone, 10 ng/ml TGF-β3, 40 μg/ml proline, 500 ng/ml BMP-6, 100 μg/ml pyruvate, 50 μg/ml ascorbate 2-phosphate, and 50 mg/ml ITS Premix (6.25 ng/ml selenous acid, 5.35 mg/ml linoleic acid, 6.25 μg/ml insulin, 1.25 mg/ml bovine serum albumin, 6.25 μg/ml transferrin; Becton Dickinson, Franklin Lakes, NJ, USA). The medium is replaced every 3 days for 21 days. The pellets are then fixed in 10% formalin solution, embedded in paraffin, and sectioned. Finally, toluidine blue is used to stain the sections.

4.4.2.2.2. Cutaneous Wound Repair

In adult mammals, scar tissue develops due to the lack of epidermal appendages and an abundance of collagen and has been a major challenge to orthopedic surgeons for many years. Skin wound healing is a complicated process involving the coordination of a cascade of cellular responses to injury, including epithelialization, inflammation, angiogenesis, and fibroproliferation [376]. Conventional treatment methods of chronic wounds in the clinical setting include topical antibiotics, debridement with or without grafting, and bandages, and advanced treatment methods such as application of bioengineered skin substitutes and growth factors. No effective treatment is available to prevent scar formation in deep dermal wounds. Obviously, there is an urgent need for new treatments for both deep dermal wounds and chronic wounds that develop hypertrophic scars [377]. Ideally, MSC-based therapies used for skin wound healing will stimulate cellular responses to injury and promote wound regeneration instead of scar formation.

The application of BMSCs in skin wound healing has been widely reported. BMSC-treated skin defects showed less skin contraction and accelerated healing with increased re-epithelialization rates [378]. Ju-Won Kim *et al.* [379] showed that MSC-processed wounds exhibited rapid defect closure and cellular proliferation as well as increased collagen synthesis and revascularization. In another model, BMSC-treated wounds showed significantly accelerated wound healing. Notably, BMSCs, but not CD34(+) bone marrow cells in the wound formed glandular structures and expressed the keratinocyte-specific protein keratin, suggesting a significant contribution of BMSCs to skin regeneration. In addition, BMSCs in conditioned medium can differentiate into endothelial cell tubes. Western blot and real-time polymerase chain reaction have shown high levels of angiopoietin-1 and VEGF in BMSCs and significantly larger numbers of these proteins in BMSC-treated skin defects. Thus, studies have indicated that MSCs facilitate skin wound healing by differentiation and secretion of proangiogenic factors [380].

Stoff *et al.* [381] suggested that MSCs might be used in defect healing even when the MSCs are from genetically unrelated donors. They created full-thickness skin lesions in adult white rabbits and then treated the wounds by transplantation of human MSCs. The transplanted human MSCs did not exhibit an inflammatory response in the rabbit model. More importantly, the transplanted human MSCs significantly inhibited scar formation and enhanced the tensile strength of the wounds in the rabbit model [381].

Rapid progress has been made in past several years; however, there is still no evidence that MSC therapy can promote regenerative "scarless" healing when MSC differentiation takes place in injured skin. A large number of difficulties must be resolved. First, it is necessary to establish a comprehensive and consistent definition of MSCs before their therapeutic administration. Moreover, the transfer and transplantation of MSCs must be optimized to increase both paracrine signaling and differentiation of MSCs within the defects. The elements that promote MSC differentiation into skin phenotypes *in vivo* will play a decisive role in tailoring MSC-based therapies that avoid or reduce scarring. How MSC paracrine signaling accurately promotes defect restoration also requires further investigation.

4.4.2.3. Oral Mucosa Regeneration

Skin grafts or flaps for reconstruction of oral defects can be used to correct tissue complications such as impregnation and infection [382]. In the treatment of cleft palate, the limited availability of oral mucosa may lead to bone exposure and scar formation, which can stunt the development of the dentomaxillary complex [383]. Oral keratinocytes can be cultured and transplanted to treat oral defects [384]. However, studies on the characteristics of these cells *in vivo* are rare. Izumi *et al.* [385] reported the successful reconstruction of oral mucosal defects in 15 patients using autologous keratinocytes cultured on an a cellular dermal substitute [385]. In addition, Peramo *et al.* [386] described the use of AlloDerm® with mucosal and skin keratinocytes for reconstruction of mucocutaneous defects at the junction of the skin and lip *in vitro*. The use of such a technique for reconstruction of other mucocutaneous defects, such as vaginal or anal defects, was also considered feasible [386].

4.4.2.4. Fat Tissue Regeneration

Soft tissue defects can develop from a variety of causes in daily life, including congenital defects, infection, trauma, and tumor resection. Some studies have been performed with the purpose of enhancing adipose tissue regeneration. Numerous researchers have indicated that cell therapy is an ideal method because it has the potential to entirely regenerate healthy vascularized adipose tissue and completely fill the defect volume. ESCs and ASCs have been considered ideal candidates for adipose tissue regeneration because of the unlimited source of cells for cell therapy. The generation of ASCs opens a new avenue for cell transplantation-based regenerative medicine. The adipogenic potential of ASCs is not essentially different from that of ESCs.

4.4.2.4.1. Adipogenic Differentiation

Preadipocytes can be induced to differentiate until they reach 80% confluence. The medium is removed, 1500 μl of pre warmed PBS (containing 1% antibiotic) is added to a six-well plate to wash the cells, and the PBS is aspirated. Finally, the induction medium is added.

Adipogenic differentiation can be induced by culture medium containing 0.5 μM dexamethasone, 50 μM indomethacin, and 0.5 mM isobutylmethylxanthine. The adipocyte differentiation medium should be changed every 3 to 12 days of differentiation until mature adipocytes are obtained. The cells can then be fixed with either 70% ethanol or 10% formalin solution. The medium is aspirated, and the cells are washed with PBS and fixed with a fixative solution (70% ethanol or 10% formalin for 30 min or 1 h, respectively). Cell staining is conducted after removal of the fixative. The fixed cells are stored at −20°C or 4°C for several months and can continue to be used during this time. Notably, when using 70% ethanol as the fixative, the lipids may become eluted from the cells. Staining of the cells with 0.5% Oil Red O at room temperature should be performed for at least 60 min to detect the neutral lipids. This can be performed outside the biosafety cabinet. In addition, two-step fluorescent staining is performed using a nuclear fluorescent dye (DAPI; maximum excitation wave length, 360 nm; maximum emission, 460 nm) and a neutral lipid fluorescent dye (BODIPY; fluorescence emission, 510–665 nm). All procedures are performed at room temperature.

After removing the fixative solution from the wells, 10 μg/ml of BODIPY working fluid is added, and the cells are stained for 20 min. This staining should be performed in a dark environment. After staining, all dye liquor is removed and the cells are immediately washed several times with PBS solution. For the DAPI staining, the fixative is completely removed from the wells and the cells are dried for 1 or 2 min. The cells are then immersed in 300 nM DAPI treatment fluid within the wells and counterstained for 20 min at room temperature.

The percentage of cells that undergoes adipogenic differentiation can be calculated by microscopic examination. The number of cells with positive staining by BODIPY can be considered as a percentage relative to the total cells in the field, whose number is determined by positive nuclear staining with DAPI.

Three methods can be used to enhance adipogenesis and avoid inefficient adipogenic differentiation: (a) 5% rabbit serum can be added to enhance differentiation, (b) different peroxisome proliferator-activated receptor-γ agonists such as pioglitazone or troglitazone can be added to the medium, or (c) the differentiation medium can be added several times after a 3-day rest period; this cycle is repeated until mature adipocytes are obtained.

4.4.3. Compatibility of the Regeneration Material

Regeneration material is defined as a synthetic or natural material used to replace part of a living tissue or organ. During formation of the matrices and surface properties of tissue, an effective regeneration material should improve or promote cell adhesion, expansion, and differentiation as well as exhibit good osteoconductivity and mechanical behavior [387]. Additionally, the ideal scaffold should have good biocompatibility, a controllable degradation rate, suitable porosity, and high mechanical strength.

4.4.3.1. Evolution of Current Concepts of Biocompatibility

Biocompatibility traditionally describes devices that are implantable within an individual and intended to remain in place for a long time. Researchers who developed and used the first generation of implantable devices found it to be increasingly evident that the best biological performance was achieved by materials that induced a minimal chemical reaction. Within metallic systems, vanadium steels and plain carbon, which had problematic corrosiveness, were gradually superseded by stainless steel and then replaced by the superior passivated titanium alloys, cobalt–chromiumalloys, and platinum group metals. With respect to polymers, readily available polyesters and versatile nylons were replaced by materials more resistant to degradation, such as PMMA, polytetrafluoroethylene, silicones, and polyethylene. Consistent with this approach, the selection criteria for implantable biomaterials developed into a list of events that must be avoided. Most of these criteria were derived from events related to products that originate from release of disintegrants or contaminants or additives of the main ingredients of the biomaterial and their subsequent systemic or local biological activity. Materials were therefore chosen, or occasionally developed, on the basis that they would be nonirritant, noncarcinogenic, nontoxic, nonthrombogenic, nonimmunogenic, and so on. Such a negative list, by default, is the definition of biocompatibility.

This position initiated a re-evaluation on account of three major factors. First, it became obvious that the response to specific individual materials could vary from one application site to another. Therefore, biocompatibility does not only depend on material properties, but also the circumstances where the materials are applied. Second, a growing number of applications were needed in which the material should react specifically to the organization instead of acting as an inert material. Third, in similar cases, some applications required that the material should degrade at the appropriate time in the body instead of being retained in the body indefinitely.

Therefore, a basic standard of biocompatibility equivalent to biological safety was considered to be that the regeneration material should do no harm to patients, which is not prerequisites of materials with biocompatibility. Therefore, biocompatibility was redefined in 1987 as follows: Biocompatibility is the ability of a material to carry out a proper host response in a particular situation [388].

4.4.3.2. Growth Factors and Biologicals

The isolation and identification of BMPs [389] was initially reported in a study by Urist [338]. The study revealed that bone formation induced by demineralized bone matrix provided the framework forbio-derived growth and morphogenetic factors to enhance bone restoration. Although biological materials such as allografts, autografts, and demineralized bone matrix have a rich history in craniofacial surgery, recent technological advances and improvements in the scientific understanding of soft tissue and bone regeneration have accelerated the development of novel biotechnology-derived therapeutics for craniofacial regeneration. These therapies have surpassed the original focus on BMPs to include autogenous growth factors, morphogenetic factors, and other recombinant human growth factors. By making use of natural peptide sequences and proteins instead of synthetic chemical analogs, these therapies could make good use of the body's own biological functions and processes to promote the development of craniofacial regeneration.

In conclusion, this section has mainly discussed three major aspects of *in vitro* studies of craniofacial regeneration: (1) cell sources, (2) tissue regeneration, and (3) the compatibility of the regeneration material. This information is expected to serve as a reference for *in vivo* experiments and lay a certain foundation for the application of tissue engineering technology to craniofacial regeneration.

4.5. *IN VIVO* STUDIES OF CRANIOFACIAL REGENERATION

Healing and regeneration of defects in the craniofacial region caused by infection, trauma, tumors, or congenital disease are great challenges for orthopedic and reconstructive surgeons [389]. Although autologous tissue transplantation is the gold standard for regeneration of craniofacial defects, current treatments cannot return the tissue to its native normal state and the desirable function [390, 391]. Since the rise of tissue engineering, this technology has gradually gained the attention of oral and maxillofacial surgeons and has evolved as a valid approach to the repair and restoration of the anatomical structure and function of the damaged or diseased tissue. However, application of the experimental results from *in vitro* studies to *in vivo* use is not always available because of the differences inherent in the living environment; cell proliferation *in vitro* is very different from that *in vivo*. Therefore, it is necessary to establish reliable and validated animal models to evaluate the results of *in vitro* experiments and related surgical techniques for their application *in vivo*.

Tissue engineering requires a proper understanding of the structure–function relationships in normal and pathologic mammalian tissues based on the application of life sciences and engineering technology to develop biologic substitutes that repair, maintain, or improve the morphology and function of the damaged tissue or organs [392]. Tissue engineering is based on a triad comprising an appropriate cell source, supporting scaffold, and physiologic signaling molecules [393]. To ensure complete repair of the defect, the desired cells applied in tissue engineering should be easy to acquire, have the capacity for multidirectional differentiation, and lack immunogenicity and tumorigenicity [390, 394]. MSCs, which are self-renewable and can develop into diverse tissue types, have become the most commonly used cells in craniofacial regeneration [395].

The scaffold, which provides the microenvironment for cell adhesion, nutrient delivery, and cell growth, division, and differentiation, is the framework and core element of tissue engineering [396]. Ideally, the scaffold should have the following features: (1) suitable degradation and resorption rates to match tissue growth *in vivo*, (2) suitable porosity and pore size to facilitate the migration of cells, the delivery of nutrients, and the excretion of metabolic waste; (3) suitable surface chemical properties and microstructure for cell attachment, proliferation, and differentiation; (4) plasticity, toughness, and strength; (5) the ability to combine with growth factors to induce tissue regeneration; and (6) ease of disinfection [396]. Scaffold materials can be roughly divided into four categories: natural biodegradable materials, artificial synthetic biodegradable materials, inorganic materials, and composite materials. Each has its own advantages and disadvantages. Below, we elaborate on each material's clinical merits and shortcomings from two viewpoints: the spatial conformation and type of material.

Growth factors have a wide range of biological activities, including promotion of target cell proliferation and differentiation. They enhance the anti-infection capacity and kill efficiency, affect cell metabolism, and are critical signaling molecules [397]. These signaling molecules are the focus of this section. We will first discuss the general biological characteristics of growth factors, then summarize the application of growth factors in different scenarios of tissue regeneration. Finally, we will cover the current status and prospects of the application of growth factors.

4.5.1. Animal Models

4.5.1.1. Bone Defect Models

Calvarial Done Defect Models

The degree of bone defect healing is largely dependent upon the wound size [398]. Thus, an experimental bone wound used to assess repair should be large enough to avoid spontaneous healing. The smallest intraosseous wound that will not heal spontaneously during the lifetime of an animal is termed the CSD [399]. Various experimental animal models of bone defects have been described; however, there is little consistency in the choice of an appropriate experimental model for craniofacial bone research [399].

The biologic inertness of thick cortical bone induced by a poor blood supply and relative deficiency of bone marrow makes defect repair in such areas difficult. The cranial bone is thus a frequently researched site for measurement of bone repair [400]. However, the calvarial bone CSD in rats varies from 2 mm [401, 402] to 4 mm [403, 404]. At present, 5-mm full-thickness defects are optimal. Other models of rat skull defects have also been described. For example, Kobayashi [405] created a 10 × 10 × 1mm full-thickness defect to test the effect of insulin-like growth factor 1 on craniofacial bone healing. Raposo-Amaral [406] produced bilateral, symmetrical, full-thickness 18-mm^2 cranial defects to quantify the amount of bone formation.

The rabbit model of cranial defects has proven to be the most convenient and effective for testing bone repair methods [407]. However, a considerable discrepancy in the size of the control defects, ranging from 8 to 15 mm [408], has been reported. To confirm the cranial CSD in the rabbit model, Dodde II [407] created different cranial defects ranging from 0.5 to 15.0 mm, then used computed tomography and histologic examination to determine the status of bone healing. The author confirmed a CSD of 15 mm based on the experimental result that defects of >15 mm failed to heal spontaneously.

Pigs are usually used for experimental regeneration of skull defects because their bone regeneration rate (1.2–1.5 mm/day) is close to that of humans (1.0–1.5 mm/day) [409]. Schlegel *et al.* [410] confirmed that monocortical CSDs (10 × 10 mm) meet the requirements of CSDs. However, Lindholm *et al.* [411] reported that a 22-mm calvarial defect was a CSD.

Dogs, mice, and sheep have also been used to establish animal models of calvarial bone defects. Cui *et al.* [412] stated that a 20 × 20-mm full-thickness calvarial defect in the canine fulfills the criteria for a CSD. A 4-mm-diameter calvarial defect is reportedly the CSD in mice [413], and that in sheep is 6 × 5 cm [414].

<u>Mandibular Bone Defect Model</u>

In oncologic surgeries of the maxillofacial region, the mandible is the site that most commonly requires bone grafts for the repair of continuity and function [399]. However, reconstructing the contour and function of the mandible remains a great challenge for maxillofacial surgeons because of masticatory stresses and potential contamination with oral flora. Therefore, it is necessary to study animal models of mandibular defects.

Rat, canine, rabbit, sheep, and swine models of mandibular defects have been used to assess the potential for bone regeneration. Different sizes of mandibular defects have been created in different animals because different species have different bone healing rates. Cheng *et al.* [415] developed a novel animal model in which the first premolar was extracted and a 10-mm-diameter, full-thickness defect was created in the mandible of rabbits. This mimicked segmental jawbone defects and met the requirements of the mandibular CSD in rabbits.

In pigs, 10-cm segmental defects [416] have been used as the bone CSD for tissue engineering studies. However, Ma *et al.* [417] reported a 6-cm CSD when the periosteum was preserved and a 2-cm bone CSD when the periosteum was removed. In growing pigs, which mimic adolescent humans, Sun [418] reported a 5-cm CSD in the mandibular angle region without buccal periosteum.

Saadeh [419] and Arosarena [420] used allogenic type I collagen or bone morphogenic proteins associated with marrow cells, respectively, to repair 4-m--diameter, full-thickness CSDs in the ramus of the mandible in rats. However, Schliephake [421] defined the CSD as a nonhealing 5-mm-diameter defect.

To determine the CSD in a canine model, Huh [422] created variously sized defects ranging from 5 to 20 mm or 30 to 60 mm in the mandible with or without periosteum, respectively. The author demonstrated that the CSD in the canine mandible was 15 and 50 mm when the periosteum was removed and preserved, respectively.

Maxillary Bone Defect Model

Bone absorption in the maxillary posterior region secondary to missing teeth can result in a short distance between the alveolar ridge of the maxilla and sinus floor as well as an inadequate bone height for placement of endosseous implants. Maxillary sinus floor elevation is an effective way to solve this problem.

Asai [423] created a 5-mm-diameter defect in the rabbit nasal bone with or without ostial occlusion to evaluate the impact of sinus air pressure on bone structure. The experimental results showed that sinus air pressure can affect the structure of the augmented bone by acting on the mucous membrane for a long time. Xia [424, 425] conducted various studies in rabbits to assess the effects of different scaffolds, cells, or cytokines on the outcome of maxillary sinus floor elevation.

Compared with rabbits, the structure and physiology of sheep are more similar to those of humans, and the maxillary sinus is larger and easier to operate on. Gutwald [426] used a sheep model of maxillary sinus elevation to determine the effect of recombinant human BMP-2 on bone formation and osseointegration in different implants. Yu [427] used a canine model to compare the effects of different stem cells on the outcome of maxillary sinus augmentation.

Abarrategi [428] created a pig model of superior maxillary CSDs to mimic the poor bone tissue volume area and test the characteristics of the robocast/BMP-2 scaffold. Lyons [429] created 5 × 8 × 3mm bilateral bone defects in the maxilla of rabbits to assess the bone tissue responses to cortico cancellous bone grafts.

Alveolar Bone Defect Model

Animal models of alveolar defects are mainly used to simulate alveolar bone defects caused by periodontal and pre-implant related bone destruction. Such models provide a basis for alveolar ridge restoration and future implant placement.

Periodontitis, which causes destruction of the supporting tissue of the tooth, is the primary cause of alveolar bone defects. Animal models of periodontal bone defects can be spontaneous or experimentally induced. Artificial defect models can be divided into three categories: acute defects, chronic defects, and acute/chronic defects. These models are often used because they are cost-effective, can reduce the experimental time, and undergo spontaneous regeneration that is asymmetrical and occurs late in life [430].

In periodontal defect models, the cementum, periodontal ligament, and alveolar bone can be surgically removed to simulate the destruction of periodontal tissues caused by periodontitis. Chantarawaratit [431] created class II furcation defects in root bifurcation in maxillary and mandibular second and third premolars in dogs by removing the most coronal 5 mm of bone as well as the bone comprising two-thirds of the diameter of the tooth crown in the buccolingual direction. To clarify the effect of periodontal ligament–derived fibroblast-like cells on regeneration of fenestration defects, Dogan [432] created defects in dogs by removing a 5 × 5mm area of the buccal alveolar plate and the remaining 3 mm of marginal bone in the coronal direction of the defects on the maxillary canine and first molar.

Acute defects caused by surgical operations are not the same as bone defects caused by human periodontitis because in the former, plaque does not affect the defect. Periodontitis-induced chronic defects can be created by placing orthodontic elastics around the circumference of the teeth, which may take up to 6 months [433]. This model is not suitable for periodontal treatment research because it destroys the interproximal bone, takes a long time, and is expensive.

An acute/chronic model is the combination of the two above-described methods. Surgical procedures are usually used to create acute defects, and plaque is used to maintain a state of chronic inflammation. These models can mimic the inflammatory state of periodontitis and inflammation-induced defects. Liu [433] created this type of model by removing 3 × 7 × 5 mm bone in the mesial region of the maxillary and mandibular first molars of pigs and placing 4-0 silk around the cervical portion of the first molars.

Peri-implant bone defects can induce a large occlusal force load on the implant, causing the implant to gradually loosen and even drop. According to the positional relationship between defects and implants, peri-implant defects in models can be roughly divided into circumferential bone defects, buccal dehiscence-type defects, mesial orbital bone defects, and supra-alveolar defects. These models also include defects caused by peri-implantitis [434]. Dogs are widely used to create peri-implant defect models because their pathological characteristics are similar to those of humans.

Kinoshita [435] created 5-mm horizontal circumferential defects by removing the cementum around the root after bone loss caused by placement of silk ligatures around the crevices had progressed to half the root length. Schwarz *et al.* [436] surgically created a buccal dehiscence-type defect model in dogs; the defect was4mm high, 3mm wide, and 2mm from the buccal surface of the bone. Kim [437] surgically created three-walled intrabony defects (4 × 4 × 4 mm) at the mesial and distal aspects of the maxillary and mandibular first and third premolars, respectively. Additionally, 5-mm supra-alveolar peri-implant defects considered CSDs were created to evaluate the availability and safety of this model before clinical application [438]. Park [439] established both conventional and immediate peri-implantitis models. To assess the effectiveness of the immediate peri-implantitis model, the author extracted the mandibular third premolars in beagles and immediately placed dental implants in the distal extraction sockets; ligatures were also placed on the bottom of the healing abutment, as in the conventional method. The experimental results showed similarly shaped bone defects after 9 months in the conventional model and after 3 months in the immediate model. Thus, because of its shorter modeling time, the immediate peri-implantitis model may be an effective alternative method with which to study the regeneration of defects caused by peri-implantitis. The vertical alveolar ridge augmentation model is mainly used to test the treatment effect of different techniques and materials on deficient alveolar ridges. Modalities to augment the vertical alveolar ridge mainly include autogenous bone grafting, guided bone regeneration, tissue engineering, and distraction osteogenesis.

Autologous bone transplantation is the gold standard for the regeneration of alveolar bone [440]. However, it is not suitable for regeneration of large bone defects because it is difficult to obtain a large amount of autologous bone, and even if the bone obtained, severe damage will be induced. Guided bone regeneration involves the use of different materials as a barrier to protect the augmentation site from the invasion of surrounding soft tissues. However, the barrier film may lead to secondary infections and subsequent failure of bone regeneration [434].

To explore a feasible method of vertical alveolar ridge augmentation, Wang [441] extracted the canine bilateral mandibular premolars and molars 8 weeks before surgery, thus creating an edentulous alveolar ridge model. Based on this model, the bone regeneration effects of different regeneration strategies, such as autogenous bone grafting and BTE, can be explored. Distraction osteogenesis is a technique for augmentation of bone defects and uses the endogenous potency of tissue regeneration to form bone in response to mechanical tension caused by distracting a distractor, which is fixed on bone segments, at a certain speed [442]. Gaggl [443] established an alveolar ridge augmentation model in sheep by inserting two distraction implants into the mandibles and reducing 4 mm of alveolar bone 7 days before surgical alveolar augmentation (Fig. **6**).

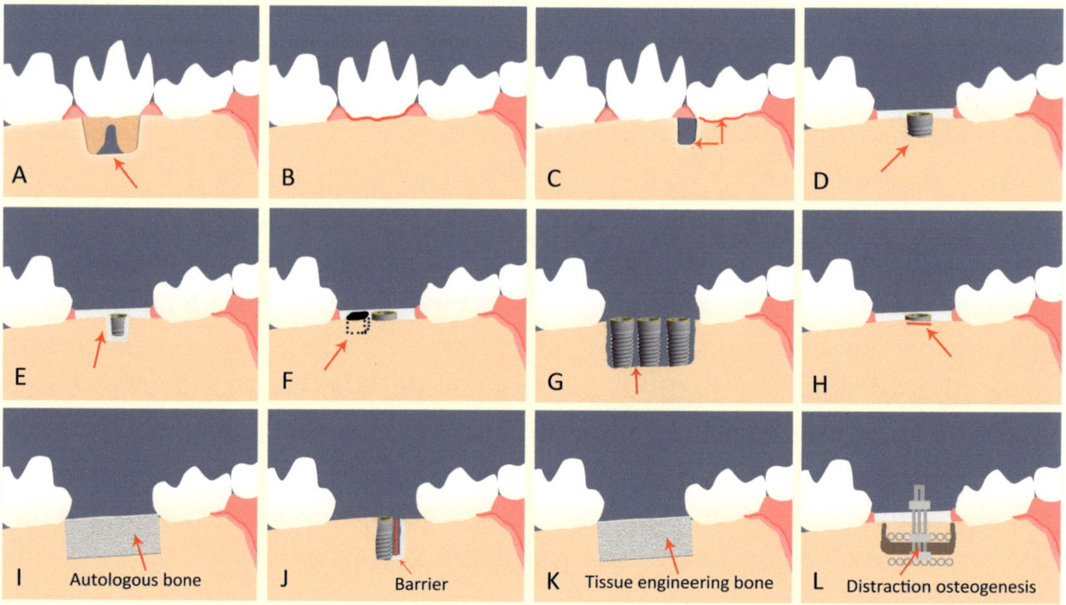

Fig. (6). Alveolar bone defect model. **(A)** Acute defect model. **(B)** Chronic defect model. **(C)** Acute/chronic defect model. **(D)** Circumferential bone defect model. **(E)** Buccal dehiscence type defect model. **(F)** Mesial/distal bone defect model. **(G)** Supra-alveolarperi-implant defect model. **(H)** Peri-implantitis model. **(I)** Autogenous bone grafting model. **(J)** Guided bone regeneration model. **(K)** Tissue engineering bone model. **(L)** Distraction osteogenesis model.

4.5.1.2. TMJ Defect Model

The TMJ is a synovial joint that plays an implant role in oral and maxillofacial function. Trauma, tumors, infection, and other factors may cause destruction of the TMJ. However, the classification of temporomandibular diseases is controversial. Different animal models, which can roughly be divided into disc repair and condylar repair according to the composition of the TMJ, have been created to mimic different types of TMJ destruction.

The TMJ disc plays an important role in rotary and translational movements and can distribute loads, absorb shocks, and lubricate the joint space. Changes in the position, contour, and structure of the disc may have serious impacts on its function. Embree [444] established a TMJ disc perforation model by creating a 2.5-mm defect using a punch biopsy in rabbits. Sharawy [445] used a double-layered flap from the synovial line of the superior and inferior recesses to repair 4- to 6-mm defects in the posterior lateral aspects of the avascular TMJ disc. The results provide a basis for repair of TMJ disc perforation using intra-articular synovial tissue. Wu [446] used a fibrin/chitosan hybrid scaffold combined with TMJ-derived synovial stem cells to repair 2-mm TMJ disc perforations in rats. The results demonstrated that the regenerative ability of TMJ synovial-derived stem cells seeded in a fibrin/chitosan scaffold could be applied to the repair of TMJ disc perforation.

A major cause of TMJ disc displacement is the presence of a pathological mechanical load [447]. Li [448] created an anterior disc displacement model in rabbits by applying different stretching forces to compare disc positions and condylar changes. The experimental results showed that a stretching force of at least 1 N successfully induced disc displacement and that a larger stretching force would induce more severe anterior disc displacement and condylar degenerative changes. Berteretche [449] created a rabbit model of partial anterior disc displacement by passing a suture around the disc over the condylar attachments and fixing two strands in the orbital cavity with a nickel-titanium spring. Legrell [450] created a non-reducing disc displacement model in growing rabbits with an intact posterior disc attachment to determine whether permanent disc displacement is a causative factor in the development of mandibular midline asymmetry.

TMJ ankylosis can occur after TMJ discectomy. Tuncel [451] used a temporalis muscle graft and fresh and cryopreserved amniotic membrane, respectively, to reconstruct the defect after discectomy and prevent TMJ ankylosis. The results showed that this method can effectively prevent TMJ ankylosis. A similar effect was obtained using autologous ASCs and polylactide discs to replace the TMJ disc in rabbits [452].

The condyle, which is covered with a layer of fibrous cartilage, is the growth center of the TMJ. If the condyle is damaged, both motor function and mandibular growth will be impaired. Animal models of condylar defects can be created by different methods, mainly surgical operations, mechanical stimuli, and drug administration [434]. Surgical operations are the most commonly used method with which to establish such animal models (*e.g.*, condylar fracture models and different degrees of condylar defect models). Impairment of the TMJ condyle can also occur after application of a mechanical stimulus that causes anterior disc displacement. Additionally, various drugs can be injected into the TMJ capsule to induce osteoarthritis, which destroys the condylar cartilage and results in remodeling of the subchondral bone. Bone/cartilage tissue engineering, distraction osteogenesis, bone grafts, and artificial succedaneum can be used in these repair models.

4.5.1.3. Cleft Lip and Palate Model

Cleft lip and palate is one of the most common types of congenital craniofacial deformities and is caused by multiple factors. Cleft lip and palate models can be divided into two categories: surgical operation models and congenital models. The surgical model only simulates the appearance of the disease, but the congenital model mimics embryogenesis. The canine unilateral alveolar lip and cleft palate model is created by removing 5 mm of bone tissue and adjacent oral and nasal mucosa between the incisor and canine teeth [453]. Inbred mice have a spontaneous tendency for the development of cleft lip, but the occurrence rate is low; it can be effectively enhanced by induction with a teratogen.

4.5.1.4. Facial Nerve Injury and Regeneration Model

Peripheral nerve injuries seriously affect both nerve function and patients' quality of life, especially when a large neural gap is present. Such injuries are usually caused by traffic accidents, natural disasters, war trauma, and tumor resection. A facial nerve injury model can be established by crushing or transection. Direct neurorrhaphy and nerve transfer can be used to repair the peripheral nerve when no nerve tissue defect is present, while nerve autografting is the gold standard method with which to bridge the neural gap [454]. However, because of the inherent disadvantages of nerve autografting, such as limited donor sources, the requirement for a second surgery area, donor site complications, and mismatch between the damaged nerve and the available nerve [455], tissue engineering has gradually become a hot research topic. Tissue-engineered nerve transplants, like other tissue engineering products, mainly comprise a scaffold, cells, and growth factors, which are the focus of the research on tissue engineering repair model. Matsumine [456] surgically created a 7-mm facial nerve buccal branch defect model and then transplanted dedifferentiated adipose cells combined with a type 1 collagen scaffold into the defect to evaluate the effects of nerve regeneration.

4.5.2. Cell Source

Healing and regenerating craniofacial region defects caused by infection, trauma, tumors, or congenital disease is a great challenge for orthopedic and reconstructive surgeons. Autologous tissue transplantation is still the gold standard for repairing defects. However, it has many disadvantages such as a limited supply, donor site morbidity, ease of infection, and mismatch between the donor and recipient site. To overcome these inherent limitations, stem cell-based regenerative medicine provides an alternative to repair and reconstruction of craniofacial defects. Stem cells can be divided into three categories based on their origin and capability of differentiation: ESCs, iPSCs, and adult stem cells [457]. ESCs are totipotent cells that can differentiate into any cell type and tissue. However, the clinical application of ESCs is limited because of immune rejection, ethical issues, and uncontrolled differentiation. Additionally, long-term trials to exclude the possibility of chromosome instability and tumorigenesis are needed [458]. iPSCs are pluripotent stem cells in an embryonic-like state and are rebuilt by importing transcription factors into somatic cells using gene transfection technology [459]. However, the use of iPSCs is less frequently recommended because of their higher incidence of tumor formation and lower differentiation ability compared with ESCs [460, 461]. Adult stem cells, especially MSCs, are often used for craniofacial reconstruction because they have the advantage of easy acquisition, good biocompatibility, a low incidence of tumor formation [462], and the potential to differentiate. In the craniofacial region, different sources of MSCs have been recognized; these include but are not limited to bone marrow, adipose tissue, peripheral blood, umbilical cord blood, and dental pulp (Fig. 7).

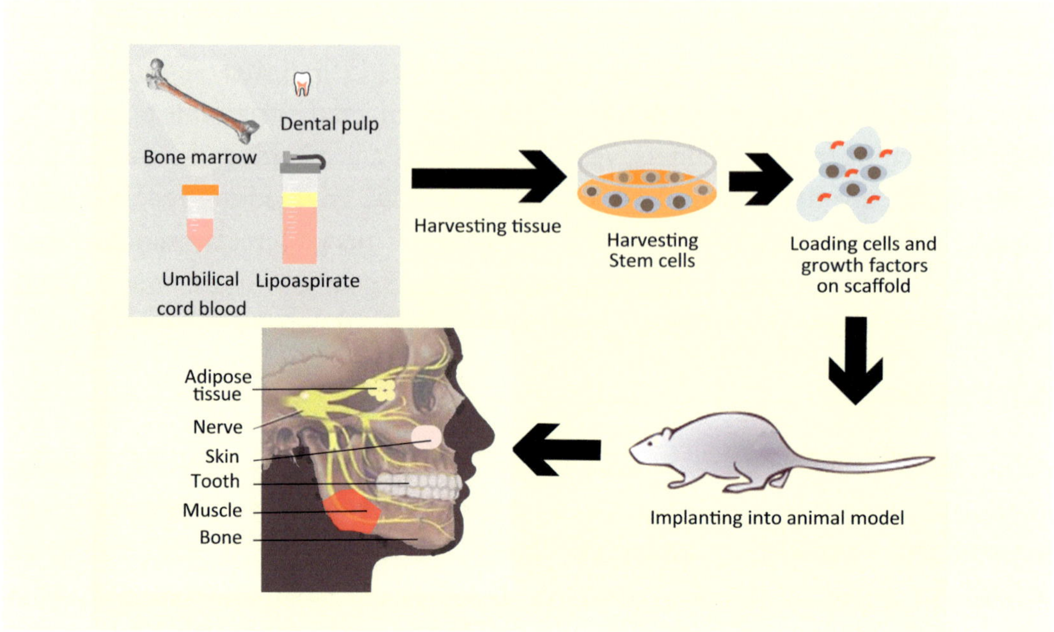

Fig. (7). *In vivo* study of craniofacial regeneration.

4.5.2.1. BMSCs

In addition to hematopoietic stem cells, BMSCs derived from bone marrow are another type of pluripotent stem cell. BMSCs have the capacity to support and regulate hematopoiesis, and exhibit multipotent differentiation into osteoblasts, chondrocytes, adipocytes, neurons, myogenic cells, and other types of cells [463]. Hence, BMSCs can be used to regenerate multiple tissue types in different craniofacial regions through the principles and practices of tissue engineering (Fig. 7).

Bone Tissue Regeneration

BMSCs can form bone *in vivo* when transplanted into immune compromised mice [464]. Additionally, defects of craniofacial bones such as the mandible, maxilla, and alveolar bone in various animal models have been successfully repaired using BMSCs [465 - 468]. The use of autologous BMSCs in the treatment of patients with different bone defects also achieved the desired results. For example, to reconstruct an obvious mandibular defect after ameloblastoma removal, Kim [469] fixed cortical block bone with plates and screws as a scaffold to form a bone tray, injected fibrin glue and differentiated autologous osteoblasts into the bone defect, and then injected autologous human BMSCs into the scaffold. After a 3-month postoperative follow-up, autologous human MSCs were injected again into the mandibular defect area to accelerate bone formation. Newly formed bone was present in the intercortical area 6 months postoperatively.

Cartilage Tissue Regeneration

Most of the cartilage of the oral and maxillofacial region is contained in the TMJ. Tissue engineering technology can be used to reconstruct cartilage destruction caused by surgery, trauma, inflammation, deformities, or TMJ disorders. BMSCs hold the capacity to give rise to chondrocytes [463]. In one study, a human-shaped mandibular condyle was generated by implanting rat MSCs encapsulated in a two-layered poly-based hydrogel into the dorsum of immunodeficient mice for 8 weeks, and microscopic analyses demonstrated stratified layers of histogenesis of cartilaginous and osseous phenotypes [470]. In another study, hyaline cartilaginous tissue formed after insertion of a gelatin β-TCP sponge loaded with BMSCs, BMP-2, and PRP into a 10-mm-diameter defect in the medial femoral condyle of horses for 16 weeks [471].

Nerve Regeneration

Facial nerve segmental defects may lead to facial muscle paralysis, which is usually found after tumor resection. *In vitro* experiments have confirmed that BMSCs can differentiate into neurons and can therefore be used as seed cells for nerve restoration in tissue engineering. One study showed that BMSCs can be induced to differentiate into Schwann cells, namely BMSC-DSCs [472]. The cells were suspended in Matrigel and shifted to 12-mm-long hollow fibers, which were then implanted into transected sciatic nerve defects in rats for 6 months. Immunohistochemical study data demonstrated that BMSC-DSCs have an enormous potential to promote peripheral nerve regeneration, and this may be an effective way to reconstruct a long-distance interval in a peripheral nerve [472]. Undifferentiated BMSCs or Schwann-like cells that had differentiated from

BMSCs within polyglycolic acid tubes were transplanted into 5-mm fragments in the marginal mandibular branch of the left facial nerve, improving facial nerve regeneration as confirmed by electromyography [473].

4.5.2.2. ASCs

The ideal stem cell for reconstructing craniofacial defects should (1) be abundant and easy to harvest, (2) be harvested through minimally invasive surgery, (3) have the capacity to differentiate into a variety of cells, and (4) be safe and effective when applied *in vivo* [474]. Although BMSCs were first used in *in vivo* studies, their inherent characteristics (low stem cell yield per harvest, necessity for a large invasive procedure, and relative difficulty in harvesting) limit the performance of *in vivo* studies and clinical application. Therefore, alternative MSCs, namely ASCs, have gradually gained the attention of researchers [475] (Fig. **7**).

Bone Regeneration

Undifferentiated human ASCs loaded onto starch-PCL scaffolds were transplanted into 4-mm calvarial CSDs in nude mice [476]. Histological and micro-computed tomography results showed new bone deposition and osseointegration in the calvarial defects [476]. Mandibular defects are larger and bear a greater force. In another study, rat mandibular CSDs were filled with ASCs transfected by lentiviral short hairpin RNA, 3D apatite-coated chitosan and chondroitin sulfate scaffolds, and BMP-2 for 8 weeks [477]. Micro-computed tomography, histology, and immunohistochemistry results suggested that the combinatorial strategy greatly contributed to mandibular regeneration and may provide an alternative method for restoration of large bone defects [477]. Sándor [478] used a similar technique to reconstruct a 10-cm anterior mandibular defect after ameloblastoma resection and inserted dental implants into the grafted site after 10 months for occlusion reconstruction.

Periodontal Tissue Regeneration

The periodontium comprises multiple tissues including cementum, the periodontal membrane, and alveolar bone. Thus, regeneration of damaged periodontium is very challenging. Tobita [479] transplanted ASCs mixed with PRP obtained from inbred rats into periodontal tissue defects of rats to determine whether ASCs promote periodontal tissue regeneration. A small quantity of alveolar bone regeneration was observed at 2 and 4 weeks after implantation, and a periodontal ligament-like structure was found 8 weeks after implantation. These experimental results showed that ASCs have the capacity to promote periodontal tissue regeneration *in vivo* and can be used as a treatment method for periodontal disease. However, clinical application will take some time because few studies have evaluated the application of ASCs in periodontal repair.

Wound Healing

ASCs can promote proliferation of dermal fibroblasts through direct contact with cells and paracrine tissues because they facilitate angiogenesis and secretion of growth factors [480]. ASCs showed properties of accelerated wound healing in animal trauma models. Kuo [481] created dorsal full-thickness skin wound defects (6 × 5 cm) in rats with streptozotocin-induced diabetes, then subcutaneously injected ASCs into the wound margin twice. Histological examination and immunohistochemical analyses of peri-wound tissue were performed after 8 weeks. The results showed that the ASCs significantly enhanced diabetic wound healing, engrafted into the local wound tissue, and entered the circulating blood. This treatment stimulated neoangiogenesis and increased tissue regeneration through paracrine and autocrine mechanisms.

Nerve Regeneration

ASCs can differentiate to Schwann-like cells originating from the ectoderm and have the ability to repair peripheral nerve injury. In an *in vivo* study, undifferentiated ASCs, differentiated ASCs, or Schwann cells were transplanted into 7mm defects of the rat facial nerves for 13 weeks [482]. Morphometric quantification analysis showed a similar nerve regeneration capability among these three different cell types. In cell-based tissue engineering, undifferentiated and differentiated ASCs can be used as a source of Schwann cells and as an alternative restorative technique for nerve regeneration [482]. With further developments in ASC differentiation and scaffold materials, the use of tissue engineering to repair nerve injury will gain broader prospects.

4.5.2.3. UCMSCs

Umbilical cord blood has the following advantages over bone marrow: rich sources, a noninvasive harvesting technique, lower immunogenicity and a lower incidence of graft-*versus*-host disease, and stronger proliferation and differentiation abilities at juvenile biological ages. Therefore, UCMSCs can be treated as a new source of cells for transplantation and gene therapy in various diseases (Fig. **7**).

Bone Regeneration

Chen [483] transplanted macroporous calcium phosphate cement scaffolds loaded with UCMSCs and BMSCs into 8-mm cranial CSDs in rats to compare the bone regeneration ability of UCMSCs and BMSCs. Micro-computed tomography and histological examination indicated that both groups had statistically similar amounts of new bone, mineral densities, and vessel densities. In short, a similar capacity for bone regeneration can be found between UCMSCs and BMSCs seeded onto calcium phosphate cement scaffolds *in vivo*. However, Todeschi [484] proposed that UCMSCs do not repair bone lesions by direct deposition of bone matrix, but by activating endogenous repair mechanisms. This conclusion was based on the fact that bone remained detectable and no new bone formation was found in an ectopic location after transplanting UCMSC-loaded scaffolds into the subcutaneous tissue for 3 weeks; however, the quantity of new blood vessels was significant increased. Additionally, orthotropic transplantation of UCMSCs into calvarial defects promoted new bone formation. However, this bone regeneration mechanism differs from that of BMSCs, which promote new bone deposition through activation of host cells.

Nerve Regeneration

MSCs can differentiate into neurons. Previous research has shown that MSCs can promote nerve regeneration and functional recovery [485]. Ryu [486] transplanted material with ASCs, BMSCs, and UCMSCs into injured canine spinal cords to identify the most suitable seed cells for nerve regeneration. After 8 weeks, all groups exhibited an increased quantity of surviving neurons and neurofilament-positive fibers within the lesions. A greater number of regenerated nerves and slightly more inflammation were found in the UCMSC group than in the other groups. However, the specific mechanism by which UCMSCs promote nerve regeneration is unclear. Paracrine action may be a key mechanism underlying the effects of human UCMSCs in peripheral nerve repair [487].

4.5.2.4. DPSCs

DPSCs can be found in the cell-rich region and core of the pulp tissue, which has a certain capacity for differentiation and proliferation. Corresponding tissue components can be found after injection of DPSCs *in vivo*, and tissue repair can be performed by various paracrine growth factors. This has important implications in tooth regeneration and dental pulp repair, especially with the development of tissue engineering technologies. DPSCs represent another choice for tissue engineering (Fig. **7**).

Bone Regeneration

DPSCs can still form mineral deposits after transplantation *in vivo* [488] and can thus be used in cell therapy and tissue engineering. Riccio [489] used fibroin scaffolds combined with osteo differentiated DPCSs to restore cranial CSDs in immune compromised rats. After 4 weeks, imaging and histological analysis showed a large amount of new bone formation, indicating that the DPSC/fibroin composites have great potential for the reconstruction of large cranial defects in animal models and may become a solution for the reconstruction of large craniofacial bone defects in humans. DPCSs can differentiate into osteoblasts and endothelial cells *in vitro*, and their transplantation into immune compromised rats regenerates an organization with an integral blood supply similar to the structure of adult bone [490]. The capability for bidirectional differentiation is important for revascularization of regenerated bone tissue.

Nerve Regeneration

DPSCs can differentiate into mature nerve cells *in vitro*. Spyridopoulos [491] transplanted collagen nerve conduits with DPSCs into intercostal nerve defects in pigs. Histological examination revealed progressive development of neural tissue and immunohistochemical expression of neuron-specific enolase, and electrophysiological examination indicated similar potentials between the regions infused with stem cells and the proximal stumps of these regions and slightly smaller potentials in the respective distal stumps. Thus, DPSCs have the potential to promote peripheral nerve regeneration.

Dental Tissue Regeneration

Composite materials are usually used to repair tooth tissue defects caused by caries, trauma, or excessive wear. However, this often leads to pulp necrosis or tooth discoloration. Tissue engineering provides a new way to resolve necrosis and discoloration. When DPSCs were transplanted into immune compromised mice, a dentin-like structure was found in the periphery of human odontoblast-like cells wrapped in medullary interstitial tissue [492]. Therefore, DPSCs have the capacity to generate a dentin/pulp-like complex. Wang [493] established a canine pulpless model mainly by performing pulpectomy and root canal preparation in young permanent incisor teeth, then transplanted expanded DPSCs with Gelfoam into the pulpless root canal. Radiographic and histologic analyses demonstrated that DPSCs generated pulp-like tissues comprising blood vessels and a dentin-like organization. These findings indicate that pulp regeneration of immature teeth is feasible using stem cells.

4.5.3. Scaffolds

The ultimate goal of tissue engineering is to form an organ or tissue with a 3D structure. However, cells only form a 2D cellular layer based on their growth and migration; this does not meet the demands of tissue engineering. Scaffolds have a 3D structure that can provide an ideal environment for cell adherence, migration, and growth. Therefore, scaffolds play a significant role in tissue regeneration.

Tissue engineering scaffolds are artificial synthesis frameworks with a 3D structure that mimics the extracellular matrix. The contour and microstructure of tissue engineering scaffolds determine the ultimate shape and structure of the reconstructed tissue. Ideal scaffolds should have specific requirements: (1) suitable degradation and resorption rates to match cell/tissue growth *in vivo*; (2) suitable porosity and pore size to facilitate the migration of cells, the delivery of nutrients, and the excretion of metabolic waste; (3) suitable surface chemical properties and microstructure for cell attachment, proliferation, and differentiation; (4) plasticity, toughness, and strength; (5) the ability to combine with growth factors to induce tissue regeneration; and (6) ease of disinfection [494] (Fig. 7).

4.5.3.1. Classification of Scaffold Materials

Natural Biodegradable Scaffolds

Natural biodegradable materials are polymer materials that can be absorbed by the body. They include polysaccharides, proteins, polynucleotides, and a variety of biofibers [495]. These materials promote cell adhesion and interaction between the scaffold and host tissue. However, they also have some disadvantages such as weak mechanical properties, poor reproducibility, and an uncontrolled degradation rate. Kim [496] designed a hybrid (core/shell) scaffold comprising external collagen and inner alginate to overcome the poor mechanical stability of scaffolds and the poor mechanical nature of collagen. Their hybrid scaffold showed good structural stability (core–shell structure), an increased Young's modulus, and excellent cell viability, similar to collagen scaffolds. Seol [497] created hybrid scaffolds comprising hydrogel for cartilage tissue and ceramic for bone tissue; the scaffolds were press-fitted into osteochondral tissue defects in rabbit knee joints. Twelve weeks after implantation, histological and immunohistochemical examination revealed that the shapes of newly generated cartilage tissues were similar to those of inherent cartilage tissues and that the newly generated cartilage tissues were smoothly linked to the surrounding inherent tissues. This indicates that hybrid scaffolds comprising hydrogel and ceramic can enhance the mechanical stability of hydrogel scaffolds and the regeneration of cartilage tissue at the defect site.

Natural hydrogel materials such as cellulose, chitosan, and alginate will transform into harder scaffolds with a certain shape and mechanical strength from the flow state when injected *in vivo*. This offers the advantages of flexibility of these materials to adapt their shape to the required form and the ability to act as delivery devices. Yao [498] injected adipocytes/collagen/alginate microspheres into the subcutaneous tissue of the head of nude mice and found that vascularized adipose tissue had formed after 4 weeks. This suggests that a functional vascular anastomosis forms between the graft and host and that the composite can act as a delivery device for cell transplantation and has the capacity to promote vascularized adipose tissue regeneration.

Synthetic Biodegradable Polymer Scaffolds

Synthetic polymers have been widely used in tissue engineering because they have good mechanical strength and controllable biodegradation rates and allow artificial modification and regulation. Such polymers do have some shortcomings: they can cause aseptic inflammation during the process of degradation, are hydrophobic, and have insufficient absorption. However, these problems can be solved by surface modification using biological molecules [499, 500]. To increase hydrophilicity, Valence [501] treated PCL vascular grafts with or without cold air plasma and subcutaneously implanted them into rats for 3 weeks. The plasma-treated scaffolds exhibited a more extended and dense cellular infiltrate, induced better cellularization of the graft wall, and did not affect the endothelialization rate or intimal hyperplasia. Therefore, plasma treatment is a useful tool with which to increase the hydrophilicity of scaffolds without compromising mechanical strength. Boron serves as a surfactant of PLGA scaffolds and can enhance diffusion, attachment, and calcium deposition of ADSCs. In bone defect models, scaffolds treated with boron can increase bone regeneration by improving the expression of osteocalcin, VEGF, and collagen type I protein [502].

Synthetic biodegradable polymeric materials mainly include polylactic acid (PLA), polyglycolic acid (PGA), and the copolymer-polylactic-co-glycolic acid (PLGA). The degradation rate of PLA is slow, while that of PGA is fast. The degradation rate and mechanical strength of implanted PLGA scaffolds can be controlled according to the proportion of the PLA and PGA. Kim [503] subcutaneously transplanted VEGF-loaded PLA porous scaffolds modified with calcium phosphate mineral in rats for 4 weeks. Histological analysis indicated excellent tissue compatibility of the scaffolds, and immunohistochemical staining showed a significant increase in new capillaries. Ren [504] repaired mandibular defects in rabbits using PLGA/MSC tissue-engineered bone, which significantly promoted cell growth and had a fairly better therapeutic effect on the mandibular defects. After 3 months, the defects were completely reconstructed.

Inorganic Scaffolds

The inorganic materials used in tissue engineering scaffolds mainly include HA, TCP, natural coral, and BAG, all of which have high compressive strength, chemical stability, and good biocompatibility and conductibility. Some shortcomings of inorganic materials include a too-slow degradation rate, low porosity, and high brittleness. Del [505] demonstrated that the creation of macropores of >100 μm can significantly improve the resorption rate of calcium phosphate cement associated with almost complete bone replacement. This scaffold material is usually used to repair bone defects in the nonloading area of the craniofacial region. Morishita [506] transplanted HA covered with autogenous BMSCs that had undergone osteogenic differentiation to fill a patient's bone defects after tumor resection. Serial plain radiographs and computed tomography showed immediate healing potential without adverse reactions. To improve the performance of inorganic materials, the scaffolds used in craniofacial defect regeneration usually involve a composite of inorganic and organic polymer materials, which can improve the biocompatibility, strength, toughness, and characteristics of the inorganic materials.

Composite Scaffolds

Generally, scaffolds fabricated by materials do not readily meet the requirements of tissue engineering. Composite scaffolds composed of two or more kinds of materials are needed to make up for the lack of various materials. Different natural biodegradable polymer materials are used to construct composite scaffolds mainly to mimic the extracellular matrix as much as possible. Whu [507] transplanted chondrocyte-seeded chitosan: gelatin (1:1) scaffolds into rabbit articular cartilage defects. After 6.5 months, the average elastic modulus of regenerative tissue was about 0.539 MPa, close to that of normal cartilage. Furthermore, the expression of type II collagen was similar to that of normal hyaline cartilage. Yan [508] transplanted silk fibroin/chondroitin sulfate/HA scaffolds onto dorsal full-thickness defects of rats to assess defect regeneration. The experimental results showed that the scaffolds improved angiogenesis and collagen deposition and promoted dermis regeneration.

Composite scaffolds that consist of both synthetic and natural biodegradable polymer materials can not only maintain mechanical strength and a proper degradation rate, but they also show improved surface bioactivity. Han [509] transplanted TGF-β1 and BMP-2 double-loaded chitosan-gelatin hydrogel/poly (l-lactide-co-glycolide) scaffolds into 5-mm-diameter defects in rabbits. After 2 months, hematoxylin and eosin staining and immunohistochemical examination showed regeneration of a hyaline cartilage-like surface and subchondral bone with good integration of the native tissues.

Composite scaffolds consisting of inorganic materials and biodegradable polymers are mainly used in BTE, and these materials can improve the strength, toughness, and biocompatibility of the scaffolds. Ben-David [510] subcutaneously transplanted a tissue-like construct formed by cultured BMSCs that had undergone osteogenic differentiation loaded into a composite comprising hydrogel microspheres and ceramic particles for 3 weeks in athymic nude mice. Significant new bone formation was present after 8 weeks. In an *in vivo* study, Koç [511] demonstrated that MC3T3-E1 cell-laden chitosan/HA was biocompatible, promoted the formation of new blood vessels and calcified matrix, and secreted osteoblast-specific protein. Thus, such composite scaffolds may be used in the regeneration and reconstruction of bone defects.

4.5.3.2. Three-Dimensional Structure

Scaffolds play a significant role in tissue engineering, especially their shape and microstructure. The structure of a scaffold defines the geometry of cell adhesion sites and thus affects cell adhesion, migration, and proliferation and the ultimate shape of the regenerated tissue [512]. The microstructure comprises the pore morphology, size, interconnectivity, and porosity, all of which influence nutrient transfer, waste disposal, and vascular development.

The shape of the scaffold is related to the scaffold surface, which is used for cell adhesion and proliferation. The greater the scaffold surface-to-volume ratio, the greater the scaffold's cellular activity and surface bioactivity [513]. Thus, an increased surface-to-volume ratio of the scaffold is beneficial to tissue regeneration. However, the shape of the scaffold should be adaptable to the tissue defect. Improvements in scaffold microstructure are needed to achieve enhanced tissue regeneration.

Pore size can affect both the mechanical properties and diffusion of a scaffold. Small pores may restrict cell infiltration and revascularization, and large pores can result in degradation of the mechanical properties [514]. Most researchers suggest that the most appropriate pore size for cell in growth, blood capillary formation, and new tissue formation should range from 100 to 500 μm and that the interval should be larger than 100 μm [515]. However, the optimal size of the scaffold is related to many factors, such as the material, porosity, location of the defect to be repaired, and other factors. Additionally, there is no standard pore size [516]. To study the influence of scaffold pore size on bone regeneration, Guda [517] designed two types of scaffolds with similar porosity and pillar thickness but with different pore sizes. One was a bilayer scaffold with a pore size of 200 μm in the outer layer and 450 μm in the inner layer; the other was a purely trabecular-like scaffold with a pore size of 340 μm. The scaffolds were transplanted into 10-mm segmental defects of rabbits for 8 weeks. Micro-computed tomography, histology, and flexural testing analysis showed that the trabecular scaffolds had more uniform new bone and greater bone regeneration. Thus, a larger uniform pore size will result in more functional bone regeneration. With respect to cartilage regeneration, Zhang [518] demonstrated that the pore size had no obvious effect on cell proliferation but exhibited different effects on cartilage regeneration because homogeneity of the cell distribution throughout the scaffolds was ensured by interconnectivity. The collagen porous scaffolds with a precisely controlled pore structure by ice particulates of 150 to 250 μm most effectively increased the formation and mechanical properties of the cartilage.

Porosity of scaffolds also plays an important role in tissue regeneration. To study the effect of porosity on bone regeneration, Aarvold [519] fabricated two kinds of scaffolds with different porosities but the same pore characteristics while maintaining mechanical stability and pore interconnectivity. One scaffold contained 45 pores per inch, and the other contained 30. The results showed that the higher porosity was associated with greater surface reactivity and a greater increase in density. Alteration in scaffold porosity positively influenced both cell proliferation and osteogenic differentiation. Horst [520] also demonstrated that increased porosity accelerated the cell proliferation and distribution *in vitro* and promoted tissue ingrowth and formation of new blood vessels *in vivo*.

The spatial arrangement of pores also affects the mechanical properties and regeneration capacity of a scaffold. The pore arrangement should ideally be consistent with that of the defect to best simulate the structure of the defect. Most scaffolds prepared for bone regeneration have a uniform isotropic, sponge-like structure to mimic the architecture of trabecular bone. Despang [521] created an anisotropic bioceramic with pores arranged in parallel, similar to the honeycomb arrangement of Haversian canals in cortical bone. Histological analysis showed that the new bone formation within the channel-like pores had the morphology of osteon-like tissues after transplantation of the scaffold into rabbit cortical femur defects.

In short, design parameters such as pore size, porosity, and shape are related to and contingent on one another. Although it is impossible to establish a uniform standard for all scaffolds, it is crucial to create scaffolds with the capacity for promoting tissue regeneration.

4.5.4. Growth Factors

Growth factors are proteins with the capacity to induce and simulate cell proliferation and promote cellular differentiation. They play an important role in tissue or organ restoration and reconstruction and have become one of the three major topics of study in the field of tissue engineering. Various growth factors are now applied in tissue engineering and are respectively described below (Fig. **7**).

4.5.4.1. BMP-2

BMP-2 can promote osteoblast differentiation and homogeneous ectopic bone formation [522]. Many studies have showed that BMP-2 can successfully repair and construct bone defects *in vivo*, and BMP-2 has been extensively examined as an osteoinductive growth factor for tissue engineering [523]. Furthermore, BMP-2 enhances the osteogenic potential of MSCs in a dose- and time-dependent manner. However, the biological half-life of growth factors is relatively short, which results in serious limitations to their clinical applications. Gene-based delivery systems and controlled-release scaffolds allow for the achievement of a sustained high concentration of growth factors within the scope of the defects [524]. Obvious bone regeneration was found in mice calvarial defects treated with as little as 0.19 μg of BMP-2 using a vitrigel delivery system 4 weeks after transplantation [525]. Additionally, BMP-2 gene-transduced human gingival fibroblasts promoted osseous healing of 8-mm cranial defects in rats, and this group had significantly greater regenerated bone area than non-gene-transduced group [526]. *In vitro*, BMP-2 can induce differentiation of MSCs into chondrocytes, promote chondrocyte proliferation, and promote synthesis of the extracellular matrix; *in vivo*, BMP-2 can promote cartilage regeneration. 3D PCL scaffolds loaded with chemically coupled BMP-2 and primary chondrocytes can promote significantly more cartilage regeneration *in vivo* than can unprocessed scaffolds and do not particularly speed up endochondral ossification even *in vivo* environment of conditions permitting and rich blood supplying [527]. As for periodontal tissue, HA particles and collagen gel with autologous canine BMP-2 and periodontal ligament stem cells were transplanted into peri-implantitis models created by ligature placement in the dental cervix. After 3 months, the results showed new bone formation and re-osseointegration in the defects caused by peri-implantitis [528].

4.5.4.2. bFGF

bFGF is an effective bone regeneration factor. It can increase the transcription level of osteoblast differentiation factor and alkaline phosphatase activity as well as induce differentiation of MSCs *in vitro* and bone regeneration *in vivo* [529]. Yang [530] transplanted Bio-Oss scaffolds seeded with BMSCs that were transfected with the bFGF gene into mandibular defects of rabbits. After 12 weeks, bone mineral density and histological examination showed that bFGF increased the mineralization of BMSCs and induced more new bone and cartilage formation compared with the scaffolds in the BMSC group. Thus, bFGF can enhance new bone regeneration.

An appropriate concentration of bFGF can maintain the activity of cultured chondrocytes, synthesize specific extracellular matrix, inhibit the activity of chondrocyte terminal differentiation and calcification *in vitro*, and promote cartilage regeneration. Takafuji [531] created an osteoarthritis model by boring a 10-mm defect on the mandibular condyle head surface, then filled the defect with lyophilized collagen containing 0.1 or 1.0 μg of FGF-2. Three weeks after the operation, initial cartilage formation was observed in the 0.1-μg group by optical microscopy, but the amount of generated cartilage was small and a 0.1 μg dose of FGF-2 induced articular cartilage regeneration. These results indicate that an optimum concentration of FGF-2 could be used in the clinical setting to restore TMJ articular cartilage defects.

Hiraoka [532] performed a similar study on the promotion of fat tissue regeneration by bFGF. Gelatin microspheres with different water contents were used to study the relationship between water content and the release rate of bFGF. The author found that the higher the water content, the faster the rate of bFGF release. After a collagen sponge scaffold loaded with microspheres containing different doses of bFGF was transplanted into rat fat pad defects, significantly more *in situ* adipose tissue formation was induced in the bFGF group than in the free bFGF group, organization accompanied by angiogenesis was observed in the scaffold implanted with the microspheres containing 1.0 μg of bFGF, and less regenerated adipose tissue was observed at the lower water content of the microspheres. Therefore, the combination of a scaffold with suitable controlled release of bFGF can better promote the regeneration of adipose tissue *in situ*.

bFGF is also considered an important angiogenic factor that can promote endothelial cell proliferation and new blood vessel formation [533]. However, bFGF placed directly in the wound region may result in loss of bioactivity. To better accelerate wound healing, Liu [534] created composite hydrogel-alginate microspheres loaded with bFGF merged into carboxymethyl chitosan–poly (vinyl alcohol) and then transplanted the composite into full-thickness burn wounds of rats. The group with the composites exhibited significantly faster recovery rates than did the control group, showed re-epithelialization and regeneration of the dermis, and had the highest density of newly formed and mature blood vessels during the 2 months of treatment.

4.5.4.3. TGF-β1

TGF-β1 acts as a cartilage-inducing factor that enhances proliferation and induces both synthetic activity of cartilage cells and MSC differentiation into chondrocytes and osteoblasts [535]. It also plays an important role in the repair and reconstruction of cartilage. Intra-articular injection of TGF-β1 can result in joints covered with a layer of well-organized fibrocartilage, an increased proteoglycan content, and maintenance of normal microarchitectural properties [536]. Because of the inherent instability of TGF-β1, high amounts of this protein cause adverse side effects with inefficient cartilage formation, limiting the clinical application of TGF-β1. Injection of TGF-β1 into a visible blue light inducible chitosan hydrogel system can result in stabilization of TGF-β1 signaling and promote cartilage restoration in challenging environments (*e.g.*, in rat models of partial-thickness cartilage defects, which exhibit a limited source of subchondral bone marrow elements). Therefore, this new injectable delivery modality can improve clinical cartilage repair [537]. TGF-β1can interact with other growth factors to produce a synergistic or antagonistic effect on tissue regeneration. In a mouse calvarial CSD model, co-delivery of TGF-β1 and stromal cell-derived factor-1β attenuated bone formation levels seen with TGF-β1 alone [538]. Tachi [539] implanted collagen sponges containing 2.5 μg of BMP-2 and 25 ng of TGF-β1 into mouse calvarial defects. The defects in the TGF-β1 group were completely filled with formed orthotopic bone, whereas those in the BMP-2 group were only partial filled. These results indicate that TGF-β1 has the capacity to enhance BMP-2 in inducing ectopic bone formation.

4.5.4.4. VEGF

Many studies have shown that VEGF can promote angiogenesis and may be used for skin and bone tissue regeneration. Yamashima [540] treated rat mandibular bone defects with VEGF, and histological and soft X-ray radiological observations after 30 days indicated that VEGF effectively induced blood vessel formation and offers potential for bone regeneration. However, high local concentrations of VEGF induce deformities and nonfunctional neovascularization. To overcome these disadvantages, Wernike [541] co-precipitated VEGF onto BCP ceramics to achieve sustained release of the growth factors. After transplanting the BCP ceramics into cranial CSDs of mice for 28 days, intravital microscopy and histomorphometric analysis showed that the extended bioavailability of low concentrations of VEGF can promote biomaterial vascularization, osseointegration, and bone formation. Feng [542] transplanted BMSCs transfected with the VEGF165 gene combined with β-TCP material into femoral defects of rabbits. Vascularization and bone volume were significantly greater than those in the control group. These findings indicate that VEGF is suitable for clinical application because it can accelerate tissue-engineered bone formation by facilitating vascularization and ossification. Thus far, the regeneration of vascularized bone tissue has been achieved using VEGF and other growth factors.

4.5.4.5. PRP

PRP contains a number of mitogenic and chemotactic growth factors such as platelet-derived growth factor, TGF-β1, insulin-like growth factor, VEGF, and epidermal growth factor. These factors can interact with one another to repair tissue defects, providing better results than with single cell factors.

Marx [543] used PRP combined with cancellous bone to repair mandibular defects of patients. Compared with the control group, the rate of bone formation in the experimental group increased by 1.62 to 2.16 times and the bone density was also greater. This proves that PRP plays a role in promoting bone formation. Since then, many scholars have carried out research on the ability of PRP to promote bone tissue repair and its application to BTE. For example, Chen [544] used PRP combined with bovine porous bone mineral (BPBM) and a bio-guide membrane (BGM) to repair rabbit circular mandible bicortical bony defects. After 4, 8, and 12 weeks postoperatively, the defects filled with PRP/BPBM/BGM showed significantly higher osteoblast activity, a higher number of mitochondria and rough endoplasmic reticulum within osteoblasts, and a higher concentration of alkaline phosphatase than in the BPBM/BGM group and control group. The entire closure rate of bone defects treated with PRP/BPBM/BGM (50.0%) was significantly higher than that of bone defects treated with BPBM/BGM (16.6%), indicating that PRP has a stronger ability to promote bone regeneration. However, the effects of different concentrations of PRP on bone regeneration vary. A too-low concentration will not achieve the best effect, but a higher concentration could inhibit bone regeneration [545]. So it is necessary to seek the optimal concentration of PRP for tissue regeneration according to different seed cells, scaffolds, defects, and so on.

Kütük [543] created bilateral TMJ defects on the condylar fibrocartilage, hyaline cartilage, and bone to induce an osteoarthritic TMJ in rabbits. PRP was applied to the right joints of the rabbits (PRP group), and the left joints were treated with physiologic saline (control group). After 4 weeks, histologic examination and scanning electron microscopy analysis showed greater regeneration of the fibrocartilage and hyaline cartilage and better ultrastructural architecture of the collagen fibrils in the PRP group. Thus, PRP may promote fibrocartilage regeneration. Mehrotra [544] used a titanium mini plate to fix preshaped HA/collagen condyles with PRP to the ramus of patients with TMJ ankylosis, and temporal fascia was placed in between. After 3 months, radiographic evaluation showed a less opaque condyle, and at 18 months, the opacity was more defined, suggesting a newly formed condyle. Thus, PRP can improve both the aesthetics and function of the condyle.

In the physiological state, a series of cellular factors are involved in wound healing. Therefore, PRP containing a multitude of factors is a better choice for soft tissue regeneration than in PRP containing a single growth factor. The use of ASCs and PRP to repair skin wounds in mice showed that a better skin regeneration effect could be achieved by the combined use of PRP and ASCs, while the effect of ASC alone was only slight. The reason for this finding is that PRP has the capacity to significantly improve ASC proliferation, stimulate the secretion of angiogenic growth factor from ASCs and surrounding mouse host cells in the wound areas, and promote neovascularization in the wound beds. PRP has a single biological effect on contained growth factors, but also affects the interactions among these growth factors. These characteristics provide new avenues for bone, cartilage, and soft tissue regeneration.

In conclusion, *in vivo* studies of craniofacial regeneration mainly focus on animal models, cell sources, scaffolds, and growth factors. The conventional approach is to establish an animal model, harvest stem cells, expand and differentiate those with the potential to form new tissue, and seed the cells on scaffolds with or without growth factors. The scaffold is then transplanted as a composite into an animal model. After a period of time, the effect of tissue regeneration is observed. This chapter has concentrated mainly on these factors with the aim of promoting further research in tissue regeneration.

CONFLICT OF INTEREST

The authors confirm that they have no conflict of interest to declare for this publication.

ACKNOWLEDGEMENTS

This work was supported by grants from the National Natural Science Foundation of China (81371125), the Program of Sichuan Provincial Science and Technology Bureau (2014JY0044), the Project of Sichuan Provincial Education Department (10ZB030), the Project of Department of Health of Sichuan Province (No. 80170), and the Key Project of Luzhou Medical College (201207).

REFERENCES

[1] Rajan A, Eubanks E, Edwards S, *et al.* Optimized cell survival and seeding efficiency for craniofacial tissue engineering using clinical stem cell therapy. Stem Cells Transl Med 2014; 3(12): 1495-503. [http://dx.doi.org/10.5966/sctm.2014-0039] [PMID: 25378653]

[2] Chen W, Liu J, Manuchehrabadi N, Weir MD, Zhu Z, Xu HH. Umbilical cord and bone marrow mesenchymal stem cell seeding on macroporous calcium phosphate for bone regeneration in rat cranial defects. Biomaterials 2013; 34(38): 9917-25. [http://dx.doi.org/10.1016/j.biomaterials.2013.09.002] [PMID: 24054499]

[3] Mikos AG, Herring SW, Ochareon P, *et al.* Engineering complex tissues. Tissue Eng 2006; 12(12): 3307-39. [http://dx.doi.org/10.1089/ten.2006.12.3307] [PMID: 17518671]

[4] Kaigler D, Krebsbach PH, Wang Z, West ER, Horger K, Mooney DJ. Transplanted endothelial cells enhance orthotopic bone regeneration. J Dent Res 2006; 85(7): 633-7. [http://dx.doi.org/10.1177/154405910608500710] [PMID: 16798864]

[5] Johnson PC, Mikos AG, Fisher JP, Jansen JA. Strategic directions in tissue engineering. Tissue Eng 2007; 13(12): 2827-37. [http://dx.doi.org/10.1089/ten.2007.0335] [PMID: 18052823]

[6] Eun SC. Facial transplantation surgery introduction. J Korean Med Sci 2015; 30(6): 669-72. [http://dx.doi.org/10.3346/jkms.2015.30.6.669] [PMID: 26028914]

[7] Phillips C, Bennett ME, Broder HL. Dentofacial disharmony: psychological status of patients seeking treatment consultation. Angle Orthod 1998; 68(6): 547-56.
[PMID: 9851353]

[8] Costello BJ, Shah G, Kumta P, Sfeir CS. Regenerative medicine for craniomaxillofacial surgery. Oral Maxillofac Surg Clin North Am 2010; 22(1): 33-42.
[http://dx.doi.org/10.1016/j.coms.2009.10.009] [PMID: 20159476]

[9] Susarla SM, Swanson E, Gordon CR. Craniomaxillofacial reconstruction using allotransplantation and tissue engineering: challenges, opportunities, and potential synergy. Ann Plast Surg 2011; 67(6): 655-61.
[http://dx.doi.org/10.1097/SAP.0b013e31822c00e6] [PMID: 21825966]

[10] Vacanti JP, Langer R. Tissue engineering: the design and fabrication of living replacement devices for surgical reconstruction and transplantation. Lancet 1999; 354 (Suppl. 1): S32-4.
[http://dx.doi.org/10.1016/S0140-6736(99)90247-7] [PMID: 10437854]

[11] Gamie Z, Tran GT, Vyzas G, *et al.* Stem cells combined with bone graft substitutes in skeletal tissue engineering. Expert Opin Biol Ther 2012; 12(6): 713-29.
[http://dx.doi.org/10.1517/14712598.2012.679652] [PMID: 22500826]

[12] Abukawa H, Papadaki M, Abulikemu M, *et al.* The engineering of craniofacial tissues in the laboratory: a review of biomaterials for scaffolds and implant coatings. Dent Clin North Am 2006; 50(2): 205-16.
[http://dx.doi.org/10.1016/j.cden.2005.11.006] [PMID: 16530058]

[13] Rodríguez-Lozano FJ, Insausti CL, Iniesta F, *et al.* Mesenchymal dental stem cells in regenerative dentistry. Med Oral Patol Oral Cir Bucal 2012; 17(6): e1062-7.
[http://dx.doi.org/10.4317/medoral.17925] [PMID: 22926467]

[14] Scheller EL, Krebsbach PH, Kohn DH. Tissue engineering: state of the art in oral rehabilitation. J Oral Rehabil 2009; 36(5): 368-89.
[http://dx.doi.org/10.1111/j.1365-2842.2009.01939.x] [PMID: 19228277]

[15] Ma J, Both SK, Yang F, *et al.* Concise review: cell-based strategies in bone tissue engineering and regenerative medicine. Stem Cells Transl Med 2014; 3(1): 98-107.
[http://dx.doi.org/10.5966/sctm.2013-0126] [PMID: 24300556]

[16] Sundelacruz S, Kaplan DL. Stem cell- and scaffold-based tissue engineering approaches to osteochondral regenerative medicine. Semin Cell Dev Biol 2009; 20(6): 646-55.
[http://dx.doi.org/10.1016/j.semcdb.2009.03.017] [PMID: 19508851]

[17] Varghese S, Hwang NS, Ferran A, *et al.* Engineering musculoskeletal tissues with human embryonic germ cell derivatives. Stem Cells 2010; 28(4): 765-74.
[http://dx.doi.org/10.1002/stem.325] [PMID: 20178108]

[18] Mao JJ, Vunjak-Novakovic G, Mikos AG, Atala A. Translational Approaches in Tissue Engineering and Regenerative Medicine. Boston, MA: Artech House 2007.

[19] Alvarez P, Hee CK, Solchaga L, *et al.* Growth factors and craniofacial surgery. J Craniofac Surg 2012; 23(1): 20-9.
[http://dx.doi.org/10.1097/SCS.0b013e318240c6a8] [PMID: 22337369]

[20] Schilephake H. Bone growth factors in maxillofacial skeletal reconstruction. Int J Oral Maxillofac Surg 2002; 31(5): 469-84.
[http://dx.doi.org/10.1054/ijom.2002.0244] [PMID: 12418561]

[21] Langer R, Vacanti JP. Tissue engineering. Science 1993; 260(5110): 920-6.
[http://dx.doi.org/10.1126/science.8493529] [PMID: 8493529]

[22] Yuan J, Cao Y, Liu W. Biomimetic scaffolds: implications for craniofacial regeneration. J Craniofac Surg 2012; 23(1): 294-7.
[http://dx.doi.org/10.1097/SCS.0b013e318241bae1] [PMID: 22337428]

[23] Brugmann SA, Cordero DR, Helms JA. Craniofacial ciliopathies: A new classification for craniofacial disorders. Am J Med Genet A 2010; 152A(12): 2995-3006.
[http://dx.doi.org/10.1002/ajmg.a.33727] [PMID: 21108387]

[24] Tang XJ, Gui L, Zhang ZY, Yin L, Pellerin P. A spectacle frame classification for rare craniofacial clefts. Plast Reconstr Surg 2012; 129(1): 195e-7e.
[http://dx.doi.org/10.1097/PRS.0b013e3182365ec3] [PMID: 22186561]

[25] Gargano F, Szymanski K, Bosman M, Podda S. Tessier 1-13 Atypical Craniofacial Cleft. Eplasty 2015; 15: ic32.
[PMID: 26171107]

[26] Koch BL. Pediatric considerations in craniofacial trauma. Neuroimaging Clin N Am 2014; 24(3): 513-29.
[http://dx.doi.org/10.1016/j.nic.2014.03.002] [PMID: 25086809]

[27] De Sousa A. Psychological issues in oral and maxillofacial reconstructive surgery. Br J Oral Maxillofac Surg 2008; 46(8): 661-4.
[http://dx.doi.org/10.1016/j.bjoms.2008.07.192] [PMID: 18771826]

[28] Stricker M. [Evolution of the craniofacial surgery in Nancy]. Bull Acad Natl Med 2003; 187(9): 1661-6.
[PMID: 15369236]

[29] Mohan R, Borsuk DE, Dorafshar AH, *et al.* Aesthetic and functional facial transplantation: a classification system and treatment algorithm. Plast Reconstr Surg 2014; 133(2): 386-97.
[http://dx.doi.org/10.1097/01.prs.0000437259.24069.35] [PMID: 24469171]

[30] Brown JS, Rogers SN, McNally DN, Boyle M. A modified classification for the maxillectomy defect. Head Neck 2000; 22(1): 17-26.
[http://dx.doi.org/10.1002/(SICI)1097-0347(200001)22:1<17::AID-HED4>3.0.CO;2-2] [PMID: 10585601]

[31] Brown JS, Shaw RJ. Reconstruction of the maxilla and midface: introducing a new classification. Lancet Oncol 2010; 11(10): 1001-8.
[http://dx.doi.org/10.1016/S1470-2045(10)70113-3] [PMID: 20932492]

[32] Ethunandan M, Mellor TK. Haemangiomas and vascular malformations of the maxillofacial region--a review. Br J Oral Maxillofac Surg 2006; 44(4): 263-72.
[http://dx.doi.org/10.1016/j.bjoms.2005.06.032] [PMID: 16107299]

[33] George A, Mani V, Noufal A. Update on the classification of hemangioma. J Oral Maxillofac Pathol 2014; 18 (Suppl. 1): S117-20.
[http://dx.doi.org/10.4103/0973-029X.141321] [PMID: 25364160]

[34] Schultz BD, Sosin M, Nam A, *et al.* Classification of mandible defects and algorithm for microvascular reconstruction. Plast Reconstr Surg 2015; 135(4): 743e-54e.
[http://dx.doi.org/10.1097/PRS.0000000000001106] [PMID: 25811586]

[35] Cordeiro PG, Santamaria E. A classification system and algorithm for reconstruction of maxillectomy and midfacial defects. Plast Reconstr Surg 2000; 105(7): 2331-46.
[http://dx.doi.org/10.1097/00006534-200006000-00004] [PMID: 10845285]

[36] Goodday RH. Management of fractures of the mandibular body and symphysis. Oral Maxillofac Surg Clin North Am 2013; 25(4): 601-16.
[http://dx.doi.org/10.1016/j.coms.2013.07.002] [PMID: 24021623]

[37] Stacey DH, Doyle JF, Mount DL, Snyder MC, Gutowski KA. Management of mandible fractures. Plast Reconstr Surg 2006; 117(3): 48e-60e.
[http://dx.doi.org/10.1097/01.prs.0000209392.85221.0b] [PMID: 16525255]

[38] Kumar V, Singh AK, Kumar P, *et al.* Blast injury face: An exemplified review of management. Natl J Maxillofac Surg 2013; 4(1): 33-9.
[http://dx.doi.org/10.4103/0975-5950.117878] [PMID: 24163550]

[39] Alvi A, Doherty T, Lewen G. Facial fractures and concomitant injuries in trauma patients. Laryngoscope 2003; 113(1): 102-6.
[http://dx.doi.org/10.1097/00005537-200301000-00019] [PMID: 12514391]

[40] Murray JM. Mandible fractures and dental trauma. Emerg Med Clin North Am 2013; 31(2): 553-73.
[http://dx.doi.org/10.1016/j.emc.2013.02.002] [PMID: 23601489]

[41] Mazzola RF, Mazzola IC. Facial clefts and facial dysplasia: revisiting the classification. J Craniofac Surg 2014; 25(1): 26-34.
[http://dx.doi.org/10.1097/SCS.0b013e3182a2ea94] [PMID: 24406554]

[42] Tanna N, Wan DC, Perry AD, Kawamoto HK, Bradley JP. Paramedian mandibular cleft: revisiting the Tessier classification. J Craniofac Surg 2012; 23(1): e38-40.
[http://dx.doi.org/10.1097/SCS.0b013e318241db58] [PMID: 22337459]

[43] Delforge A, Raoul G, Wiss A, Kerbrat JB, Ferri J. A classification of cranio-facial syndromes. Orthod Fr 2011; 82(2): 223-32.
[http://dx.doi.org/10.1051/orthodfr/2011121] [PMID: 21624342]

[44] Rychlik D, Wójcicki P, Koźlik M. Osteoplasty of the alveolar cleft defect. Adv Clin Exp Med 2012; 21(2): 255-62.
[PMID: 23214291]

[45] Sun J, Shen Y, Li J, Zhang ZY. Reconstruction of high maxillectomy defects with the fibula osteomyocutaneous flap in combination with titanium mesh or a zygomatic implant. Plast Reconstr Surg 2011; 127(1): 150-60.
[http://dx.doi.org/10.1097/PRS.0b013e3181fad2d3] [PMID: 21200209]

[46] Zan T, Li H, Gu B, *et al.* Surgical treatment of facial soft-tissue deformities in postburn patients: a proposed classification based on a retrospective study. Plast Reconstr Surg 2013; 132(6): 1001e-14e. [http://dx.doi.org/10.1097/PRS.0b013e3182a97e81] [PMID: 24281605]

[47] Fearon JA. Rare craniofacial clefts: a surgical classification. J Craniofac Surg 2008; 19(1): 110-2. [PMID: 18216674]

[48] Allam KA, Wan DC, Kawamoto HK, Bradley JP, Sedano HO, Saied S. The spectrum of median craniofacial dysplasia. Plast Reconstr Surg 2011; 127(2): 812-21. [http://dx.doi.org/10.1097/PRS.0b013e318200aa08] [PMID: 21285785]

[49] Patel NP, Tantri MD. Median cleft of the upper lip: a rare case. Cleft Palate Craniofac J 2010; 47(6): 642-4. [http://dx.doi.org/10.1597/09-065] [PMID: 20500055]

[50] Ichida M, Komuro Y, Yanai A. Consideration of median cleft lip with frenulum labii superior. J Craniofac Surg 2009; 20(5): 1370-4. [http://dx.doi.org/10.1097/SCS.0b013e3181ae42bf] [PMID: 19816259]

[51] Kayser MR. Surgical flaps. Selected Readings in Plastic Surgery 1999; 9: 1-63.

[52] McGregor IA, Morgan G. Axial and random pattern flaps. Br J Plast Surg 1973; 26(3): 202-13. [http://dx.doi.org/10.1016/0007-1226(73)90003-9] [PMID: 4580012]

[53] Bak M, Jacobson AS, Buchbinder D, Urken ML. Contemporary reconstruction of the mandible. Oral Oncol 2010; 46(2): 71-6. [http://dx.doi.org/10.1016/j.oraloncology.2009.11.006] [PMID: 20036611]

[54] Hayden RE, Mullin DP, Patel AK. Reconstruction of the segmental mandibular defect: current state of the art. Curr Opin Otolaryngol Head Neck Surg 2012; 20(4): 231-6. [http://dx.doi.org/10.1097/MOO.0b013e328355d0f3] [PMID: 22894990]

[55] Klotch DW, Gump J, Kuhn L. Reconstruction of mandibular defects in irradiated patients. Am J Surg 1990; 160(4): 396-8. [http://dx.doi.org/10.1016/S0002-9610(05)80551-3] [PMID: 2221241]

[56] Petersen K, Hayes DK, Blice JP, Hale RG. Prevention and management of infections associated with combat-related head and neck injuries. J Trauma 2008; 64(3) (Suppl.): S265-76. [http://dx.doi.org/10.1097/TA.0b013e318163d2a6] [PMID: 18316971]

[57] Foster RD, Anthony JP, Sharma A, Pogrel MA. Vascularized bone flaps versus nonvascularized bone grafts for mandibular reconstruction: an outcome analysis of primary bony union and endosseous implant success. Head Neck 1999; 21(1): 66-71. [http://dx.doi.org/10.1002/(SICI)1097-0347(199901)21:1<66::AID-HED9>3.0.CO;2-Z] [PMID: 9890353]

[58] Cordeiro PG, Disa JJ, Hidalgo DA, Hu QY. Reconstruction of the mandible with osseous free flaps: a 10-year experience with 150 consecutive patients. Plast Reconstr Surg 1999; 104(5): 1314-20. [http://dx.doi.org/10.1097/00006534-199910000-00011] [PMID: 10513911]

[59] Kroll SS, Schusterman MA, Reece GP, *et al.* Choice of flap and incidence of free flap success. Plast Reconstr Surg 1996; 98(3): 459-63. [http://dx.doi.org/10.1097/00006534-199609000-00015] [PMID: 8700982]

[60] Tatara AM, Wong ME, Mikos AG. *In vivo* bioreactors for mandibular reconstruction. J Dent Res 2014; 93(12): 1196-202.
[http://dx.doi.org/10.1177/0022034514547763] [PMID: 25139360]

[61] Heliotis M, Lavery KM, Ripamonti U, Tsiridis E, di Silvio L. Transformation of a prefabricated hydroxyapatite/osteogenic protein-1 implant into a vascularised pedicled bone flap in the human chest. Int J Oral Maxillofac Surg 2006; 35(3): 265-9.
[http://dx.doi.org/10.1016/j.ijom.2005.07.013] [PMID: 16257511]

[62] Warnke PH, Springer IN, Wiltfang J, *et al.* Growth and transplantation of a custom vascularised bone graft in a man. Lancet 2004; 364(9436): 766-70.
[http://dx.doi.org/10.1016/S0140-6736(04)16935-3] [PMID: 15337402]

[63] Orringer JS, Shaw WW, Borud LJ, Freymiller EG, Wang SA, Markowitz BL. Total mandibular and lower lip reconstruction with a prefabricated osteocutaneous free flap. Plast Reconstr Surg 1999; 104(3): 793-7.
[http://dx.doi.org/10.1097/00006534-199909010-00028] [PMID: 10456533]

[64] Boraiah S, Paul O, Hawkes D, Wickham M, Lorich DG. Complications of recombinant human BMP-2 for treating complex tibial plateau fractures: a preliminary report. Clin Orthop Relat Res 2009; 467(12): 3257-62.
[http://dx.doi.org/10.1007/s11999-009-1039-8] [PMID: 19693635]

[65] Miller MJ, Goldberg DP, Yasko AW, *et al.* Guided bone growth in sheep: a model for tissue-engineered bone flaps. Tissue Eng 1996; 2(1): 51-9.
[http://dx.doi.org/10.1089/ten.1996.2.51] [PMID: 19877951]

[66] Cheng MH, Brey EM, Allori A, *et al.* Ovine model for engineering bone segments. Tissue Eng 2005; 11(1-2): 214-25.
[http://dx.doi.org/10.1089/ten.2005.11.214] [PMID: 15738676]

[67] Brey EM, Cheng MH, Allori A, *et al.* Comparison of guided bone formation from periosteum and muscle fascia. Plast Reconstr Surg 2007; 119(4): 1216-22.
[http://dx.doi.org/10.1097/01.prs.0000254361.74614.bb] [PMID: 17496593]

[68] Thomson RC, Mikos AG, Beahm E, *et al.* Guided tissue fabrication from periosteum using preformed biodegradable polymer scaffolds. Biomaterials 1999; 20(21): 2007-18.
[http://dx.doi.org/10.1016/S0142-9612(99)00103-9] [PMID: 10535812]

[69] Cheng MH, Brey EM, Allori AC, *et al.* Periosteum-guided prefabrication of vascularized bone of clinical shape and volume. Plast Reconstr Surg 2009; 124(3): 787-95.
[http://dx.doi.org/10.1097/PRS.0b013e3181b17a91] [PMID: 19730297]

[70] Cheng MH, Brey EM, Ulusal BG, Wei FC. Mandible augmentation for osseointegrated implants using tissue engineering strategies. Plast Reconstr Surg 2006; 118(1): 1e-4e.
[http://dx.doi.org/10.1097/01.prs.0000221120.11128.1a] [PMID: 16816662]

[71] Moore WR, Graves SE, Bain GI. Synthetic bone graft substitutes. ANZ J Surg 2001; 71(6): 354-61.
[http://dx.doi.org/10.1046/j.1440-1622.2001.02128.x] [PMID: 11409021]

[72] Kagami H, Agata H, Inoue M, *et al.* The use of bone marrow stromal cells (bone marrow-derived multipotent mesenchymal stromal cells) for alveolar bone tissue engineering: basic science to clinical

translation. Tissue Eng Part B Rev 2014; 20(3): 229-32. [http://dx.doi.org/10.1089/ten.teb.2013.0578] [PMID: 24494719]

[73] Wakimoto M, Ueno T, Hirata A, Iida S, Aghaloo T, Moy PK. Histologic evaluation of human alveolar sockets treated with an artificial bone substitute material. J Craniofac Surg 2011; 22(2): 490-3. [http://dx.doi.org/10.1097/SCS.0b013e318208bacf] [PMID: 21415629]

[74] LeGeros RZ, Lin S, Rohanizadeh R, Mijares D, LeGeros JP. Biphasic calcium phosphate bioceramics: preparation, properties and applications. J Mater Sci Mater Med 2003; 14(3): 201-9. [http://dx.doi.org/10.1023/A:1022872421333] [PMID: 15348465]

[75] Hollinger JO, Kleinschmidt JC. The critical size defect as an experimental model to test bone repair materials. J Craniofac Surg 1990; 1(1): 60-8. [http://dx.doi.org/10.1097/00001665-199001000-00011] [PMID: 1965154]

[76] Lu M, Rabie AB. The effect of demineralized intramembranous bone matrix and basic fibroblast growth factor on the healing of allogeneic intramembranous bone grafts in the rabbit. Arch Oral Biol 2002; 47(12): 831-41. [http://dx.doi.org/10.1016/S0003-9969(02)00119-X] [PMID: 12450514]

[77] Miloro M, Haralson DJ, Desa V. Bone healing in a rabbit mandibular defect using platelet-rich plasma. J Oral Maxillofac Surg 2010; 68(6): 1225-30. [http://dx.doi.org/10.1016/j.joms.2009.09.090] [PMID: 20227807]

[78] Busuttil Naudi K, Ayoub A, McMahon J, *et al.* Mandibular reconstruction in the rabbit using beta-tricalcium phosphate (β-TCP) scaffolding and recombinant bone morphogenetic protein 7 (rhBMP-7) - histological, radiographic and mechanical evaluations. J Craniomaxillofac Surg 2012; 40(8): e461-9. [http://dx.doi.org/10.1016/j.jcms.2012.03.005] [PMID: 22507295]

[79] Kneser U, Schaefer DJ, Polykandriotis E, Horch RE. Tissue engineering of bone: the reconstructive surgeon's point of view. J Cell Mol Med 2006; 10(1): 7-19. [http://dx.doi.org/10.1111/j.1582-4934.2006.tb00287.x] [PMID: 16563218]

[80] Baqain ZH, Anabtawi M, Karaky AA, Malkawi Z. Morbidity from anterior iliac crest bone harvesting for secondary alveolar bone grafting: an outcome assessment study. J Oral Maxillofac Surg 2009; 67(3): 570-5. [http://dx.doi.org/10.1016/j.joms.2008.09.023] [PMID: 19231782]

[81] Goudy S, Lott D, Burton R, Wheeler J, Canady J. Secondary alveolar bone grafting: outcomes, revisions, and new applications. Cleft Palate Craniofac J 2009; 46(6): 610-2. [http://dx.doi.org/10.1597/08-126.1] [PMID: 19860505]

[82] Szpalski C, Wetterau M, Barr J, Warren SM. Bone tissue engineering: current strategies and techniques--part I: Scaffolds. Tissue Eng Part B Rev 2012; 18(4): 246-57. [http://dx.doi.org/10.1089/ten.teb.2011.0427] [PMID: 22029448]

[83] Szpalski C, Barbaro M, Sagebin F, Warren SM. Bone tissue engineering: current strategies and techniques--part II: Cell types. Tissue Eng Part B Rev 2012; 18(4): 258-69. [http://dx.doi.org/10.1089/ten.teb.2011.0440] [PMID: 22224439]

[84] Vacanti CA, Vacanti JP. The science of tissue engineering. Orthop Clin North Am 2000; 31(3): 351-6. [http://dx.doi.org/10.1016/S0030-5898(05)70155-3] [PMID: 10882462]

[85] Vacanti JP. Editorial: tissue engineering: a 20-year personal perspective. Tissue Eng 2007; 13(2): 231-2.
[http://dx.doi.org/10.1089/ten.2006.0351] [PMID: 17504059]

[86] Vacanti JP, Langer R. Tissue engineering: the design and fabrication of living replacement devices for surgical reconstruction and transplantation. Lancet 1999; 354 (Suppl. 1): S32-4.
[http://dx.doi.org/10.1016/S0140-6736(99)90247-7] [PMID: 10437854]

[87] Nolff MC, Gellrich NC, Hauschild G, *et al.* Comparison of two beta-tricalcium phosphate composite grafts used for reconstruction of mandibular critical size bone defects. Vet Comp Orthop Traumatol 2009; 22(2): 96-102.
[PMID: 19290389]

[88] Pinto JG, Primo BT, Gassen HT, *et al.* Alpha-tricalcium phosphate cement in the reconstruction of bone defects in rats. Acta Cir Bras 2011; 26(2): 135-9.
[http://dx.doi.org/10.1590/S0102-86502011000200011] [PMID: 21445477]

[89] Rojbani H, Nyan M, Ohya K, Kasugai S. Evaluation of the osteoconductivity of α-tricalcium phosphate, β-tricalcium phosphate, and hydroxyapatite combined with or without simvastatin in rat calvarial defect. J Biomed Mater Res A 2011; 98(4): 488-98.
[http://dx.doi.org/10.1002/jbm.a.33117] [PMID: 21681941]

[90] Kamitakahara M, Ohtsuki C, Miyazaki T. Review paper: behavior of ceramic biomaterials derived from tricalcium phosphate in physiological condition. J Biomater Appl 2008; 23(3): 197-212.
[http://dx.doi.org/10.1177/0885328208096798] [PMID: 18996965]

[91] Mokbel N, Bou Serhal C, Matni G, Naaman N. Healing patterns of critical size bony defects in rat following bone graft. Oral Maxillofac Surg 2008; 12(2): 73-8.
[http://dx.doi.org/10.1007/s10006-008-0107-7] [PMID: 18618164]

[92] Payer M, Lohberger B, Stadelmeyer E, Bartmann C, Windhager R, Jakse N. Behaviour of multipotent maxillary bone-derived cells on beta-tricalcium phosphate and highly porous bovine bone mineral. Clin Oral Implants Res 2010; 21(7): 699-708.
[http://dx.doi.org/10.1111/j.1600-0501.2009.01856.x] [PMID: 20412093]

[93] Schwartz Z, Weesner T, van Dijk S, *et al.* Ability of deproteinized cancellous bovine bone to induce new bone formation. J Periodontol 2000; 71(8): 1258-69.
[http://dx.doi.org/10.1902/jop.2000.71.8.1258] [PMID: 10972641]

[94] Tapety FI, Amizuka N, Uoshima K, Nomura S, Maeda T. A histological evaluation of the involvement of Bio-Oss in osteoblastic differentiation and matrix synthesis. Clin Oral Implants Res 2004; 15(3): 315-24.
[http://dx.doi.org/10.1111/j.1600-0501.2004.01012.x] [PMID: 15142094]

[95] Dominici M, Le Blanc K, Mueller I, *et al.* Minimal criteria for defining multipotent mesenchymal stromal cells. The International Society for Cellular Therapy position statement. Cytotherapy 2006; 8(4): 315-7.
[http://dx.doi.org/10.1080/14653240600855905] [PMID: 16923606]

[96] Horwitz EM, Le Blanc K, Dominici M, *et al.* Clarification of the nomenclature for MSC: the international society for cellular therapy position statement. Cytotherapy 2005; 7(5): 393-5.

[http://dx.doi.org/10.1080/14653240500319234] [PMID: 16236628]

[97] Bueno DF, Kerkis I, Costa AM, *et al.* New source of muscle-derived stem cells with potential for alveolar bone reconstruction in cleft lip and/or palate patients. Tissue Eng Part A 2009; 15(2): 427-35. [http://dx.doi.org/10.1089/ten.tea.2007.0417] [PMID: 18816169]

[98] Takano-Yamamoto T, Kawakami M, Sakuda M. Defects of the rat premaxilla as a model of alveolar clefts for testing bone-inductive agents. J Oral Maxillofac Surg 1993; 51(8): 887-91. [http://dx.doi.org/10.1016/S0278-2391(10)80110-7] [PMID: 8336226]

[99] Raposo-Amaral CE, Kobayashi GS, Almeida AB, *et al.* Alveolar osseous defect in rat for cell therapy: preliminary report. Acta Cir Bras 2010; 25(4): 313-7. [http://dx.doi.org/10.1590/S0102-86502010000400002] [PMID: 20676486]

[100] Asko Seljavaara S. Abstracts of the Seventh Congress of the International Society of Reconstructive Microsurgery. New York, NY. 1983; pp. 19-30.

[101] Saint-Cyr M, Wong C, Schaverien M, Mojallal A, Rohrich RJ. The perforasome theory: vascular anatomy and clinical implications. Plast Reconstr Surg 2009; 124(5): 1529-44. [http://dx.doi.org/10.1097/PRS.0b013e3181b98a6c] [PMID: 20009839]

[102] Taylor GI, Palmer JH. The vascular territories (angiosomes) of the body: experimental study and clinical applications. Br J Plast Surg 1987; 40(2): 113-41. [http://dx.doi.org/10.1016/0007-1226(87)90185-8] [PMID: 3567445]

[103] Lecours C, Saint-Cyr M, Wong C, *et al.* Freestyle pedicle perforator flaps: clinical results and vascular anatomy. Plast Reconstr Surg 2010; 126(5): 1589-603. [http://dx.doi.org/10.1097/PRS.0b013e3181f02ee3] [PMID: 21042115]

[104] D'Arpa S, Cordova A, Pignatti M, Moschella F. Freestyle pedicled perforator flaps: safety, prevention of complications, and management based on 85 consecutive cases. Plast Reconstr Surg 2011; 128(4): 892-906. [http://dx.doi.org/10.1097/PRS.0b013e3182268c83] [PMID: 21921765]

[105] Gir P, Cheng A, Oni G, Mojallal A, Saint-Cyr M. Pedicled-perforator (propeller) flaps in lower extremity defects: a systematic review. J Reconstr Microsurg 2012; 28(9): 595-601. [http://dx.doi.org/10.1055/s-0032-1315786] [PMID: 22715046]

[106] Nelson JA, Fischer JP, Brazio PS, Kovach SJ, Rosson GD, Rad AN. A review of propeller flaps for distal lower extremity soft tissue reconstruction: Is flap loss too high? Microsurgery 2013; 33(7): 578-86. [http://dx.doi.org/10.1002/micr.22134] [PMID: 23861186]

[107] Innocenti M, Menichini G, Baldrighi C, Delcroix L, Vignini L, Tos P. Are there risk factors for complications of perforator-based propeller flaps for lower-extremity reconstruction? Clin Orthop Relat Res 2014; 472(7): 2276-86. [http://dx.doi.org/10.1007/s11999-014-3537-6] [PMID: 24706021]

[108] Lazzeri D, Huemer GM, Nicoli F, *et al.* Indications, outcomes, and complications of pedicled propeller perforator flaps for upper body defects: a systematic review. Arch Plast Surg 2013; 40(1): 44-50. [http://dx.doi.org/10.5999/aps.2013.40.1.44] [PMID: 23362479]

[109] Innocenti M, Baldrighi C, Delcroix L, Adani R. Local perforator flaps in soft tissue reconstruction of

the upper limb. Handchir Mikrochir Plast Chir 2009; 41(6): 315-21.
[http://dx.doi.org/10.1055/s-0029-1237357] [PMID: 20024864]

[110] Pribaz JJ, Chan RK. Where do perforator flaps fit in our armamentarium? Clin Plast Surg 2010; 37(4): 571-9.
[http://dx.doi.org/10.1016/j.cps.2010.06.007] [PMID: 20816513]

[111] Hanasono MM, Lee JC, Yang JS, Skoracki RJ, Reece GP, Esmaeli B. An algorithmic approach to reconstructive surgery and prosthetic rehabilitation after orbital exenteration. Plast Reconstr Surg 2009; 123(1): 98-105.
[http://dx.doi.org/10.1097/PRS.0b013e3181904b95] [PMID: 19116542]

[112] Neligan PC, Mulholland S, Irish J, *et al.* Flap selection in cranial base reconstruction. Plast Reconstr Surg 1996; 98(7): 1159-66.
[http://dx.doi.org/10.1097/00006534-199612000-00005] [PMID: 8942901]

[113] Gliklich RE, Rounds MF, Cheney ML, Varvares MA. Combining free flap reconstruction and craniofacial prosthetic technique for orbit, scalp, and temporal defects. Laryngoscope 1998; 108(4): 482-7.
[http://dx.doi.org/10.1097/00005537-199804000-00004] [PMID: 9546256]

[114] Federspil PA. [Craniofacial prostheses for facial defects]. HNO 2010; 58(6): 621-31.
[http://dx.doi.org/10.1007/s00106-010-2087-9] [PMID: 20393689]

[115] Archibald S, Jackson S, Thoma A. Paranasal sinus and midfacial reconstruction. Clin Plast Surg 2005; 32(3): 309-25.
[http://dx.doi.org/10.1016/j.cps.2005.01.001] [PMID: 15979471]

[116] Menick FJ. Artistry in aesthetic surgery. Aesthetic perception and the subunit principle. Clin Plast Surg 1987; 14(4): 723-35.
[PMID: 3652614]

[117] McGregor IA, Morgan G. Axial and random pattern flaps. Br J Plast Surg 1973; 26(3): 202-13.
[http://dx.doi.org/10.1016/0007-1226(73)90003-9] [PMID: 4580012]

[118] Manson PN, Hoopes JE, Chambers RG, Jaques DA. Algorithm for nasal reconstruction. Am J Surg 1979; 138(4): 528-32.
[http://dx.doi.org/10.1016/0002-9610(79)90413-6] [PMID: 384819]

[119] Adams C, Ratner D. Composite and free cartilage grafting. Dermatol Clin 2005; 23(1): 129-40.
[http://dx.doi.org/10.1016/j.det.2004.08.005] [PMID: 15620624]

[120] Ayhan M, Gorgu M, Erdogan B, *et al.* Various applications of chondrocutaneous composite grafts in secondary cleft lip nose patients. J Craniofac Surg 2006; 17(6): 1065-71.
[http://dx.doi.org/10.1097/01.scs.0000231628.10511.bb] [PMID: 17119406]

[121] Jackson IT. Local flaps of the head and neck. Quality Medical Pub 2010; p. 2.

[122] Thornton JF, Weathers WM. Nasolabial flap for nasal tip reconstruction. Plast Reconstr Surg 2008; 122(3): 775-81.
[http://dx.doi.org/10.1097/PRS.0b013e318182370b] [PMID: 18766040]

[123] Jackson IT. Midline forehead flaps in nasal reconstruction. Eur J Plast Surg 2004; 27(3): 105-13.
[http://dx.doi.org/10.1007/s00238-004-0632-z]

[124] Rohrich RJ, Griffin JR, Ansari M, Beran SJ, Potter JK. Nasal reconstruction--beyond aesthetic subunits: a 15-year review of 1334 cases. Plast Reconstr Surg 2004; 114(6): 1405-16. [http://dx.doi.org/10.1097/01.PRS.0000138596.57393.05] [PMID: 15509926]

[125] Cöloğlu H, Koçer U, Kankaya Y, Sungur N, Oruç M. Lower eyelid orbicularis oculi musculocutaneous flap for reconstruction of nasal tip and supratip defects. Plast Reconstr Surg 2006; 117(1): 239-46. [http://dx.doi.org/10.1097/01.prs.0000195078.56771.97] [PMID: 16404274]

[126] Walton RL, Burget GC, Beahm EK. Microsurgical reconstruction of the nasal lining. Plast Reconstr Surg 2005; 115(7): 1813-29. [http://dx.doi.org/10.1097/01.PRS.0000164683.34767.2F] [PMID: 15923823]

[127] Li K, Tjong SC. Mechanical, thermal and bioactive behaviors of polyamide 6/hydroxyapatite nanocomposites. J Nanosci Nanotechnol 2011; 11(12): 10644-8. [http://dx.doi.org/10.1166/jnn.2011.3955] [PMID: 22408965]

[128] Liao CZ, Li K, Wong HM, Tong WY, Yeung KW, Tjong SC. Novel polypropylene biocomposites reinforced with carbon nanotubes and hydroxyapatite nanorods for bone replacements. Mater Sci Eng C Mater Biol Appl 2013; 33(3): 1380-8. [http://dx.doi.org/10.1016/j.msec.2012.12.039] [PMID: 23827585]

[129] McGregor IA, Jackson IT. The groin flap. Br J Plast Surg 1972; 25(1): 3-16. [http://dx.doi.org/10.1016/S0007-1226(72)80003-1] [PMID: 4550433]

[130] Smith PJ, Foley B, McGregor IA, Jackson IT. The anatomical basis of the groin flap. Plast Reconstr Surg 1972; 49(1): 41-7. [http://dx.doi.org/10.1097/00006534-197201000-00008] [PMID: 4550070]

[131] Kroll SS, Rosenfield L. Perforator-based flaps for low posterior midline defects. Plast Reconstr Surg 1988; 81(4): 561-6. [http://dx.doi.org/10.1097/00006534-198804000-00012] [PMID: 3279442]

[132] Koshima I, Soeda S. Inferior epigastric artery skin flaps without rectus abdominis muscle. Br J Plast Surg 1989; 42(6): 645-8. [http://dx.doi.org/10.1016/0007-1226(89)90075-1] [PMID: 2605399]

[133] Taylor GI. The angiosomes of the body and their supply to perforator flaps. Clin Plast Surg 2003; 30(3): 331-42. [http://dx.doi.org/10.1016/S0094-1298(03)00034-8] [PMID: 12916590]

[134] Taylor GI, Palmer JH. The vascular territories (angiosomes) of the body: experimental study and clinical applications. Br J Plast Surg 1987; 40(2): 113-41. [http://dx.doi.org/10.1016/0007-1226(87)90185-8] [PMID: 3567445]

[135] Saint-Cyr M, Wong C, Schaverien M, Mojallal A, Rohrich RJ. The perforasome theory: vascular anatomy and clinical implications. Plast Reconstr Surg 2009; 124(5): 1529-44. [http://dx.doi.org/10.1097/PRS.0b013e3181b98a6c] [PMID: 20009839]

[136] Gunnarsson GL, Jackson IT, Thomsen JB. Freestyle facial perforator flaps-a safe reconstructive option for moderate-sized facial defects. Eur J Plast Surg 2014; 37(6): 315-8. [http://dx.doi.org/10.1007/s00238-014-0936-6] [PMID: 24899784]

[137] Kudo K, Fujioka Y. Clinical course and evaluation of shape and function after reconstruction using various pedicle flaps for oral and maxillofacial defects. Int J Oral Maxillofac Surg 1987; 16(5): 529-39.
[http://dx.doi.org/10.1016/S0901-5027(87)80102-9] [PMID: 3116107]

[138] Wei FC, Mardini S. Free-style free flaps. Plast Reconstr Surg 2004; 114(4): 910-6.
[http://dx.doi.org/10.1097/01.PRS.0000133171.65075.81] [PMID: 15468398]

[139] Wei FC, Jain V, Suominen S, Chen HC. Confusion among perforator flaps: what is a true perforator flap? Plast Reconstr Surg 2001; 107(3): 874-6.
[http://dx.doi.org/10.1097/00006534-200103000-00037] [PMID: 11304620]

[140] Chang YM, Santamaria E, Wei FC, *et al.* Primary insertion of osseointegrated dental implants into fibula osteoseptocutaneous free flap for mandible reconstruction. Plast Reconstr Surg 1998; 102(3): 680-8.
[http://dx.doi.org/10.1097/00006534-199809010-00010] [PMID: 9727431]

[141] Yim KK, Wei FC. Fibula osteoseptocutaneous flap for mandible reconstruction. Microsurgery 1994; 15(4): 245-9.
[http://dx.doi.org/10.1002/micr.1920150405] [PMID: 8035669]

[142] Tessier P, Kawamoto H, Matthews D, *et al.* Autogenous bone grafts and bone substitute—tools and techniques: I A 20000-case Fig. 7 Patient, 2 years after reconstruction (patient 2). Good functional and esthetic outcome is seen/experience in maxillofacial and craniofacial surgery. Plast Reconstr Surg Suppl 2005; 116(5): 6-24.
[http://dx.doi.org/10.1097/01.prs.0000173862.20563.12]

[143] Lenz JH, Henkel KO, Hingst V, von Versen R, Gundlach KK. Reconstruction of the frontal calvarian continuity in a child using a freeze-preserved autogenous bone graft. J Craniomaxillofac Surg 2003; 31(3): 154-8.
[http://dx.doi.org/10.1016/S1010-5182(03)00024-6] [PMID: 12818600]

[144] Rodrigues M, O'malley BW Jr, Staecker H, Tamargo R. Extended pericranial flap and bone graft reconstruction in anterior skull base surgery. Otolaryngol Head Neck Surg 2004; 131(1): 69-76.
[http://dx.doi.org/10.1016/j.otohns.2004.02.033] [PMID: 15243560]

[145] Ahlmann E, Patzakis M, Roidis N, Shepherd L, Holtom P. Comparison of anterior and posterior iliac crest bone grafts in terms of harvest-site morbidity and functional outcomes. J Bone Joint Surg Am 2002; 84-A(5): 716-20.
[PMID: 12004011]

[146] Eppley BL. Alloplastic implantation. Plast Reconstr Surg 1999; 104(6): 1761-83.
[http://dx.doi.org/10.1097/00006534-199911000-00025] [PMID: 10541182]

[147] Ducic Y. Titanium mesh and hydroxyapatite cement cranioplasty: a report of 20 cases. J Oral Maxillofac Surg 2002; 60(3): 272-6.
[http://dx.doi.org/10.1053/joms.2002.30575] [PMID: 11887138]

[148] Eppley BL, Kilgo M, Coleman JJ III. Cranial reconstruction with computer-generated hard-tissue replacement patient-matched implants: indications, surgical technique, and long-term follow-up. Plast Reconstr Surg 2002; 109(3): 864-71.

[http://dx.doi.org/10.1097/00006534-200203000-00005] [PMID: 11884798]

[149] Roeder RK, Converse GL, Kane RJ, Yue W. Hydroxyapatitereinforced polymer biocomposites for synthetic bone substitutes. JOM 2008; 60(3): 38-45.
[http://dx.doi.org/10.1007/s11837-008-0030-2]

[150] Ngiam M, Liao S, Patil AJ, Cheng Z, Chan CK, Ramakrishna S. The fabrication of nano-hydroxyapatite on PLGA and PLGA/collagen nanofibrous composite scaffolds and their effects in osteoblastic behavior for bone tissue engineering. Bone 2009; 45(1): 4-16.
[http://dx.doi.org/10.1016/j.bone.2009.03.674] [PMID: 19358900]

[151] Huang J, Di Silvio L, Wang M, Tanner KE, Bonfield W. *In vitro* mechanical and biological assessment of hydroxyapatite-reinforced polyethylene composite. J Mater Sci Mater Med 1997; 8(12): 775-9.
[http://dx.doi.org/10.1023/A:1018516813604] [PMID: 15348789]

[152] Guild FJ, Bonfield W. Predictive modelling of the mechanical properties and failure processes in hydroxyapatite- polyethylene (Hapex) composite. J Mater Sci Mater Med 1998; 9(9): 497-502.
[http://dx.doi.org/10.1023/A:1008831720198] [PMID: 15348846]

[153] Wang M, Joseph R. Bonf ield W. Hydroxyapatite-high density polyethylene composites: effect of ceramic particle size and morphology. Biomaterials 1998; 19(24): 2357-66.
[http://dx.doi.org/10.1016/S0142-9612(98)00154-9] [PMID: 9884050]

[154] Abu Bakar MS, Cheang P, Khor KA. Mechanical properties of injection molded hydroxyapatite-polyetheretherketone biocomposites. Compos Sci Technol 2003; 63(3–4): 421-5.
[http://dx.doi.org/10.1016/S0266-3538(02)00230-0]

[155] Tjong SC, Chen H. Nanocrystalline materials and coatings. Mater Sci Eng Rep 2004; 45(1–2): 1-88.
[http://dx.doi.org/10.1016/j.mser.2004.07.001]

[156] Yang L, Zhang LJ, Webster TJ. Nanobiomaterials: state of the art and future trends. Adv Eng Mater 2011; 13(6): B197-217.
[http://dx.doi.org/10.1002/adem.201080140]

[157] Zhang LJ, Webster TJ. Nanotechnology and nanomaterials: promises for improved tissue regeneration. Nano Today 2009; 4(1): 66-80.
[http://dx.doi.org/10.1016/j.nantod.2008.10.014]

[158] Ergun C, Liu H, Halloran JW, Webster TJ. Increased osteoblast adhesion on nanograined hydroxyapatite and tricalcium phosphate containing calcium titanate. J Biomed Mater Res A 2007; 80(4): 990-7.
[http://dx.doi.org/10.1002/jbm.a.30923] [PMID: 17120201]

[159] Price RL, Waid MC, Haberstroh KM, Webster TJ. Selective bone cell adhesion on formulations containing carbon nanofibers. Biomaterials 2003; 24(11): 1877-87.
[http://dx.doi.org/10.1016/S0142-9612(02)00609-9] [PMID: 12615478]

[160] Li X, Gao H, Uo M, *et al.* Effect of carbon nanotubes on cellular functions *in vitro.* J Biomed Mater Res A 2009; 91(1): 132-9.
[http://dx.doi.org/10.1002/jbm.a.32203] [PMID: 18767064]

[161] Newman P, Minett A, Ellis-Behnke R, Zreiqat H. Carbon nanotubes: their potential and pitfalls for

bone tissue regeneration and engineering. Nanomedicine (Lond) 2013; 9(8): 1139-58. [PMID: 23770067]

[162] Li K, Yeung Yeung CY, Yeung KW, Tjong SC. Sintered hydroxyapatite polyetheretherketone nanocomposites: mechanical behavior and biocompatibility. Adv Eng Mater 2012; 14(4): B155-65. [http://dx.doi.org/10.1002/adem.201080145]

[163] Li X, Xu Z. Controllable synthesis of helical, straight, hollow and nitrogen-doped carbon nanofibers and their magnetic properties. Mater Res Bull 2012; 47(12): 4383-91. [http://dx.doi.org/10.1016/j.materresbull.2012.09.025]

[164] Jian X, Jiang M, Zhou Z, *et al.* Gas-induced formation of Cu nanoparticle as catalyst for high-purity straight and helical carbon nanofibers. ACS Nano 2012; 6(10): 8611-9. [http://dx.doi.org/10.1021/nn301880w] [PMID: 22963353]

[165] Stelnicki EJ, Ousterhout DK. Prevention of thermal tissue injury induced by the application of polymethylmethacrylate to the calvarium. J Craniofac Surg 1996; 7(3): 192-5. [http://dx.doi.org/10.1097/00001665-199605000-00006] [PMID: 9086884]

[166] Neovius E, Engstrand T. Craniofacial reconstruction with bone and biomaterials: review over the last 11 years. J Plast Reconstr Aesthet Surg 2010; 63(10): 1615-23. [http://dx.doi.org/10.1016/j.bjps.2009.06.003] [PMID: 19577527]

[167] Lo KW, Ulery BD, Ashe KM, Laurencin CT. Studies of bone morphogenetic protein-based surgical repair. Adv Drug Deliv Rev 2012; 64(12): 1277-91. [http://dx.doi.org/10.1016/j.addr.2012.03.014] [PMID: 22512928]

[168] Liao CZ, Wong HM, Yeung KW, Tjong SC. The development, fabrication, and material characterization of polypropylene composites reinforced with carbon nanofiber and hydroxyapatite nanorod hybrid fillers. Int J Nanomedicine 2014; 9: 1299-310. [PMID: 24648729]

[169] Yuan H, Fernandes H, Habibovic P, *et al.* Osteoinductive ceramics as a synthetic alternative to autologous bone grafting. Proc Natl Acad Sci USA 2010; 107(31): 13614-9. [http://dx.doi.org/10.1073/pnas.1003600107] [PMID: 20643969]

[170] Quarto R, Mastrogiacomo M, Cancedda R, *et al.* Repair of large bone defects with the use of autologous bone marrow stromal cells. N Engl J Med 2001; 344(5): 385-6. [http://dx.doi.org/10.1056/NEJM200102013440516] [PMID: 11195802]

[171] Taylor BC, French BG, Fowler TT, Russell J, Poka A. Induced membrane technique for reconstruction to manage bone loss. J Am Acad Orthop Surg 2012; 20(3): 142-50. [http://dx.doi.org/10.5435/JAAOS-20-03-142] [PMID: 22382286]

[172] Dorozhkin SV. Bioceramics of calcium orthophosphates. Biomaterials 2010; 31(7): 1465-85. [http://dx.doi.org/10.1016/j.biomaterials.2009.11.050] [PMID: 19969343]

[173] Aberg J, Brisby H, Henriksson HB, Lindahl A, Thomsen P, Engqvist H. Premixed acidic calcium phosphate cement: characterization of strength and microstructure. J Biomed Mater Res B Appl Biomater 2010; 93(2): 436-41. [http://dx.doi.org/10.1002/jbm.b.31600] [PMID: 20127991]

[174] Apelt D, Theiss F, El-Warrak AO, *et al.* *In vivo* behavior of three different injectable hydraulic

calcium phosphate cements. Biomaterials 2004; 25(7-8): 1439-51.
[http://dx.doi.org/10.1016/j.biomaterials.2003.08.073] [PMID: 14643619]

[175] Lee JH, Chang BS, Jeung UO, Park KW, Kim MS, Lee CK. The first clinical trial of beta-calcium pyrophosphate as a novel bone graft extender in instrumented posterolateral lumbar fusion. Clin Orthop Surg 2011; 3(3): 238-44.
[http://dx.doi.org/10.4055/cios.2011.3.3.238] [PMID: 21909472]

[176] LeGeros RZ. Properties of osteoconductive biomaterials: calcium phosphates. Clin Orthop Relat Res 2002; 395(395): 81-98.
[http://dx.doi.org/10.1097/00003086-200202000-00009] [PMID: 11937868]

[177] Peltola M, Aitasalo K, Suonpää J. Varpula M, Yli-Urpo A.Bioactive glassS53P4 in frontal sinus obliteration: a long-term clinical experience. Head Neck 2006; 28(9): 834-41.
[http://dx.doi.org/10.1002/hed.20436] [PMID: 16823870]

[178] Turunen T, Peltola J, Yli-Urpo A, Happonen RP. Bioactive glass granules as a bone adjunctive material in maxillary sinus floor augmentation. Clin Oral Implants Res 2004; 15(2): 135-41.
[http://dx.doi.org/10.1111/j.1600-0501.2004.00989.x] [PMID: 15008925]

[179] Lindfors NC, Heikkilä JT, Koski I, Mattila K, Aho AJ. Bioactive glass and autogenous bone as bone graft substitutes in benign bone tumors. J Biomed Mater Res B Appl Biomater 2009; 90(1): 131-6.
[PMID: 18988277]

[180] Leppäranta O, Vaahtio M, Peltola T, *et al.* Antibacterial effect of bioactive glasses on clinically important anaerobic bacteria *in vitro.* J Mater Sci Mater Med 2008; 19(2): 547-51.
[http://dx.doi.org/10.1007/s10856-007-3018-5] [PMID: 17619981]

[181] Munukka E, Leppäranta O, Korkeamäki M, *et al.* Bactericidal effects of bioactive glasses on clinically important aerobic bacteria. J Mater Sci Mater Med 2008; 19(1): 27-32.
[http://dx.doi.org/10.1007/s10856-007-3143-1] [PMID: 17569007]

[182] Susarla SM, Swanson E, Gordon CR. Craniomaxillofacial reconstruction using allotransplantation and tissue engineering: challenges, opportunities, and potential synergy. Ann Plast Surg 2011; 67(6): 655-61.
[http://dx.doi.org/10.1097/SAP.0b013e31822c00e6] [PMID: 21825966]

[183] Kim RY, Fasi AC, Feinberg SE. Soft tissue engineering in craniomaxillofacial surgery. Ann Maxillofac Surg 2014; 4(1): 4-8.
[http://dx.doi.org/10.4103/2231-0746.133064] [PMID: 24987591]

[184] Susarla SM, Swanson E, Gordon CR. Craniomaxillofacial reconstruction using allotransplantation and tissue engineering: challenges, opportunities, and potential synergy. Ann Plast Surg 2011; 67(6): 655-61.
[http://dx.doi.org/10.1097/SAP.0b013e31822c00e6] [PMID: 21825966]

[185] Langer R, Vacanti JP. Tissue engineering. Science 1993; 260(5110): 920-6.
[http://dx.doi.org/10.1126/science.8493529] [PMID: 8493529]

[186] Payne KF, Balasundaram I, Deb S, Di Silvio L, Fan KF. Tissue engineering technology and its possible applications in oral and maxillofacial surgery. Br J Oral Maxillofac Surg 2014; 52(1): 7-15.
[http://dx.doi.org/10.1016/j.bjoms.2013.03.005] [PMID: 23601833]

[187] Yang M, Zhang H, Gangolli R. Advances of mesenchymal stem cells derived from bone marrow and dental tissue in craniofacial tissue engineering. Curr Stem Cell Res Ther 2014; 9(3): 150-61. [http://dx.doi.org/10.2174/1574888X09666140213142258] [PMID: 24524798]

[188] Polak JM, Bishop AE. Stem cells and tissue engineering: past, present, and future. Ann N Y Acad Sci 2006; 1068(1068): 352-66. [http://dx.doi.org/10.1196/annals.1346.001] [PMID: 16831937]

[189] Zaidi N, Nixon AJ. Stem cell therapy in bone repair and regeneration. Ann N Y Acad Sci 2007; 1117(1): 62-72. [http://dx.doi.org/10.1196/annals.1402.074] [PMID: 18056038]

[190] Bajada S, Mazakova I, Richardson JB, Ashammakhi N. Updates on stem cells and their applications in regenerative medicine. J Tissue Eng Regen Med 2008; 2(4): 169-83. [http://dx.doi.org/10.1002/term.83] [PMID: 18493906]

[191] Alwattar BJ, Schwarzkopf R, Kirsch T. Stem cells in orthopaedics and fracture healing. Bull NYU Hosp Jt Dis 2011; 69(1): 6-10. [PMID: 21332433]

[192] Zhang L, Feng G, Wei X, *et al.* The effects of mesenchymal stem cells in craniofacial tissue engineering. Curr Stem Cell Res Ther 2014; 9(3): 280-9. [http://dx.doi.org/10.2174/1574888X09666140213204202] [PMID: 24524786]

[193] Evans MJ, Kaufman MH. Establishment in culture of pluripotential cells from mouse embryos. Nature 1981; 292(5819): 154-6. [http://dx.doi.org/10.1038/292154a0] [PMID: 7242681]

[194] Thomson JA, Itskovitz-Eldor J, Shapiro SS, *et al.* Embryonic stem cell lines derived from human blastocysts. Science 1998; 282(5391): 1145-7. [http://dx.doi.org/10.1126/science.282.5391.1145] [PMID: 9804556]

[195] Rai S, Kaur M, Kaur S. Applications of stem cells in interdisciplinary dentistry and beyond: an overview. Ann Med Health Sci Res 2013; 3(2): 245-54. [http://dx.doi.org/10.4103/2141-9248.113670] [PMID: 23919198]

[196] Huang GT, Gronthos S, Shi S. Mesenchymal stem cells derived from dental tissues *vs.* those from other sources: their biology and role in regenerative medicine. J Dent Res 2009; 88(9): 792-806. [http://dx.doi.org/10.1177/0022034509340867] [PMID: 19767575]

[197] Mao JJ. Stem cells and the future of dental care. N Y State Dent J 2008; 74(2): 20-4. [PMID: 18450184]

[198] Reznick JB. Continuing education: stem cells: emerging medical and dental therapies for the dental professional Dentaltown magazine 2008; 42-53.

[199] Nedel F, André DdeA, de Oliveira IO, *et al.* Stem cells: therapeutic potential in dentistry. J Contemp Dent Pract 2009; 10(4): 90-6. [PMID: 19575059]

[200] Buttery LD, Bourne S, Xynos JD, *et al.* Differentiation of osteoblasts and *in vitro* bone formation from murine embryonic stem cells. Tissue Eng 2001; 7(1): 89-99. [http://dx.doi.org/10.1089/107632700300003323] [PMID: 11224927]

[201] Jukes JM, van Blitterswijk CA, de Boer J. Skeletal tissue engineering using embryonic stem cells. J Tissue Eng Regen Med 2010; 4(3): 165-80.
[http://dx.doi.org/10.1002/term.234] [PMID: 19967745]

[202] Assady S, Maor G, Amit M, Itskovitz-Eldor J, Skorecki KL, Tzukerman M. Insulin production by human embryonic stem cells. Diabetes 2001; 50(8): 1691-7.
[http://dx.doi.org/10.2337/diabetes.50.8.1691] [PMID: 11473026]

[203] Schwartz SD, Hubschman JP, Heilwell G, *et al.* Embryonic stem cell trials for macular degeneration: a preliminary report. Lancet 2012; 379(9817): 713-20.
[http://dx.doi.org/10.1016/S0140-6736(12)60028-2] [PMID: 22281388]

[204] Kang HK, Roh S, Lee G, Hong SD, Kang H, Min BM. Osteogenic potential of embryonic stem cells in tooth sockets. Int J Mol Med 2008; 21(5): 539-44.
[PMID: 18425344]

[205] Inanç B, Elçin AE, Elçin YM. *In vitro* differentiation and attachment of human embryonic stem cells on periodontal tooth root surfaces. Tissue Eng Part A 2009; 15(11): 3427-35.
[http://dx.doi.org/10.1089/ten.tea.2008.0380] [PMID: 19405785]

[206] Shamis Y, Hewitt KJ, Carlson MW, *et al.* Fibroblasts derived from human embryonic stem cells direct development and repair of 3D human skin equivalents. Stem Cell Res Ther 2011; 2(1): 10.
[http://dx.doi.org/10.1186/scrt51] [PMID: 21338517]

[207] Abe K, Niwa H, Iwase K, *et al.* Endoderm-specific gene expression in embryonic stem cells differentiated to embryoid bodies. Exp Cell Res 1996; 229(1): 27-34.
[http://dx.doi.org/10.1006/excr.1996.0340] [PMID: 8940246]

[208] Ohazama A, Modino SA, Miletich I, Sharpe PT. Stem-cell-based tissue engineering of murine teeth. J Dent Res 2004; 83(7): 518-22.
[http://dx.doi.org/10.1177/154405910408300702] [PMID: 15218039]

[209] Leahy A, Xiong JW, Kuhnert F, Stuhlmann H. Use of developmental marker genes to define temporal and spatial patterns of differentiation during embryoid body formation. J Exp Zool 1999; 284(1): 67-81.
[http://dx.doi.org/10.1002/(SICI)1097-010X(19990615)284:1<67::AID-JEZ10>3.0.CO;2-O] [PMID: 10368935]

[210] Itskovitz-Eldor J, Schuldiner M, Karsenti D, *et al.* Differentiation of human embryonic stem cells into embryoid bodies compromising the three embryonic germ layers. Mol Med 2000; 6(2): 88-95.
[PMID: 10859025]

[211] Rippon HJ, Bishop AE. Embryonic stem cells. Cell Prolif 2004; 37(1): 23-34.
[http://dx.doi.org/10.1111/j.1365-2184.2004.00298.x] [PMID: 14871235]

[212] Takahashi K, Tanabe K, Ohnuki M, *et al.* Induction of pluripotent stem cells from adult human fibroblasts by defined factors. Cell 2007; 131(5): 861-72.
[http://dx.doi.org/10.1016/j.cell.2007.11.019] [PMID: 18035408]

[213] Takahashi K, Yamanaka S. Induction of pluripotent stem cells from mouse embryonic and adult fibroblast cultures by defined factors. Cell 2006; 126(4): 663-76.
[http://dx.doi.org/10.1016/j.cell.2006.07.024] [PMID: 16904174]

[214] Yu J, Vodyanik MA, Smuga-Otto K, *et al.* Induced pluripotent stem cell lines derived from human somatic cells. Science 2007; 318(5858): 1917-20.
[http://dx.doi.org/10.1126/science.1151526] [PMID: 18029452]

[215] Illich DJ, Demir N, Stojković M, *et al.* Concise review: induced pluripotent stem cells and lineage reprogramming: prospects for bone regeneration. Stem Cells 2011; 29(4): 555-63.
[http://dx.doi.org/10.1002/stem.611] [PMID: 21308867]

[216] Duan X, Tu Q, Zhang J, *et al.* Application of induced pluripotent stem (iPS) cells in periodontal tissue regeneration. J Cell Physiol 2011; 226(1): 150-7.
[http://dx.doi.org/10.1002/jcp.22316] [PMID: 20658533]

[217] Ye JH, Xu YJ, Gao J, *et al.* Critical-size calvarial bone defects healing in a mouse model with silk scaffolds and SATB2-modified iPSCs. Biomaterials 2011; 32(22): 5065-76.
[http://dx.doi.org/10.1016/j.biomaterials.2011.03.053] [PMID: 21492931]

[218] Takahashi K, Yamanaka S. Induction of pluripotent stem cells from mouse embryonic and adult fibroblast cultures by defined factors. Cell 2006; 126(4): 663-76.
[http://dx.doi.org/10.1016/j.cell.2006.07.024] [PMID: 16904174]

[219] Ebert AD, Liang P, Wu JC. Induced pluripotent stem cells as a disease modeling and drug screening platform. J Cardiovasc Pharmacol 2012; 60(4): 408-16.
[http://dx.doi.org/10.1097/FJC.0b013e318247f642] [PMID: 22240913]

[220] Miyoshi K, Tsuji D, Kudoh K, *et al.* Generation of human induced pluripotent stem cells from oral mucosa. J Biosci Bioeng 2010; 110(3): 345-50.
[http://dx.doi.org/10.1016/j.jbiosc.2010.03.004] [PMID: 20547351]

[221] Oda Y, Yoshimura Y, Ohnishi H, *et al.* Induction of pluripotent stem cells from human third molar mesenchymal stromal cells. J Biol Chem 2010; 285(38): 29270-8.
[http://dx.doi.org/10.1074/jbc.M109.055889] [PMID: 20595386]

[222] Wada N, Wang B, Lin NH, Laslett AL, Gronthos S, Bartold PM. Induced pluripotent stem cell lines derived from human gingival fibroblasts and periodontal ligament fibroblasts. J Periodontal Res 2011; 46(4): 438-47.
[http://dx.doi.org/10.1111/j.1600-0765.2011.01358.x] [PMID: 21443752]

[223] Tamaoki N, Takahashi K, Tanaka T, *et al.* Dental pulp cells for induced pluripotent stem cell banking. J Dent Res 2010; 89(8): 773-8.
[http://dx.doi.org/10.1177/0022034510366846] [PMID: 20554890]

[224] Otsu K, Kumakami-Sakano M, Fujiwara N, *et al.* Stem cell sources for tooth regeneration: current status and future prospects. Front Physiol 2014; 5: 36.
[http://dx.doi.org/10.3389/fphys.2014.00036] [PMID: 24550845]

[225] Di KQ, Gao S, Cui LF, *et al.* Generation of fully pluripotent female murine-induced pluripotent stem cells. Biol Reprod 2015; 92(5): 123.
[http://dx.doi.org/10.1095/biolreprod.114.124958] [PMID: 25788660]

[226] Li G, Chunxu Y, Guisheng L. Efficient p53 gene targeting by homologous recombination in rat-induced pluripotent stem cells. Cell Prolif 2013; 46(1): 1-9.
[http://dx.doi.org/10.1111/cpr.12000] [PMID: 23278941]

[227] Kaji K, Norrby K, Paca A, Mileikovsky M, Mohseni P, Woltjen K. Virus-free induction of pluripotency and subsequent excision of reprogramming factors. Nature 2009; 458(7239): 771-5. [http://dx.doi.org/10.1038/nature07864] [PMID: 19252477]

[228] Eggenschwiler R, Cantz T. Induced pluripotent stem cells generated without viral integration. Hepatology 2009; 49(3): 1048-9. [http://dx.doi.org/10.1002/hep.22827] [PMID: 19242974]

[229] Woltjen K, Michael IP, Mohseni P, *et al.* piggyBac transposition reprograms fibroblasts to induced pluripotent stem cells. Nature 2009; 458(7239): 766-70. [http://dx.doi.org/10.1038/nature07863] [PMID: 19252478]

[230] Judson RL, Babiarz JE, Venere M, Blelloch R. Embryonic stem cell-specific microRNAs promote induced pluripotency. Nat Biotechnol 2009; 27(5): 459-61. [http://dx.doi.org/10.1038/nbt.1535] [PMID: 19363475]

[231] Li W, Wei W, Zhu S, *et al.* Generation of rat and human induced pluripotent stem cells by combining genetic reprogramming and chemical inhibitors. Cell Stem Cell 2009; 4(1): 16-9. [http://dx.doi.org/10.1016/j.stem.2008.11.014] [PMID: 19097958]

[232] Rybak A, Fuchs H, Smirnova L, *et al.* A feedback loop comprising lin-28 and let-7 controls pre-let-7 maturation during neural stem-cell commitment. Nat Cell Biol 2008; 10(8): 987-93. [http://dx.doi.org/10.1038/ncb1759] [PMID: 18604195]

[233] Pompilio G, Cannata A, Peccatori F, *et al.* Autologous peripheral blood stem cell transplantation for myocardial regeneration: a novel strategy for cell collection and surgical injection. Ann Thorac Surg 2004; 78(5): 1808-12. [http://dx.doi.org/10.1016/j.athoracsur.2003.09.084] [PMID: 15511478]

[234] De Ugarte DA, Morizono K, Elbarbary A, *et al.* Comparison of multi-lineage cells from human adipose tissue and bone marrow. Cells Tissues Organs (Print) 2003; 174(3): 101-9. [http://dx.doi.org/10.1159/000071150] [PMID: 12835573]

[235] Laughlin MJ. Umbilical cord blood for allogeneic transplantation in children and adults. Bone Marrow Transplant 2001; 27(1): 1-6. [http://dx.doi.org/10.1038/sj.bmt.1702740] [PMID: 11244431]

[236] Gang EJ, Jeong JA, Hong SH, *et al.* Skeletal myogenic differentiation of mesenchymal stem cells isolated from human umbilical cord blood. Stem Cells 2004; 22(4): 617-24. [http://dx.doi.org/10.1634/stemcells.22-4-617] [PMID: 15277707]

[237] De Gemmis P, Lapucci C, Bertelli M, *et al.* A real-time PCR approach to evaluate adipogenic potential of amniotic fluid-derived human mesenchymal stem cells. Stem Cells Dev 2006; 15(5): 719-28. [http://dx.doi.org/10.1089/scd.2006.15.719] [PMID: 17105407]

[238] Caplan AI. Mesenchymal stem cells. J Orthop Res 1991; 9(5): 641-50. [http://dx.doi.org/10.1002/jor.1100090504] [PMID: 1870029]

[239] Bluteau G, Luder HU, De Bari C, Mitsiadis TA. Stem cells for tooth engineering. Eur Cell Mater 2008; 16: 1-9. [PMID: 18671204]

[240] Bianco P, Riminucci M, Gronthos S, Robey PG. Bone marrow stromal stem cells: nature, biology, and potential applications. Stem Cells 2001; 19(3): 180-92.
[http://dx.doi.org/10.1634/stemcells.19-3-180] [PMID: 11359943]

[241] Friedenstein AJ, Piatetzky-Shapiro II, Petrakova KV. Osteogenesis in transplants of bone marrow cells. J Embryol Exp Morphol 1966; 16(3): 381-90.
[PMID: 5336210]

[242] Owen M, Friedenstein AJ. Stromal stem cells: marrow-derived osteogenic precursors. Ciba Found Symp 1988; 136: 42-60.
[PMID: 3068016]

[243] Abukawa H, Shin M, Williams WB, Vacanti JP, Kaban LB, Troulis MJ. Reconstruction of mandibular defects with autologous tissue-engineered bone. J Oral Maxillofac Surg 2004; 62(5): 601-6.
[http://dx.doi.org/10.1016/j.joms.2003.11.010] [PMID: 15122567]

[244] Mendonça JJ, Juiz-Lopez P. Regenerative facial reconstruction of terminal stage osteoradionecrosis and other advanced craniofacial diseases with adult cultured stem and progenitor cells. Plast Reconstr Surg 2010; 126(5): 1699-709.
[http://dx.doi.org/10.1097/PRS.0b013e3181f24164] [PMID: 21042127]

[245] Kaku M, Tai M, Kawata T, *et al.* Mesenchymal stem cell-induced cranial suture-like gap in rats. Plast Reconstr Surg 2011; 127(1): 69-77.
[http://dx.doi.org/10.1097/PRS.0b013e3181f95849] [PMID: 21200202]

[246] Chen J, Chen H, Li P, *et al.* Simultaneous regeneration of articular cartilage and subchondral bone *in vivo* using MSCs induced by a spatially controlled gene delivery system in bilayered integrated scaffolds. Biomaterials 2011; 32(21): 4793-805.
[http://dx.doi.org/10.1016/j.biomaterials.2011.03.041] [PMID: 21489619]

[247] Hay ED. An overview of epithelio-mesenchymal transformation. Acta Anat (Basel) 1995; 154(1): 8-20.
[http://dx.doi.org/10.1159/000147748] [PMID: 8714286]

[248] Battula VL, Evans KW, Hollier BG, *et al.* Epithelial-mesenchymal transition-derived cells exhibit multilineage differentiation potential similar to mesenchymal stem cells. Stem Cells 2010; 28(8): 1435-45.
[http://dx.doi.org/10.1002/stem.467] [PMID: 20572012]

[249] Rubio D, Garcia S, De la Cueva T, *et al.* Human mesenchymal stem cell transformation is associated with a mesenchymal-epithelial transition. Exp Cell Res 2008; 314(4): 691-8.
[http://dx.doi.org/10.1016/j.yexcr.2007.11.017] [PMID: 18201695]

[250] Limbert C, Ebert R, Schilling T, *et al.* Functional signature of human islet-derived precursor cells compared to bone marrow-derived mesenchymal stem cells. Stem Cells Dev 2010; 19(5): 679-91.
[http://dx.doi.org/10.1089/scd.2009.0241] [PMID: 19895235]

[251] Franklin WA, Christison WH, Colley M, Montag AG, Stephens JK, Hart CE. *In situ* distribution of the beta-subunit of platelet-derived growth factor receptor in nonneoplastic tissue and in soft tissue tumors. Cancer Res 1990; 50(19): 6344-8.
[PMID: 2169345]

[252] Schwab KE, Gargett CE. Co-expression of two perivascular cell markers isolates mesenchymal stem-like cells from human endometrium. Hum Reprod 2007; 22(11): 2903-11. [http://dx.doi.org/10.1093/humrep/dem265] [PMID: 17872908]

[253] Khan WS, Adesida AB, Tew SR, Lowe ET, Hardingham TE. Bone marrow-derived mesenchymal stem cells express the pericyte marker 3G5 in culture and show enhanced chondrogenesis in hypoxic conditions. J Orthop Res 2010; 28(6): 834-40. [PMID: 20058274]

[254] Sacchetti B, Funari A, Michienzi S, *et al.* Self-renewing osteoprogenitors in bone marrow sinusoids can organize a hematopoietic microenvironment. Cell 2007; 131(2): 324-36. [http://dx.doi.org/10.1016/j.cell.2007.08.025] [PMID: 17956733]

[255] Crisan M, Yap S, Casteilla L, *et al.* A perivascular origin for mesenchymal stem cells in multiple human organs. Cell Stem Cell 2008; 3(3): 301-13. [http://dx.doi.org/10.1016/j.stem.2008.07.003] [PMID: 18786417]

[256] Corselli M, Chen CW, Crisan M, Lazzari L, Péault B. Perivascular ancestors of adult multipotent stem cells. Arterioscler Thromb Vasc Biol 2010; 30(6): 1104-9. [http://dx.doi.org/10.1161/ATVBAHA.109.191643] [PMID: 20453168]

[257] Cai X, Lin Y, Friedrich CC, *et al.* Bone marrow derived pluripotent cells are pericytes which contribute to vascularization. Stem Cell Rev 2009; 5(4): 437-45. [http://dx.doi.org/10.1007/s12015-009-9097-6] [PMID: 20058207]

[258] Cai X, Lin Y, Hauschka PV, Grottkau BE. Adipose stem cells originate from perivascular cells. Biol Cell 2011; 103(9): 435-47. [http://dx.doi.org/10.1042/BC20110033] [PMID: 21679159]

[259] Dore-Duffy P, Katychev A, Wang X, Van Buren E. CNS microvascular pericytes exhibit multipotential stem cell activity. J Cereb Blood Flow Metab 2006; 26(5): 613-24. [http://dx.doi.org/10.1038/sj.jcbfm.9600272] [PMID: 16421511]

[260] Feng J, Mantesso A, De Bari C, Nishiyama A, Sharpe PT. Dual origin of mesenchymal stem cells contributing to organ growth and repair. Proc Natl Acad Sci USA 2011; 108(16): 6503-8. [http://dx.doi.org/10.1073/pnas.1015449108] [PMID: 21464310]

[261] Risbud MV, Guttapalli A, Tsai TT, *et al.* Evidence for skeletal progenitor cells in the degenerate human intervertebral disc. Spine 2007; 32(23): 2537-44. [http://dx.doi.org/10.1097/BRS.0b013e318158dea6] [PMID: 17978651]

[262] Henriksson H, Thornemo M, Karlsson C, *et al.* Identification of cell proliferation zones, progenitor cells and a potential stem cell niche in the intervertebral disc region: a study in four species. Spine 2009; 34(21): 2278-87. [http://dx.doi.org/10.1097/BRS.0b013e3181a95ad2] [PMID: 19755937]

[263] Hoshino A, Chiba H, Nagai K, Ishii G, Ochiai A. Human vascular adventitial fibroblasts contain mesenchymal stem/progenitor cells. Biochem Biophys Res Commun 2008; 368(2): 305-10. [http://dx.doi.org/10.1016/j.bbrc.2008.01.090] [PMID: 18230345]

[264] Corselli M, Chen CW, Sun B, Yap S, Rubin JP, Péault B. The tunica adventitia of human arteries and veins as a source of mesenchymal stem cells. Stem Cells Dev 2012; 21(8): 1299-308.

[http://dx.doi.org/10.1089/scd.2011.0200] [PMID: 21861688]

[265] Hilkens P, Gervois P, Fanton Y, *et al.* Effect of isolation methodology on stem cell properties and multilineage differentiation potential of human dental pulp stem cells. Cell Tissue Res 2013; 353(1): 65-78.
[http://dx.doi.org/10.1007/s00441-013-1630-x] [PMID: 23715720]

[266] Mahmood A, Lu D, Lu M, Chopp M. Treatment of traumatic brain injury in adult rats with intravenous administration of human bone marrow stromal cells. Neurosurgery 2003; 53(3): 697-702.
[http://dx.doi.org/10.1227/01.NEU.0000079333.61863.AA] [PMID: 12943585]

[267] Horwitz EM, Gordon PL, Koo WK, *et al.* Isolated allogeneic bone marrow-derived mesenchymal cells engraft and stimulate growth in children with osteogenesis imperfecta: Implications for cell therapy of bone. Proc Natl Acad Sci USA 2002; 99(13): 8932-7.
[http://dx.doi.org/10.1073/pnas.132252399] [PMID: 12084934]

[268] Rüster B, Göttig S, Ludwig RJ, *et al.* Mesenchymal stem cells display coordinated rolling and adhesion behavior on endothelial cells. Blood 2006; 108(12): 3938-44.
[http://dx.doi.org/10.1182/blood-2006-05-025098] [PMID: 16896152]

[269] Belema-Bedada F, Uchida S, Martire A, Kostin S, Braun T. Efficient homing of multipotent adult mesenchymal stem cells depends on FROUNT-mediated clustering of CCR2. Cell Stem Cell 2008; 2(6): 566-75.
[http://dx.doi.org/10.1016/j.stem.2008.03.003] [PMID: 18522849]

[270] Wynn RF, Hart CA, Corradi-Perini C, *et al.* A small proportion of mesenchymal stem cells strongly expresses functionally active CXCR4 receptor capable of promoting migration to bone marrow. Blood 2004; 104(9): 2643-5.
[http://dx.doi.org/10.1182/blood-2004-02-0526] [PMID: 15251986]

[271] Ren G, Zhao X, Zhang L, *et al.* Inflammatory cytokine-induced intercellular adhesion molecule-1 and vascular cell adhesion molecule-1 in mesenchymal stem cells are critical for immunosuppression. J Immunol 2010; 184(5): 2321-8.
[http://dx.doi.org/10.4049/jimmunol.0902023] [PMID: 20130212]

[272] Shi M, Li J, Liao L, *et al.* Regulation of CXCR4 expression in human mesenchymal stem cells by cytokine treatment: role in homing efficiency in NOD/SCID mice. Haematologica 2007; 92(7): 897-904.
[http://dx.doi.org/10.3324/haematol.10669] [PMID: 17606439]

[273] Leung AY, Mendenhall EM, Kwan TT, *et al.* Characterization of expanded intermediate cell mass in zebrafish chordin morphant embryos. Dev Biol 2005; 277(1): 235-54.
[http://dx.doi.org/10.1016/j.ydbio.2004.09.032] [PMID: 15572152]

[274] Ferrari G, Cusella-De Angelis G, Coletta M, *et al.* Muscle regeneration by bone marrow-derived myogenic progenitors. Science 1998; 279(5356): 1528-30.
[http://dx.doi.org/10.1126/science.279.5356.1528] [PMID: 9488650]

[275] Grottkau BE, Purudappa PP, Lin YF. Multilineage differentiation of dental pulp stem cells from green fluorescent protein transgenic mice. Int J Oral Sci 2010; 2(1): 21-7.
[http://dx.doi.org/10.4248/IJOS10015] [PMID: 20690415]

[276] Pittenger MF, Mackay AM, Beck SC, *et al.* Multilineage potential of adult human mesenchymal stem cells. Science 1999; 284(5411): 143-7.
[http://dx.doi.org/10.1126/science.284.5411.143] [PMID: 10102814]

[277] Prockop DJ. Marrow stromal cells as stem cells for nonhematopoietic tissues. Science 1997; 276(5309): 71-4.
[http://dx.doi.org/10.1126/science.276.5309.71] [PMID: 9082988]

[278] Castro-Malaspina H, Gay RE, Resnick G, *et al.* Characterization of human bone marrow fibroblast colony-forming cells (CFU-F) and their progeny. Blood 1980; 56(2): 289-301.
[PMID: 6994839]

[279] Tomita M, Yamada H, Adachi Y, *et al.* Choroidal neovascularization is provided by bone marrow cells. Stem Cells 2004; 22(1): 21-6.
[http://dx.doi.org/10.1634/stemcells.22-1-21] [PMID: 14688388]

[280] Jones EA, Kinsey SE, English A, *et al.* Isolation and characterization of bone marrow multipotential mesenchymal progenitor cells. Arthritis Rheum 2002; 46(12): 3349-60.
[http://dx.doi.org/10.1002/art.10696] [PMID: 12483742]

[281] Krause DS, Theise ND, Collector MI, *et al.* Multi-organ, multi-lineage engraftment by a single bone marrow-derived stem cell. Cell 2001; 105(3): 369-77.
[http://dx.doi.org/10.1016/S0092-8674(01)00328-2] [PMID: 11348593]

[282] Eglitis MA, Mezey E. Hematopoietic cells differentiate into both microglia and macroglia in the brains of adult mice. Proc Natl Acad Sci USA 1997; 94(8): 4080-5.
[http://dx.doi.org/10.1073/pnas.94.8.4080] [PMID: 9108108]

[283] Makino S, Fukuda K, Miyoshi S, *et al.* Cardiomyocytes can be generated from marrow stromal cells *in vitro.* J Clin Invest 1999; 103(5): 697-705.
[http://dx.doi.org/10.1172/JCI5298] [PMID: 10074487]

[284] Petersen BE, Bowen WC, Patrene KD, *et al.* Bone marrow as a potential source of hepatic oval cells. Science 1999; 284(5417): 1168-70.
[http://dx.doi.org/10.1126/science.284.5417.1168] [PMID: 10325227]

[285] Jiang Y, Jahagirdar BN, Reinhardt RL, *et al.* Pluripotency of mesenchymal stem cells derived from adult marrow. Nature 2002; 418(6893): 41-9.
[http://dx.doi.org/10.1038/nature00870] [PMID: 12077603]

[286] Graf T. Historical origins of transdifferentiation and reprogramming. Cell Stem Cell 2011; 9(6): 504-16.
[http://dx.doi.org/10.1016/j.stem.2011.11.012] [PMID: 22136926]

[287] Ying QL, Nichols J, Evans EP, Smith AG. Changing potency by spontaneous fusion. Nature 2002; 416(6880): 545-8.
[http://dx.doi.org/10.1038/nature729] [PMID: 11932748]

[288] Terada N, Hamazaki T, Oka M, *et al.* Bone marrow cells adopt the phenotype of other cells by spontaneous cell fusion. Nature 2002; 416(6880): 542-5.
[http://dx.doi.org/10.1038/nature730] [PMID: 11932747]

[289] Jo S, Kim S, Cho TH, Shin E, Hwang SJ, Noh I. Effects of recombinant human bone morphogenic protein-2 and human bone marrow-derived stromal cells on *in vivo* bone regeneration of chitosan-poly(ethylene oxide) hydrogel. J Biomed Mater Res A 2013; 101(3): 892-901.
[http://dx.doi.org/10.1002/jbm.a.34354] [PMID: 23023810]

[290] Pelled G, Snedeker JG, Ben-Arav A, *et al.* Smad8/BMP2-engineered mesenchymal stem cells induce accelerated recovery of the biomechanical properties of the Achilles tendon. J Orthop Res 2012; 30(12): 1932-9.
[http://dx.doi.org/10.1002/jor.22167] [PMID: 22696396]

[291] Németh K, Leelahavanichkul A, Yuen PS, *et al.* Bone marrow stromal cells attenuate sepsis *via* prostaglandin E(2)-dependent reprogramming of host macrophages to increase their interleukin-10 production. Nat Med 2009; 15(1): 42-9.
[http://dx.doi.org/10.1038/nm.1905] [PMID: 19098906]

[292] Uccelli A, Prockop DJ. Why should mesenchymal stem cells (MSCs) cure autoimmune diseases? Curr Opin Immunol 2010; 22(6): 768-74.
[http://dx.doi.org/10.1016/j.coi.2010.10.012] [PMID: 21093239]

[293] Ren G, Zhang L, Zhao X, *et al.* Mesenchymal stem cell-mediated immunosuppression occurs *via* concerted action of chemokines and nitric oxide. Cell Stem Cell 2008; 2(2): 141-50.
[http://dx.doi.org/10.1016/j.stem.2007.11.014] [PMID: 18371435]

[294] Crisostomo PR, Wang Y, Markel TA, Wang M, Lahm T, Meldrum DR. Human mesenchymal stem cells stimulated by TNF-alpha, LPS, or hypoxia produce growth factors by an NF kappa B- but not JNK-dependent mechanism. Am J Physiol Cell Physiol 2008; 294(3): C675-82.
[http://dx.doi.org/10.1152/ajpcell.00437.2007] [PMID: 18234850]

[295] Glennie S, Soeiro I, Dyson PJ, Lam EW, Dazzi F. Bone marrow mesenchymal stem cells induce division arrest anergy of activated T cells. Blood 2005; 105(7): 2821-7.
[http://dx.doi.org/10.1182/blood-2004-09-3696] [PMID: 15591115]

[296] English K, French A, Wood KJ. Mesenchymal stromal cells: facilitators of successful transplantation? Cell Stem Cell 2010; 7(4): 431-42.
[http://dx.doi.org/10.1016/j.stem.2010.09.009] [PMID: 20887949]

[297] Yamout B, Hourani R, Salti H, *et al.* Bone marrow mesenchymal stem cell transplantation in patients with multiple sclerosis: a pilot study. J Neuroimmunol 2010; 227(1-2): 185-9.
[http://dx.doi.org/10.1016/j.jneuroim.2010.07.013] [PMID: 20728948]

[298] Zhou B, Yuan J, Zhou Y, *et al.* Administering human adipose-derived mesenchymal stem cells to prevent and treat experimental arthritis. Clin Immunol 2011; 141(3): 328-37.
[http://dx.doi.org/10.1016/j.clim.2011.08.014] [PMID: 21944669]

[299] Liang J, Zhang H, Hua B, *et al.* Allogenic mesenchymal stem cells transplantation in refractory systemic lupus erythematosus: a pilot clinical study. Ann Rheum Dis 2010; 69(8): 1423-9.
[http://dx.doi.org/10.1136/ard.2009.123463] [PMID: 20650877]

[300] Xu G, Zhang Y, Zhang L, Ren G, Shi Y. The role of IL-6 in inhibition of lymphocyte apoptosis by mesenchymal stem cells. Biochem Biophys Res Commun 2007; 361(3): 745-50.
[http://dx.doi.org/10.1016/j.bbrc.2007.07.052] [PMID: 17678624]

[301] Osugi M, Katagiri W, Yoshimi R, Inukai T, Hibi H, Ueda M. Conditioned media from mesenchymal stem cells enhanced bone regeneration in rat calvarial bone defects. Tissue Eng Part A 2012; 18(13-14): 1479-89.
[http://dx.doi.org/10.1089/ten.tea.2011.0325] [PMID: 22443121]

[302] Bray GA. Medical consequences of obesity. J Clin Endocrinol Metab 2004; 89(6): 2583-9.
[http://dx.doi.org/10.1210/jc.2004-0535] [PMID: 15181027]

[303] Katz AJ, Llull R, Hedrick MH, Futrell JW. Emerging approaches to the tissue engineering of fat. Clin Plast Surg 1999; 26(4): 587-603.
[PMID: 10553215]

[304] Fajas L, Fruchart JC, Auwerx J. Transcriptional control of adipogenesis. Curr Opin Cell Biol 1998; 10(2): 165-73.
[http://dx.doi.org/10.1016/S0955-0674(98)80138-5] [PMID: 9561840]

[305] Deslex S, Negrel R, Vannier C, Etienne J, Ailhaud G. Differentiation of human adipocyte precursors in a chemically defined serum-free medium. Int J Obes 1987; 11(1): 19-27.
[PMID: 3570635]

[306] Hauner H, Entenmann G, Wabitsch M, *et al.* Promoting effect of glucocorticoids on the differentiation of human adipocyte precursor cells cultured in a chemically defined medium. J Clin Invest 1989; 84(5): 1663-70.
[http://dx.doi.org/10.1172/JCI114345] [PMID: 2681273]

[307] Halvorsen YD, Bond A, Sen A, *et al.* Thiazolidinediones and glucocorticoids synergistically induce differentiation of human adipose tissue stromal cells: biochemical, cellular, and molecular analysis. Metabolism 2001; 50(4): 407-13.
[http://dx.doi.org/10.1053/meta.2001.21690] [PMID: 11288034]

[308] Zuk PA, Zhu M, Ashjian P, *et al.* Human adipose tissue is a source of multipotent stem cells. Mol Biol Cell 2002; 13(12): 4279-95.
[http://dx.doi.org/10.1091/mbc.E02-02-0105] [PMID: 12475952]

[309] Guilak F, Lott KE, Awad HA, *et al.* Clonal analysis of the differentiation potential of human adipose-derived adult stem cells. J Cell Physiol 2006; 206(1): 229-37.
[http://dx.doi.org/10.1002/jcp.20463] [PMID: 16021633]

[310] Gimble JM, Guilak F. Differentiation potential of adipose derived adult stem (ADAS) cells. Curr Top Dev Biol 2003; 58(2): 137-60.
[http://dx.doi.org/10.1016/S0070-2153(03)58005-X] [PMID: 14711015]

[311] Rodbell M. Metabolism of isolated fat cells. II. The similar effects of phospholipase C (Clostridium perfringens alpha toxin) and of insulin on glucose and amino acid metabolism. J Biol Chem 1966; 241(1): 130-9.
[PMID: 4379054]

[312] Rodbell M. The metabolism of isolated fat cells. IV. Regulation of release of protein by lipolytic hormones and insulin. J Biol Chem 1966; 241(17): 3909-17.
[PMID: 4288359]

[313] Van RL, Bayliss CE, Roncari DA. Cytological and enzymological characterization of adult human

adipocyte precursors in culture. J Clin Invest 1976; 58(3): 699-704.
[http://dx.doi.org/10.1172/JCI108516] [PMID: 956396]

[314] Björntorp P, Karlsson M, Pertoft H, Pettersson P, Sjöström L, Smith U. Isolation and characterization of cells from rat adipose tissue developing into adipocytes. J Lipid Res 1978; 19(3): 316-24.
[PMID: 206638]

[315] Deslex S, Negrel R, Vannier C, Etienne J, Ailhaud G. Differentiation of human adipocyte precursors in a chemically defined serum-free medium. Int J Obes 1987; 11(1): 19-27.
[PMID: 3570635]

[316] Hauner H, Entenmann G, Wabitsch M, *et al.* Promoting effect of glucocorticoids on the differentiation of human adipocyte precursor cells cultured in a chemically defined medium. J Clin Invest 1989; 84(5): 1663-70.
[http://dx.doi.org/10.1172/JCI114345] [PMID: 2681273]

[317] Hauner H, Wabitsch M, Pfeiffer EF. Differentiation of adipocyte precursor cells from obese and nonobese adult women and from different adipose tissue sites. Horm Metab Res Suppl 1988; 19: 35-9.
[PMID: 3235057]

[318] Illouz YG. Body contouring by lipolysis: a 5-year experience with over 3000 cases. Plast Reconstr Surg 1983; 72(5): 591-7.
[http://dx.doi.org/10.1097/00006534-198311000-00001] [PMID: 6622564]

[319] Moore JH Jr, Kolaczynski JW, Morales LM, *et al.* Viability of fat obtained by syringe suction lipectomy: effects of local anesthesia with lidocaine. Aesthetic Plast Surg 1995; 19(4): 335-9.
[http://dx.doi.org/10.1007/BF00451659] [PMID: 7484471]

[320] Oedayrajsingh-Varma MJ, van Ham SM, Knippenberg M, *et al.* Adipose tissue-derived mesenchymal stem cell yield and growth characteristics are affected by the tissue-harvesting procedure. Cytotherapy 2006; 8(2): 166-77.
[http://dx.doi.org/10.1080/14653240600621125] [PMID: 16698690]

[321] Yoshimura K, Shigeura T, Matsumoto D, *et al.* Characterization of freshly isolated and cultured cells derived from the fatty and fluid portions of liposuction aspirates. J Cell Physiol 2006; 208(1): 64-76.
[http://dx.doi.org/10.1002/jcp.20636] [PMID: 16557516]

[322] Dominici M, Le Blanc K, Mueller I, *et al.* Minimal criteria for defining multipotent mesenchymal stromal cells. The International Society for Cellular Therapy position statement. Cytotherapy 2006; 8(4): 315-7.
[http://dx.doi.org/10.1080/14653240600855905] [PMID: 16923606]

[323] Zimmerlin L, Donnenberg VS, Pfeifer ME, *et al.* Stromal vascular progenitors in adult human adipose tissue. Cytometry A 2010; 77(1): 22-30.
[PMID: 19852056]

[324] Corselli M, Chen CW, Sun B, Yap S, Rubin JP, Péault B. The tunica adventitia of human arteries and veins as a source of mesenchymal stem cells. Stem Cells Dev 2012; 21(8): 1299-308.
[http://dx.doi.org/10.1089/scd.2011.0200] [PMID: 21861688]

[325] Zimmerlin L, Donnenberg VS, Donnenberg AD. Pericytes: a universal adult tissue stem cell? Cytometry A 2012; 81(1): 12-4.

[http://dx.doi.org/10.1002/cyto.a.21168] [PMID: 22069300]

[326] Zimmerlin L, Donnenberg VS, Rubin JP, Donnenberg AD. Mesenchymal markers on human adipose stem/progenitor cells. Cytometry A 2013; 83(1): 134-40. [http://dx.doi.org/10.1002/cyto.a.22227] [PMID: 23184564]

[327] Mitchell JB, McIntosh K, Zvonic S, *et al.* Immunophenotype of human adipose-derived cells: temporal changes in stromal-associated and stem cell-associated markers. Stem Cells 2006; 24(2): 376-85. [http://dx.doi.org/10.1634/stemcells.2005-0234] [PMID: 16322640]

[328] Gronthos S, Franklin DM, Leddy HA, Robey PG, Storms RW, Gimble JM. Surface protein characterization of human adipose tissue-derived stromal cells. J Cell Physiol 2001; 189(1): 54-63. [http://dx.doi.org/10.1002/jcp.1138] [PMID: 11573204]

[329] Suga H, Matsumoto D, Eto H, *et al.* Functional implications of CD34 expression in human adipose-derived stem/progenitor cells. Stem Cells Dev 2009; 18(8): 1201-10. [http://dx.doi.org/10.1089/scd.2009.0003] [PMID: 19226222]

[330] Bailey AM, Kapur S, Katz AJ. Characterization of adipose-derived stem cells: an update. Curr Stem Cell Res Ther 2010; 5(2): 95-102. [http://dx.doi.org/10.2174/157488810791268555] [PMID: 19941461]

[331] Bourin P, Bunnell BA, Casteilla L, *et al.* Stromal cells from the adipose tissue-derived stromal vascular fraction and culture expanded adipose tissue-derived stromal/stem cells: a joint statement of the International Federation for Adipose Therapeutics and Science (IFATS) and the International Society for Cellular Therapy (ISCT). Cytotherapy 2013; 15(6): 641-8. [http://dx.doi.org/10.1016/j.jcyt.2013.02.006] [PMID: 23570660]

[332] Lattanzi W, Bernardini C. Genes and molecular pathways of the osteogenic process. Intech 2012; pp. 21-56. [http://dx.doi.org/10.5772/34022]

[333] Albertson KS, Medoff RJ, Mitsunaga MM. The use of periosteally vascularized autografts to augment the fixation of large segmental allografts. Clin Orthop Relat Res 1991; 269(269): 113-9. [PMID: 1864028]

[334] Enneking WF, Mindell ER. Observations on massive retrieved human allografts. J Bone Joint Surg Am 1991; 73(8): 1123-42. [PMID: 1890115]

[335] Borden M, Attawia M, Khan Y, El-Amin SF, Laurencin CT. Tissue-engineered bone formation *in vivo* using a novel sintered polymeric microsphere matrix. J Bone Joint Surg Br 2004; 86(8): 1200-8. [http://dx.doi.org/10.1302/0301-620X.86B8.14267] [PMID: 15568538]

[336] Parrilla C, Lattanzi W, Rita Fetoni A, Bussu F, Pola E, Paludetti G. *Ex vivo* gene therapy using autologous dermal fibroblasts expressing hLMP3 for rat mandibular bone regeneration. Head Neck 2010; 32(3): 310-8. [PMID: 19626643]

[337] Urist MR. Bone: formation by autoinduction. Science 1965; 150(3698): 893-9. [http://dx.doi.org/10.1126/science.150.3698.893] [PMID: 5319761]

[338] Di Silvio L. Bone tissue engineering and biomineralization Tissue engineering using ceramics and

polymers. Boca Raton: CRC Press 2007; pp. 319-34.
[http://dx.doi.org/10.1533/9781845693817.2.319]

[339] Seo S, Na K. Mesenchymal stem cell-based tissue engineering for chondrogenesis J Biomed Biotechnol 2011; 2011: 806-91.

[340] Scaglione S, Quarto R. Stem cells and tissue scaffolds for bone repair. Cellular Response to Biomaterials 2008; p. 291.

[341] Zhang Z. Bone regeneration by stem cell and tissue engineering in oral and maxillofacial region. Front Med 2011; 5(4): 401-13.
[http://dx.doi.org/10.1007/s11684-011-0161-7] [PMID: 22198752]

[342] Lindroos B, Suuronen R, Miettinen S. The potential of adipose stem cells in regenerative medicine. Stem Cell Rev 2011; 7(2): 269-91.
[http://dx.doi.org/10.1007/s12015-010-9193-7] [PMID: 20853072]

[343] Lattanzi W, Pola E, Pecorini G, Logroscino CA, Robbins PD. Gene therapy for *in vivo* bone formation: recent advances. Eur Rev Med Pharmacol Sci 2005; 9(3): 167-74.
[PMID: 16080636]

[344] Kapur SK, Katz AJ. Review of the adipose derived stem cell secretome. Biochimie 2013; 95(12): 2222-8.
[http://dx.doi.org/10.1016/j.biochi.2013.06.001] [PMID: 23770442]

[345] Kilroy GE, Foster SJ, Wu X, *et al.* Cytokine profile of human adipose-derived stem cells: expression of angiogenic, hematopoietic, and pro-inflammatory factors. J Cell Physiol 2007; 212(3): 702-9.
[http://dx.doi.org/10.1002/jcp.21068] [PMID: 17477371]

[346] He J, Duan H, Xiong Y, *et al.* Participation of CD34-enriched mouse adipose cells in hair morphogenesis. Mol Med Rep 2013; 7(4): 1111-6.
[PMID: 23404453]

[347] De Francesco F, Tirino V, Desiderio V, *et al.* Human CD34/CD90 ASCs are capable of growing as sphere clusters, producing high levels of VEGF and forming capillaries. PLoS One 2009; 4(8): e6537.
[http://dx.doi.org/10.1371/journal.pone.0006537] [PMID: 19657392]

[348] Ducy P. The role of osteocalcin in the endocrine cross-talk between bone remodelling and energy metabolism. Diabetologia 2011; 54(6): 1291-7.
[http://dx.doi.org/10.1007/s00125-011-2155-z] [PMID: 21503740]

[349] Colnot C. Cellular and molecular interactions regulating skeletogenesis. J Cell Biochem 2005; 95(4): 688-97.
[http://dx.doi.org/10.1002/jcb.20449] [PMID: 15880692]

[350] Colnot C. Cellular and molecular interactions regulating skeletogenesis. J Cell Biochem 2005; 95(4): 688-97.
[http://dx.doi.org/10.1002/jcb.20449] [PMID: 15880692]

[351] Ferrara N. Vascular endothelial growth factor: basic science and clinical progress. Endocr Rev 2004; 25(4): 581-611.
[http://dx.doi.org/10.1210/er.2003-0027] [PMID: 15294883]

[352] Martinetti R, Dolcini L, Belpassi A, *et al.* Inspired porosity for cells and tissues. Key Eng Mater 2004; 254: 1095-8.
[http://dx.doi.org/10.4028/www.scientific.net/KEM.254-256.1095]

[353] Tampieri A, Celotti G, Landi E. From biomimetic apatites to biologically inspired composites. Anal Bioanal Chem 2005; 381(3): 568-76.
[http://dx.doi.org/10.1007/s00216-004-2943-0] [PMID: 15696277]

[354] Schmitz JP, Hollinger JO, Milam SB. Reconstruction of bone using calcium phosphate bone cements: a critical review. J Oral Maxillofac Surg 1999; 57(9): 1122-6.
[http://dx.doi.org/10.1016/S0278-2391(99)90338-5] [PMID: 10484115]

[355] Dorozhkin SV, Epple M. Biological and medical significance of calcium phosphates. Angew Chem Int Ed Engl 2002; 41(17): 3130-46.
[http://dx.doi.org/10.1002/1521-3773(20020902)41:17<3130::AID-ANIE3130>3.0.CO;2-1] [PMID: 12207375]

[356] Hing KA, Wilson LF, Buckland T. Comparative performance of three ceramic bone graft substitutes. Spine J 2007; 7(4): 475-90.
[http://dx.doi.org/10.1016/j.spinee.2006.07.017] [PMID: 17630146]

[357] Liu Q, Cen L, Yin S, *et al.* A comparative study of proliferation and osteogenic differentiation of adipose-derived stem cells on akermanite and beta-TCP ceramics. Biomaterials 2008; 29(36): 4792-9.
[http://dx.doi.org/10.1016/j.biomaterials.2008.08.039] [PMID: 18823660]

[358] Kruyt MC, Wilson CE, de Bruijn JD, *et al.* The effect of cell-based bone tissue engineering in a goat transverse process model. Biomaterials 2006; 27(29): 5099-106.
[http://dx.doi.org/10.1016/j.biomaterials.2006.05.048] [PMID: 16782186]

[359] Petite H, Viateau V, Bensaïd W, *et al.* Tissue-engineered bone regeneration. Nat Biotechnol 2000; 18(9): 959-63.
[http://dx.doi.org/10.1038/79449] [PMID: 10973216]

[360] Liao HT, Lee MY, Tsai WW, Wang HC, Lu WC. Osteogenesis of adipose-derived stem cells on polycaprolactone-β-tricalcium phosphate scaffold fabricated *via* selective laser sintering and surface coating with collagen type I. J Tissue Eng Regen Med 2013.
[http://dx.doi.org/10.1002/term.1811] [PMID: 23955935]

[361] Marino G, Rosso F, Cafiero G, *et al.* Beta-tricalcium phosphate 3D scaffold promote alone osteogenic differentiation of human adipose stem cells: *in vitro* study. J Mater Sci Mater Med 2010; 21(1): 353-63.
[http://dx.doi.org/10.1007/s10856-009-3840-z] [PMID: 19655233]

[362] Arrigoni E, de Girolamo L, Di Giancamillo A, *et al.* Adipose-derived stem cells and rabbit bone regeneration: histomorphometric, immunohistochemical and mechanical characterization. J Orthop Sci 2013; 18(2): 331-9.
[http://dx.doi.org/10.1007/s00776-012-0349-y] [PMID: 23344932]

[363] Zanetti AS, McCandless GT, Chan JY, Gimble JM, Hayes DJ. Characterization of novel akermanite:poly-ϵ-caprolactone scaffolds for human adipose-derived stem cells bone tissue engineering. J Tissue Eng Regen Med 2015; 9(4): 389-404.

[http://dx.doi.org/10.1002/term.1646] [PMID: 23166107]

[364] Zanetti AS, McCandless GT, Chan JY, Gimble JM, Hayes DJ. *In vitro* human adipose-derived stromal/stem cells osteogenesis in akermanite:poly-ε-caprolactone scaffolds. J Biomater Appl 2014; 28(7): 998-1007.
[http://dx.doi.org/10.1177/0885328213490974] [PMID: 23796629]

[365] Wongtriratanachai P, Pruksakorn D, Pothacharoen P, *et al.* Autologous chondrocyte implantation for cartilage injury treatment in Chiang Mai University Hospital: a case report. J Med Assoc Thai 2013; 96(11): 1518-22.
[PMID: 24428104]

[366] Itoh M, Kiuru M, Cairo MS, Christiano AM. Generation of keratinocytes from normal and recessive dystrophic epidermolysis bullosa-induced pluripotent stem cells. Proc Natl Acad Sci USA 2011; 108(21): 8797-802.
[http://dx.doi.org/10.1073/pnas.1100332108] [PMID: 21555586]

[367] Medvedev SP, Grigor'eva EV, Shevchenko AI, *et al.* Human induced pluripotent stem cells derived from fetal neural stem cells successfully undergo directed differentiation into cartilage. Stem Cells Dev 2011; 20(6): 1099-112.
[http://dx.doi.org/10.1089/scd.2010.0249] [PMID: 20846027]

[368] Ahmed N, Taylor DW, Wunder J, Nagy A, Gross AE, Kandel RA. Passaged human chondrocytes accumulate extracellular matrix when induced by bovine chondrocytes. J Tissue Eng Regen Med 2010; 4(3): 233-41.
[http://dx.doi.org/10.1002/term.235] [PMID: 19998328]

[369] Wei Y, Zeng W, Wan R, *et al.* Chondrogenic differentiation of induced pluripotent stem cells from osteoarthritic chondrocytes in alginate matrix. Eur Cell Mater 2012; 23(2): 1-12.
[PMID: 22241609]

[370] Kim JB, Zaehres H, Wu G, *et al.* Pluripotent stem cells induced from adult neural stem cells by reprogramming with two factors. Nature 2008; 454(7204): 646-50.
[http://dx.doi.org/10.1038/nature07061] [PMID: 18594515]

[371] Kuboth S, Kramer J, Rohwedel J. Chondrogenic differentiation *in vitro* of murine two-factor induced pluripotent stem cells is comparable to murine embryonic stem cells. Cells Tissues Organs (Print) 2012; 196(6): 481-9.
[http://dx.doi.org/10.1159/000338527] [PMID: 22797361]

[372] Guzzo RM, Gibson J, Xu RH, Lee FY, Drissi H. Efficient differentiation of human iPSC-derived mesenchymal stem cells to chondroprogenitor cells. J Cell Biochem 2013; 114(2): 480-90.
[http://dx.doi.org/10.1002/jcb.24388] [PMID: 22961870]

[373] Yoshida Y, Yamanaka S. Recent stem cell advances: induced pluripotent stem cells for disease modeling and stem cell-based regeneration. Circulation 2010; 122(1): 80-7.
[http://dx.doi.org/10.1161/CIRCULATIONAHA.109.881433] [PMID: 20606130]

[374] Diekman BO, Christoforou N, Willard VP, *et al.* Cartilage tissue engineering using differentiated and purified induced pluripotent stem cells. Proc Natl Acad Sci USA 2012; 109(47): 19172-7.
[http://dx.doi.org/10.1073/pnas.1210422109] [PMID: 23115336]

[375] Gurtner GC, Werner S, Barrandon Y, Longaker MT. Wound repair and regeneration. Nature 2008; 453(7193): 314-21.
[http://dx.doi.org/10.1038/nature07039] [PMID: 18480812]

[376] Hocking AM, Gibran NS. Mesenchymal stem cells: paracrine signaling and differentiation during cutaneous wound repair. Exp Cell Res 2010; 316(14): 2213-9.
[http://dx.doi.org/10.1016/j.yexcr.2010.05.009] [PMID: 20471978]

[377] Huang S, Lu G, Wu Y, *et al.* Mesenchymal stem cells delivered in a microsphere-based engineered skin contribute to cutaneous wound healing and sweat gland repair. J Dermatol Sci 2012; 66(1): 29-36.
[http://dx.doi.org/10.1016/j.jdermsci.2012.02.002] [PMID: 22398148]

[378] Kim JW, Lee JH, Lyoo YS, Jung DI, Park HM. The effects of topical mesenchymal stem cell transplantation in canine experimental cutaneous wounds. Vet Dermatol 2013; 24(2): 242-e53.
[http://dx.doi.org/10.1111/vde.12011] [PMID: 23432413]

[379] Wu Y, Chen L, Scott PG, Tredget EE. Mesenchymal stem cells enhance wound healing through differentiation and angiogenesis. Stem Cells 2007; 25(10): 2648-59.
[http://dx.doi.org/10.1634/stemcells.2007-0226] [PMID: 17615264]

[380] Stoff A, Rivera AA, Sanjib Banerjee N, *et al.* Promotion of incisional wound repair by human mesenchymal stem cell transplantation. Exp Dermatol 2009; 18(4): 362-9.
[http://dx.doi.org/10.1111/j.1600-0625.2008.00792.x] [PMID: 18803656]

[381] Liu J, Bian Z, Kuijpers-Jagtman AM, Von den Hoff JW. Skin and oral mucosa equivalents: construction and performance. Orthod Craniofac Res 2010; 13(1): 11-20.
[http://dx.doi.org/10.1111/j.1601-6343.2009.01475.x] [PMID: 20078790]

[382] Mølsted K. Treatment outcome in cleft lip and palate: issues and perspectives. Crit Rev Oral Biol Med 1999; 10(2): 225-39.
[http://dx.doi.org/10.1177/10454411990100020801] [PMID: 10759424]

[383] Sauerbier S, Gutwald R, Wiedmann-Al-Ahmad M, Lauer G, Schmelzeisen R. Clinical application of tissue-engineered transplants. Part I: mucosa. Clin Oral Implants Res 2006; 17(6): 625-32.
[http://dx.doi.org/10.1111/j.1600-0501.2006.01229.x] [PMID: 17092219]

[384] Izumi K, Feinberg SE, Iida A, Yoshizawa M. Intraoral grafting of an *ex vivo* produced oral mucosa equivalent: a preliminary report. Int J Oral Maxillofac Surg 2003; 32(2): 188-97.
[http://dx.doi.org/10.1054/ijom.2002.0365] [PMID: 12729781]

[385] Peramo A, Marcelo CL, Feinberg SE. Tissue engineering of lips and muco-cutaneous junctions: *in vitro* development of tissue engineered constructs of oral mucosa and skin for lip reconstruction. Tissue Eng Part C Methods 2012; 18(4): 273-82.
[http://dx.doi.org/10.1089/ten.tec.2011.0406] [PMID: 22067042]

[386] Eldesoqi K, Seebach C, Nguyen Ngoc C, *et al.* High calcium bioglass enhances differentiation and survival of endothelial progenitor cells, inducing early vascularization in critical size bone defects. PLoS One 2013; 8(11): e79058.
[http://dx.doi.org/10.1371/journal.pone.0079058] [PMID: 24244419]

[387] Canalis E, Economides AN, Gazzerro E. Bone morphogenetic proteins, their antagonists, and the skeleton. Endocr Rev 2003; 24(2): 218-35.

[http://dx.doi.org/10.1210/er.2002-0023] [PMID: 12700180]

[388] Williams DF. On the mechanisms of biocompatibility. Biomaterials 2008; 29(20): 2941-53. [http://dx.doi.org/10.1016/j.biomaterials.2008.04.023] [PMID: 18440630]

[389] Liu N, Lyu X, Fan H, Shi J, Hu J, Luo E. Animal models for craniofacial reconstruction by stem/stromal cells. Curr Stem Cell Res Ther 2014; 9(3): 174-86. [http://dx.doi.org/10.2174/1574888X09666140213150811] [PMID: 24524796]

[390] Payne KF, Balasundaram I, Deb S, Di Silvio L, Fan KF. Tissue engineering technology and its possible applications in oral and maxillofacial surgery. Br J Oral Maxillofac Surg 2014; 52(1): 7-15. [http://dx.doi.org/10.1016/j.bjoms.2013.03.005] [PMID: 23601833]

[391] Ferreira JR, Padilla R, Urkasemsin G, *et al.* Titanium-enriched hydroxyapatite-gelatin scaffolds with osteogenically differentiated progenitor cell aggregates for calvaria bone regeneration. Tissue Eng Part A 2013; 19(15-16): 1803-16. [http://dx.doi.org/10.1089/ten.tea.2012.0520] [PMID: 23495972]

[392] Kim RY, Fasi AC, Feinberg SE. Soft tissue engineering in craniomaxillofacial surgery. Ann Maxillofac Surg 2014; 4(1): 4-8. [http://dx.doi.org/10.4103/2231-0746.133064] [PMID: 24987591]

[393] Murphy CM, O'Brien FJ, Little DG, Schindeler A. Cell-scaffold interactions in the bone tissue engineering triad. Eur Cell Mater 2013; 26: 120-32. [PMID: 24052425]

[394] Seo S, Na K. Mesenchymal stem cell-based tissue engineering for chondrogenesis. J Biomed Biotechnol 2011; 2011(1): 806891. [PMID: 22007146]

[395] Tatullo M, Marrelli M, Paduano F. The regenerative medicine in oral and maxillofacial surgery: the most important innovations in the clinical application of mesenchymal stem cells. Int J Med Sci 2015; 12(1): 72-7. [http://dx.doi.org/10.7150/ijms.10706] [PMID: 25552921]

[396] Hutmacher DW. Scaffolds in tissue engineering bone and cartilage. Biomaterials 2000; 21(24): 2529-43. [http://dx.doi.org/10.1016/S0142-9612(00)00121-6] [PMID: 11071603]

[397] Lee K, Silva EA, Mooney DJ. Growth factor delivery-based tissue engineering: general approaches and a review of recent developments. J R Soc Interface 2011; 8(55): 153-70. [http://dx.doi.org/10.1098/rsif.2010.0223] [PMID: 20719768]

[398] Hjørting-Hansen E, Andreasen JO. Incomplete bone healing of experimental cavities in dog mandibles. Br J Oral Surg 1971; 9(1): 33-40. [http://dx.doi.org/10.1016/S0007-117X(71)80006-9] [PMID: 5286078]

[399] Schmitz JP, Hollinger JO. The critical size defect as an experimental model for craniomandibulofacial nonunions. Clin Orthop Relat Res 1986; 205(205): 299-308. [PMID: 3084153]

[400] Prolo DJ, Pedrotti PW, Burres KP, Oklund S. Superior osteogenesis in transplanted allogeneic canine skull following chemical sterilization. Clin Orthop Relat Res 1982; 168(168): 230-42.

[PMID: 7049486]

[401] Freeman E, Turnbull RS. The value of osseous coagulum as a graft material. J Periodontal Res 1973; 8(4): 229-36.
[http://dx.doi.org/10.1111/j.1600-0765.1973.tb00762.x] [PMID: 4269589]

[402] Turnbull RS, Freeman E. Use of wounds in the parietal bone of the rat for evaluating bone marrow for grafting into periodontal defects. J Periodontal Res 1974; 9(1): 39-43.
[http://dx.doi.org/10.1111/j.1600-0765.1974.tb00651.x] [PMID: 4277754]

[403] Mulliken JB, Glowacki J. Induced osteogenesis for repair and construction in the craniofacial region. Plast Reconstr Surg 1980; 65(5): 553-60.
[http://dx.doi.org/10.1097/00006534-198005000-00001] [PMID: 6988853]

[404] Shirasu N, Ueno T, Hirata Y, *et al.* Bone formation in a rat calvarial defect model after transplanting autogenous bone marrow with beta-tricalcium phosphate. Acta Histochem 2010; 112(3): 270-7.
[http://dx.doi.org/10.1016/j.acthis.2009.01.003] [PMID: 19403161]

[405] Kobayashi K, Agrawal K, Jackson IT, Vega JB. The effect of insulin-like growth factor 1 on craniofacial bone healing. Plast Reconstr Surg 1996; 97(6): 1129-35.
[http://dx.doi.org/10.1097/00006534-199605000-00006] [PMID: 8628795]

[406] Raposo-Amaral CE, Almeida AB, Paschoal G, *et al.* Histological and radiological changes in cranial bone in the presence of bone wax. Acta Cir Bras 2011; 26(4): 274-8.
[http://dx.doi.org/10.1590/S0102-86502011000400005] [PMID: 21808839]

[407] Dodde R II, Yavuzer R, Bier UC, Alkadri A, Jackson IT. Spontaneous bone healing in the rabbit. J Craniofac Surg 2000; 11(4): 346-9.
[http://dx.doi.org/10.1097/00001665-200011040-00013] [PMID: 11314382]

[408] Song JY, Kim SG, Lee JW, *et al.* Accelerated healing with the use of a silk fibroin membrane for the guided bone regeneration technique. Oral Surg Oral Med Oral Pathol Oral Radiol Endod 2011; 112(6): e26-33.
[http://dx.doi.org/10.1016/j.tripleo.2011.05.002] [PMID: 21856192]

[409] Wiltfang J, Kloss FR, Kessler P, *et al.* Effects of platelet-rich plasma on bone healing in combination with autogenous bone and bone substitutes in critical-size defects. An animal experiment. Clin Oral Implants Res 2004; 15(2): 187-93.
[http://dx.doi.org/10.1111/j.1600-0501.2004.00980.x] [PMID: 15008930]

[410] Schlegel KA, Lang FJ, Donath K, Kulow JT, Wiltfang J. The monocortical critical size bone defect as an alternative experimental model in testing bone substitute materials. Oral Surg Oral Med Oral Pathol Oral Radiol Endod 2006; 102(1): 7-13.
[http://dx.doi.org/10.1016/j.tripleo.2005.09.011] [PMID: 16831666]

[411] Lindholm TC, Lindholm TS, Marttinen A, Urist MR. Bovine bone morphogenetic protein (bBMP/NCP)-induced repair of skull trephine defects in pigs. Clin Orthop Relat Res 1994; (301): 263-70.
[PMID: 8156687]

[412] Cui L, Liu B, Liu G, *et al.* Repair of cranial bone defects with adipose derived stem cells and coral scaffold in a canine model. Biomaterials 2007; 28(36): 5477-86.

[http://dx.doi.org/10.1016/j.biomaterials.2007.08.042] [PMID: 17888508]

[413] Gupta DM, Kwan MD, Slater BJ, Wan DC, Longaker MT. Applications of an athymic nude mouse model of nonhealing critical-sized calvarial defects. J Craniofac Surg 2008; 19(1): 192-7. [PMID: 18216688]

[414] Kramer FJ, Mueller M, Rahmstorf M, Swennen GR, Dempf R, Schierle H. Ortho- and heterotopic bone grafts in bifocal transport osteogenesis for craniofacial reconstruction--an experimental study in sheep. Int J Oral Maxillofac Surg 2004; 33(6): 575-83. [http://dx.doi.org/10.1016/j.ijom.2004.01.003] [PMID: 15308258]

[415] Cheng G, Li Z, Wan Q, *et al.* A novel animal model treated with tooth extraction to repair the full-thickness defects in the mandible of rabbits. J Surg Res 2015; 194(2): 706-16. [http://dx.doi.org/10.1016/j.jss.2014.11.010] [PMID: 25491176]

[416] Carstens MH, Chin M, Li XJ. *In situ* osteogenesis: regeneration of 10-cm mandibular defect in porcine model using recombinant human bone morphogenetic protein-2 (rhBMP-2) and Helistat absorbable collagen sponge. J Craniofac Surg 2005; 16(6): 1033-42. [http://dx.doi.org/10.1097/01.scs.0000186307.09171.20] [PMID: 16327551]

[417] Ma JL, Pan JL, Tan BS, Cui FZ. Determination of critical size defect of minipig mandible. J Tissue Eng Regen Med 2009; 3(8): 615-22. [http://dx.doi.org/10.1002/term.203] [PMID: 19731258]

[418] Sun Z, Kennedy KS, Tee BC, Damron JB, Allen MJ. Establishing a critical-size mandibular defect model in growing pigs: characterization of spontaneous healing. J Oral Maxillofac Surg 2014; 72(9): 1852-68. [http://dx.doi.org/10.1016/j.joms.2014.02.024] [PMID: 24815793]

[419] Saadeh PB, Khosla RK, Mehrara BJ, *et al.* Repair of a critical size defect in the rat mandible using allogenic type I collagen. J Craniofac Surg 2001; 12(6): 573-9. [http://dx.doi.org/10.1097/00001665-200111000-00015] [PMID: 11711826]

[420] Arosarena OA, Falk A, Malmgren L, *et al.* Defect repair in the rat mandible with bone morphogenic proteins and marrow cells. Arch Facial Plast Surg 2003; 5(1): 103-8. [http://dx.doi.org/10.1001/archfaci.5.1.103] [PMID: 12533151]

[421] Schliephake H, Zghoul N, Jäger V, *et al.* Bone formation in trabecular bone cell seeded scaffolds used for reconstruction of the rat mandible. Int J Oral Maxillofac Surg 2009; 38(2): 166-72. [http://dx.doi.org/10.1016/j.ijom.2008.11.018] [PMID: 19121923]

[422] Huh JY, Choi BH, Kim BY, Lee SH, Zhu SJ, Jung JH. Critical size defect in the canine mandible. Oral Surg Oral Med Oral Pathol Oral Radiol Endod 2005; 100(3): 296-301. [http://dx.doi.org/10.1016/j.tripleo.2004.12.015] [PMID: 16122656]

[423] Asai S, Shimizu Y, Ooya K. Maxillary sinus augmentation model in rabbits: effect of occluded nasal ostium on new bone formation. Clin Oral Implants Res 2002; 13(4): 405-9. [http://dx.doi.org/10.1034/j.1600-0501.2002.130409.x] [PMID: 12175378]

[424] Xia L, Xu Y, Wei J, *et al.* Maxillary sinus floor elevation using a tissue-engineered bone with rhBMP-2-loaded porous calcium phosphate cement scaffold and bone marrow stromal cells in rabbits. Cells Tissues Organs (Print) 2011; 194(6): 481-93.

[http://dx.doi.org/10.1159/000323918] [PMID: 21494013]

[425] Xia L, Xu Y, Chang Q, *et al.* Maxillary sinus floor elevation using BMP-2 and Nell-1 gene-modified bone marrow stromal cells and TCP in rabbits. Calcif Tissue Int 2011; 89(1): 53-64. [http://dx.doi.org/10.1007/s00223-011-9493-1] [PMID: 21584647]

[426] Gutwald R, Haberstroh J, Stricker A, *et al.* Influence of rhBMP-2 on bone formation and osseointegration in different implant systems after sinus-floor elevation. An *in vivo* study on sheep. J Craniomaxillofac Surg 2010; 38(8): 571-9. [http://dx.doi.org/10.1016/j.jcms.2010.02.010] [PMID: 20381369]

[427] Yu BH, Zhou Q, Wang ZL. Comparison of tissue-engineered bone from different stem cell sources for maxillary sinus floor augmentation: a study in a canine model. J Oral Maxillofac Surg 2014; 72(6): 1084-92. [http://dx.doi.org/10.1016/j.joms.2013.12.024] [PMID: 24576438]

[428] Abarrategi A, Moreno-Vicente C, Martínez-Vázquez FJ, *et al.* Biological properties of solid free form designed ceramic scaffolds with BMP-2: *in vitro* and *in vivo* evaluation. PLoS One 2012; 7(3): e34117. [http://dx.doi.org/10.1371/journal.pone.0034117] [PMID: 22470527]

[429] Lyons KM, Beumer J III, Caputo AA. Abutment load transfer by removable partial denture obturator frameworks in different acquired maxillary defects. J Prosthet Dent 2005; 94(3): 281-8. [http://dx.doi.org/10.1016/j.prosdent.2005.06.005] [PMID: 16126081]

[430] Caton J, Mota L, Gandini L, Laskaris B. Non-human primate models for testing the efficacy and safety of periodontal regeneration procedures. J Periodontol 1994; 65(12): 1143-50. [http://dx.doi.org/10.1902/jop.1994.65.12.1143] [PMID: 7877087]

[431] Chantarawaratit P, Sangvanich P, Banlunara W, Soontornvipart K, Thunyakitpisal P. Acemannan sponges stimulate alveolar bone, cementum and periodontal ligament regeneration in a canine class II furcation defect model. J Periodontal Res 2014; 49(2): 164-78. [http://dx.doi.org/10.1111/jre.12090] [PMID: 23710575]

[432] Doğan A, Özdemir A, Kubar A, Oygür T. Healing of artificial fenestration defects by seeding of fibroblast-like cells derived from regenerated periodontal ligament in a dog: a preliminary study. Tissue Eng 2003; 9(6): 1189-96. [http://dx.doi.org/10.1089/10763270360728099] [PMID: 14670106]

[433] Liu Y, Zheng Y, Ding G, *et al.* Periodontal ligament stem cell-mediated treatment for periodontitis in miniature swine. Stem Cells 2008; 26(4): 1065-73. [http://dx.doi.org/10.1634/stemcells.2007-0734] [PMID: 18238856]

[434] Liu N, Lyu X, Fan H, Shi J, Hu J, Luo E. Animal models for craniofacial reconstruction by stem/stromal cells. Curr Stem Cell Res Ther 2014; 9(3): 174-86. [http://dx.doi.org/10.2174/1574888X09666140213150811] [PMID: 24524796]

[435] Kinoshita A, Oda S, Takahashi K, Yokota S, Ishikawa I. Periodontal regeneration by application of recombinant human bone morphogenetic protein-2 to horizontal circumferential defects created by experimental periodontitis in beagle dogs. J Periodontol 1997; 68(2): 103-9. [http://dx.doi.org/10.1902/jop.1997.68.2.103] [PMID: 9058326]

[436] Schwarz F, Sager M, Kadelka I, Ferrari D, Becker J. Influence of titanium implant surface characteristics on bone regeneration in dehiscence-type defects: an experimental study in dogs. J Clin Periodontol 2010; 37(5): 466-73.
[http://dx.doi.org/10.1111/j.1600-051X.2010.01533.x] [PMID: 20507369]

[437] Kim CK, Kim HY, Chai JK, *et al.* Effect of a calcium sulfate implant with calcium sulfate barrier on periodontal healing in 3-wall intrabony defects in dogs. J Periodontol 1998; 69(9): 982-8.
[http://dx.doi.org/10.1902/jop.1998.69.9.982] [PMID: 9776026]

[438] Wikesjö UM, Susin C, Qahash M, *et al.* The critical-size supraalveolar peri-implant defect model: characteristics and use. J Clin Periodontol 2006; 33(11): 846-54.
[http://dx.doi.org/10.1111/j.1600-051X.2006.00985.x] [PMID: 16965525]

[439] Park SY, Kim KH, Rhee SH, *et al.* An immediate peri-implantitis induction model to study regenerative peri-implantitis treatments. Clin Oral Implants Res 2015.
[http://dx.doi.org/10.1111/clr.12611] [PMID: 25958979]

[440] Misch CM, Misch CE, Resnik RR, Ismail YH. Reconstruction of maxillary alveolar defects with mandibular symphysis grafts for dental implants: a preliminary procedural report. Int J Oral Maxillofac Implants 1992; 7(3): 360-6.
[PMID: 1289262]

[441] Wang S, Zhang Z, Zhao J, *et al.* Vertical alveolar ridge augmentation with β-tricalcium phosphate and autologous osteoblasts in canine mandible. Biomaterials 2009; 30(13): 2489-98.
[http://dx.doi.org/10.1016/j.biomaterials.2008.12.067] [PMID: 19147220]

[442] Goldwaser BR, Magill J, Papadaki ME, *et al.* Continuous mandibular distraction osteogenesis: novel device and preliminary results in minipigs. J Oral Maxillofac Surg 2013; 71(4): e168-77.
[http://dx.doi.org/10.1016/j.joms.2012.09.012] [PMID: 23507324]

[443] Gaggl A, Schultes G, Regauer S, Kärcher H. Healing process after alveolar ridge distraction in sheep. Oral Surg Oral Med Oral Pathol Oral Radiol Endod 2000; 90(4): 420-9.
[http://dx.doi.org/10.1067/moe.2000.109159] [PMID: 11027377]

[444] Embree MC, Iwaoka GM, Kong D, *et al.* Soft tissue ossification and condylar cartilage degeneration following TMJ disc perforation in a rabbit pilot study. Osteoarthritis Cartilage 2015; 23(4): 629-39.
[http://dx.doi.org/10.1016/j.joca.2014.12.015] [PMID: 25573797]

[445] Sharawy MM, Helmy ES, Bays RA, Larke VB. Repair of temporomandibular joint disc perforation using a synovial membrane flap in Macaca fascicularis monkeys: light and electron microscopy studies. J Oral Maxillofac Surg 1994; 52(3): 259-70.
[http://dx.doi.org/10.1016/0278-2391(94)90296-8] [PMID: 8308624]

[446] Wu Y, Gong Z, Li J, Meng Q, Fang W, Long X. The pilot study of fibrin with temporomandibular joint derived synovial stem cells in repairing TMJ disc perforation. BioMed Res Int 2014; 2014(2): 454021.
[PMID: 24822210]

[447] Milam SB. Pathogenesis of degenerative temporomandibular joint arthritides. Odontology 2005; 93(1): 7-15.
[http://dx.doi.org/10.1007/s10266-005-0056-7] [PMID: 16170470]

[448] Li H, Cai X, Wang S, Yang C, Song H, Huang L. Disc positions and condylar changes induced by different stretching forces in the model for anterior disc displacement of temporomandibular joint. J Craniofac Surg 2014; 25(6): 2112-6.
[PMID: 25304145]

[449] Berteretche MV, Foucart JM, Meunier A, Carpentier P. Histologic changes associated with experimental partial anterior disc displacement in the rabbit temporomandibular joint. J Orofac Pain 2001; 15(4): 306-19.
[PMID: 12400399]

[450] Legrell PE, Isberg A. Mandibular length and midline asymmetry after experimentally induced temporomandibular joint disk displacement in rabbits. Am J Orthod Dentofacial Orthop 1999; 115(3): 247-53.
[http://dx.doi.org/10.1016/S0889-5406(99)70325-2] [PMID: 10066971]

[451] Tuncel U, Kostakoglu N, Turan A, Markoç F, Gokçe E, Erkorkmaz U. The use of temporalis muscle graft, fresh and cryopreserved amniotic membrane in preventing temporomandibular joint ankylosis after discectomy in rabbits. J Craniomaxillofac Surg 2014; 42(8): 1868-76.
[http://dx.doi.org/10.1016/j.jcms.2014.07.005] [PMID: 25218148]

[452] Ahtiainen K, Mauno J, Ellä V, *et al.* Autologous adipose stem cells and polylactide discs in the replacement of the rabbit temporomandibular joint disc. J R Soc Interface 2013; 10(85): 20130287.
[http://dx.doi.org/10.1098/rsif.2013.0287] [PMID: 23720535]

[453] Kawamoto T, Motohashi N, Kitamura A, *et al.* A histological study on experimental tooth movement into bone induced by recombinant human bone morphogenetic protein-2 in beagle dogs. Cleft Palate Craniofac J 2002; 39(4): 439-48.
[http://dx.doi.org/10.1597/1545-1569(2002)039<0439:AHSOET>2.0.CO;2] [PMID: 12071792]

[454] Gu X, Ding F, Yang Y, Liu J. Construction of tissue engineered nerve grafts and their application in peripheral nerve regeneration. Prog Neurobiol 2011; 93(2): 204-30.
[http://dx.doi.org/10.1016/j.pneurobio.2010.11.002] [PMID: 21130136]

[455] Mackinnon SE, Hudson AR. Clinical application of peripheral nerve transplantation. Plast Reconstr Surg 1992; 90(4): 695-9.
[http://dx.doi.org/10.1097/00006534-199210000-00024] [PMID: 1410009]

[456] Matsumine H, Takeuchi Y, Sasaki R, *et al.* Adipocyte-derived and dedifferentiated fat cells promoting facial nerve regeneration in a rat model. Plast Reconstr Surg 2014; 134(4): 686-97.
[http://dx.doi.org/10.1097/PRS.0000000000000537] [PMID: 25357029]

[457] Alwattar BJ, Schwarzkopf R, Kirsch T. Stem cells in orthopaedics and fracture healing. Bull NYU Hosp Jt Dis 2011; 69(1): 6-10.
[PMID: 21332433]

[458] Miura M, Miura Y, Sonoyama W, Yamaza T, Gronthos S, Shi S. Bone marrow-derived mesenchymal stem cells for regenerative medicine in craniofacial region. Oral Dis 2006; 12(6): 514-22.
[http://dx.doi.org/10.1111/j.1601-0825.2006.01300.x] [PMID: 17054762]

[459] Takahashi K, Yamanaka S. Induction of pluripotent stem cells from mouse embryonic and adult fibroblast cultures by defined factors. Cell 2006; 126(4): 663-76.

[http://dx.doi.org/10.1016/j.cell.2006.07.024] [PMID: 16904174]

[460] Miura K, Okada Y, Aoi T, *et al.* Variation in the safety of induced pluripotent stem cell lines. Nat Biotechnol 2009; 27(8): 743-5.
[http://dx.doi.org/10.1038/nbt.1554] [PMID: 19590502]

[461] Hu BY, Weick JP, Yu J, *et al.* Neural differentiation of human induced pluripotent stem cells follows developmental principles but with variable potency. Proc Natl Acad Sci USA 2010; 107(9): 4335-40.
[http://dx.doi.org/10.1073/pnas.0910012107] [PMID: 20160098]

[462] Rosenthal N. Prometheus's vulture and the stem-cell promise. N Engl J Med 2003; 349(3): 267-74.
[http://dx.doi.org/10.1056/NEJMra020849] [PMID: 12867611]

[463] Bianco P, Riminucci M, Gronthos S, Robey PG. Bone marrow stromal stem cells: nature, biology, and potential applications. Stem Cells 2001; 19(3): 180-92.
[http://dx.doi.org/10.1634/stemcells.19-3-180] [PMID: 11359943]

[464] Gronthos S, Zannettino AC, Hay SJ, *et al.* Molecular and cellular characterisation of highly purified stromal stem cells derived from human bone marrow. J Cell Sci 2003; 116(Pt 9): 1827-35.
[http://dx.doi.org/10.1242/jcs.00369] [PMID: 12665563]

[465] Akita S, Fukui M, Nakagawa H, Fujii T, Akino K. Cranial bone defect healing is accelerated by mesenchymal stem cells induced by coadministration of bone morphogenetic protein-2 and basic fibroblast growth factor. Wound Repair Regen 2004; 12(2): 252-9.
[http://dx.doi.org/10.1111/j.1067-1927.2004.012118.x] [PMID: 15086777]

[466] Yuan J, Cui L, Zhang WJ, Liu W, Cao Y. Repair of canine mandibular bone defects with bone marrow stromal cells and porous β-tricalcium phosphate. Biomaterials 2007; 28(6): 1005-13.
[http://dx.doi.org/10.1016/j.biomaterials.2006.10.015] [PMID: 17092556]

[467] Chang SC, Chuang HL, Chen YR, *et al. Ex vivo* gene therapy in autologous bone marrow stromal stem cells for tissue-engineered maxillofacial bone regeneration. Gene Ther 2003; 10(24): 2013-9.
[http://dx.doi.org/10.1038/sj.gt.3302106] [PMID: 14566360]

[468] Weng Y, Wang M, Liu W, *et al.* Repair of experimental alveolar bone defects by tissue-engineered bone. Tissue Eng 2006; 12(6): 1503-13.
[http://dx.doi.org/10.1089/ten.2006.12.1503] [PMID: 16846347]

[469] Kim BC, Yoon JH, Choi B, Lee J. Mandibular reconstruction with autologous human bone marrow stem cells and autogenous bone graft in a patient with plexiform ameloblastoma. J Craniofac Surg 2013; 24(4): e409-11.
[http://dx.doi.org/10.1097/SCS.0b013e318292c91d] [PMID: 23851884]

[470] Alhadlaq A, Mao JJ. Tissue-engineered neogenesis of human-shaped mandibular condyle from rat mesenchymal stem cells. J Dent Res 2003; 82(12): 951-6.
[http://dx.doi.org/10.1177/154405910308201203] [PMID: 14630893]

[471] Tsuzuki N, Seo JP, Yamada K, *et al.* The effect of a gelatin β-tricalcium phosphate sponge loaded with mesenchymal stem cells (MSC), bone morphogenic protein-2, and platelet-rich plasma (PRP) on equine articular cartilage defect. Can Vet J 2013; 54(6): 573-80.
[PMID: 24155448]

[472] Mimura T, Dezawa M, Kanno H, Sawada H, Yamamoto I. Peripheral nerve regeneration by

transplantation of bone marrow stromal cell-derived Schwann cells in adult rats. J Neurosurg 2004; 101(5): 806-12.
[http://dx.doi.org/10.3171/jns.2004.101.5.0806] [PMID: 15540919]

[473] Costa HJ, Ferreira Bento R, Salomone R, *et al.* Mesenchymal bone marrow stem cells within polyglycolic acid tube observed *in vivo* after six weeks enhance facial nerve regeneration. Brain Res 2013; 1510: 10-21.
[http://dx.doi.org/10.1016/j.brainres.2013.03.025] [PMID: 23542586]

[474] Cornelius P, MacDougald OA, Lane MD. Regulation of adipocyte development. Annu Rev Nutr 1994; 14(1): 99-129.
[http://dx.doi.org/10.1146/annurev.nu.14.070194.000531] [PMID: 7946535]

[475] Marra KG, Rubin JP. The potential of adipose-derived stem cells in craniofacial repair and regeneration. Birth Defects Res C Embryo Today 2012; 96(1): 95-7.
[http://dx.doi.org/10.1002/bdrc.21001] [PMID: 22457180]

[476] Carvalho PP, Leonor IB, Smith BJ, *et al.* Undifferentiated human adipose-derived stromal/stem cells loaded onto wet-spun starch-polycaprolactone scaffolds enhance bone regeneration: nude mice calvarial defect *in vivo* study. J Biomed Mater Res A 2014; 102(9): 3102-11.
[http://dx.doi.org/10.1002/jbm.a.34983] [PMID: 24123913]

[477] Fan J, Park H, Lee MK, *et al.* Adipose-derived stem cells and BMP-2 delivery in chitosan-based 3D constructs to enhance bone regeneration in a rat mandibular defect model. Tissue Eng Part A 2014; 20(15-16): 2169-79.
[http://dx.doi.org/10.1089/ten.tea.2013.0523] [PMID: 24524819]

[478] Sándor GK, Tuovinen VJ, Wolff J, *et al.* Adipose stem cell tissue-engineered construct used to treat large anterior mandibular defect: a case report and review of the clinical application of good manufacturing practice-level adipose stem cells for bone regeneration. J Oral Maxillofac Surg 2013; 71(5): 938-50.
[http://dx.doi.org/10.1016/j.joms.2012.11.014] [PMID: 23375899]

[479] Tobita M, Uysal AC, Ogawa R, Hyakusoku H, Mizuno H. Periodontal tissue regeneration with adipose-derived stem cells. Tissue Eng Part A 2008; 14(6): 945-53.
[http://dx.doi.org/10.1089/ten.tea.2007.0048] [PMID: 18558814]

[480] Kim WS, Park BS, Sung JH, *et al.* Wound healing effect of adipose-derived stem cells: a critical role of secretory factors on human dermal fibroblasts. J Dermatol Sci 2007; 48(1): 15-24.
[http://dx.doi.org/10.1016/j.jdermsci.2007.05.018] [PMID: 17643966]

[481] Kuo YR, Wang CT, Cheng JT, *et al.* Adipose-derived Stem Cells Accelerate Diabetic Wound Healing through the Induction of Autocrine and Paracrine Effects. Cell Transplant 2015.
[PMID: 25853951]

[482] Watanabe Y, Sasaki R, Matsumine H, Yamato M, Okano T. Undifferentiated and differentiated adipose-derived stem cells improve nerve regeneration in a rat model of facial nerve defect. J Tissue Eng Regen Med 2014.
[http://dx.doi.org/10.1002/term.1919] [PMID: 24889763]

[483] Chen W, Liu J, Manuchehrabadi N, Weir MD, Zhu Z, Xu HH. Umbilical cord and bone marrow mesenchymal stem cell seeding on macroporous calcium phosphate for bone regeneration in rat cranial

defects. Biomaterials 2013; 34(38): 9917-25.
[http://dx.doi.org/10.1016/j.biomaterials.2013.09.002] [PMID: 24054499]

[484] Todeschi MR, El Backly R, Capelli C, *et al.* Transplanted Umbilical Cord Mesenchymal Stem Cells Modify the *In Vivo* Microenvironment Enhancing Angiogenesis and Leading to Bone Regeneration. Stem Cells Dev 2015; 24(13): 1570-81.
[http://dx.doi.org/10.1089/scd.2014.0490] [PMID: 25685989]

[485] Sasaki M, Radtke C, Tan AM, *et al.* BDNF-hypersecreting human mesenchymal stem cells promote functional recovery, axonal sprouting, and protection of corticospinal neurons after spinal cord injury. J Neurosci 2009; 29(47): 14932-41.
[http://dx.doi.org/10.1523/JNEUROSCI.2769-09.2009] [PMID: 19940189]

[486] Ryu HH, Kang BJ, Park SS, *et al.* Comparison of mesenchymal stem cells derived from fat, bone marrow, Wharton's jelly, and umbilical cord blood for treating spinal cord injuries in dogs. J Vet Med Sci 2012; 74(12): 1617-30.
[http://dx.doi.org/10.1292/jvms.12-0065] [PMID: 22878503]

[487] Guo ZY, Sun X, Xu XL, Zhao Q, Peng J, Wang Y. Human umbilical cord mesenchymal stem cells promote peripheral nerve repair *via* paracrine mechanisms. Neural Regen Res 2015; 10(4): 651-8.
[http://dx.doi.org/10.4103/1673-5374.155442] [PMID: 26170829]

[488] Lei M, Li K, Li B, Gao LN, Chen FM, Jin Y. Mesenchymal stem cell characteristics of dental pulp and periodontal ligament stem cells after *in vivo* transplantation. Biomaterials 2014; 35(24): 6332-43.
[http://dx.doi.org/10.1016/j.biomaterials.2014.04.071] [PMID: 24824581]

[489] Riccio M, Maraldi T, Pisciotta A, *et al.* Fibroin scaffold repairs critical-size bone defects *in vivo* supported by human amniotic fluid and dental pulp stem cells. Tissue Eng Part A 2012; 18(9-10): 1006-13.
[http://dx.doi.org/10.1089/ten.tea.2011.0542] [PMID: 22166080]

[490] d'Aquino R, Graziano A, Sampaolesi M, *et al.* Human postnatal dental pulp cells co-differentiate into osteoblasts and endothelial cysts: a pivotal synergy leading to adult bone tissue formation. Cell Death Differ 2007; 14(6): 1162-71.
[http://dx.doi.org/10.1038/sj.cdd.4402121] [PMID: 17347663]

[491] Spyridopoulos T, Lambropoulou M, Pagonopoulou O, *et al.* Regenerated Nerve Defects with a Nerve Conduit Containing Dental Pulp Stem Cells in Pigs: An Immunohistochemical and Electrophysiological Evaluation. J Reconstr Microsurg 2015; 31(7): 516-26.
[http://dx.doi.org/10.1055/s-0035-1555751] [PMID: 26125150]

[492] Gronthos S, Mankani M, Brahim J, Robey PG, Shi S. Postnatal human dental pulp stem cells (DPSCs) *in vitro* and *in vivo.* Proc Natl Acad Sci USA 2000; 97(25): 13625-30.
[http://dx.doi.org/10.1073/pnas.240309797] [PMID: 11087820]

[493] Wang Y, Zhao Y, Jia W, Yang J, Ge L. Preliminary study on dental pulp stem cell-mediated pulp regeneration in canine immature permanent teeth. J Endod 2013; 39(2): 195-201.
[http://dx.doi.org/10.1016/j.joen.2012.10.002] [PMID: 23321230]

[494] Hutmacher DW. Scaffolds in tissue engineering bone and cartilage. Biomaterials 2000; 21(24): 2529-43.
[http://dx.doi.org/10.1016/S0142-9612(00)00121-6] [PMID: 11071603]

[495] Asti A, Gioglio L. Natural and synthetic biodegradable polymers: different scaffolds for cell expansion and tissue formation. Int J Artif Organs 2014; 37(3): 187-205.
[PMID: 24744164]

[496] Kim GH, Ahn S, Kim YY, *et al.* Coaxial structured collagen–alginate scaffolds: fabrication, physical properties, and biomedical application for skin tissue regeneration. J Mater Chem 2011; 21(17): 6165-72.
[http://dx.doi.org/10.1039/c0jm03452e]

[497] Seol YJ, Park JY, Jeong W, Kim TH, Kim SY, Cho DW. Development of hybrid scaffolds using ceramic and hydrogel for articular cartilage tissue regeneration. J Biomed Mater Res A 2015; 103(4): 1404-13.
[http://dx.doi.org/10.1002/jbm.a.35276] [PMID: 25044835]

[498] Yao R, Zhang R, Lin F, Luan J. Injectable cell/hydrogel microspheres induce the formation of fat lobule-like microtissues and vascularized adipose tissue regeneration. Biofabrication 2012; 4(4): 045003.
[http://dx.doi.org/10.1088/1758-5082/4/4/045003] [PMID: 23075755]

[499] Oh SH, Lee JH. Hydrophilization of synthetic biodegradable polymer scaffolds for improved cell/tissue compatibility. Biomed Mater 2013; 8(1): 014101.
[http://dx.doi.org/10.1088/1748-6041/8/1/014101] [PMID: 23472257]

[500] Shin H, Jo S, Mikos AG. Biomimetic materials for tissue engineering. Biomaterials 2003; 24(24): 4353-64.
[http://dx.doi.org/10.1016/S0142-9612(03)00339-9] [PMID: 12922148]

[501] Valence Sd, Tille JC, Chaabane C, *et al.* Plasma treatment for improving cell biocompatibility of a biodegradable polymer scaffold for vascular graft applications. Eur J Pharm Biopharm 2013; 85(1): 78-86.
[http://dx.doi.org/10.1016/j.ejpb.2013.06.012] [PMID: 23958319]

[502] Doğan A, Demirci S, Bayir Y, *et al.* Boron containing poly-(lactide-co-glycolide)(PLGA) scaffolds for bone tissue engineering Mater Sci Eng C Mater Biol Appl 2014; 44: 246-53.

[503] Kim JH, Kim TH, Jin GZ, *et al.* Mineralized poly(lactic acid) scaffolds loading vascular endothelial growth factor and the *in vivo* performance in rat subcutaneous model. J Biomed Mater Res A 2013; 101(5): 1447-55.
[http://dx.doi.org/10.1002/jbm.a.34446] [PMID: 23114998]

[504] Ren T, Ren J, Jia X, Pan K. The bone formation *in vitro* and mandibular defect repair using PLGA porous scaffolds. J Biomed Mater Res A 2005; 74(4): 562-9.
[http://dx.doi.org/10.1002/jbm.a.30324] [PMID: 16025492]

[505] del Real RP, Ooms E, Wolke JG, Vallet-Regí M, Jansen JA. *In vivo* bone response to porous calcium phosphate cement. J Biomed Mater Res A 2003; 65(1): 30-6.
[http://dx.doi.org/10.1002/jbm.a.10432] [PMID: 12635151]

[506] Morishita T, Honoki K, Ohgushi H, Kotobuki N, Matsushima A, Takakura Y. Tissue engineering approach to the treatment of bone tumors: three cases of cultured bone grafts derived from patients' mesenchymal stem cells. Artif Organs 2006; 30(2): 115-8.

[http://dx.doi.org/10.1111/j.1525-1594.2006.00190.x] [PMID: 16433845]

[507] Whu SW, Hung KC, Hsieh KH, Chen CH, Tsai CL, Hsu SH. *In vitro* and *in vivo* evaluation of chitosan-gelatin scaffolds for cartilage tissue engineering. Mater Sci Eng C Mater Biol Appl 2013; 33(5): 2855-63.
[http://dx.doi.org/10.1016/j.msec.2013.03.003] [PMID: 23623106]

[508] Yan S, Zhang Q, Wang J, *et al.* Silk fibroin/chondroitin sulfate/hyaluronic acid ternary scaffolds for dermal tissue reconstruction. Acta Biomater 2013; 9(6): 6771-82.
[http://dx.doi.org/10.1016/j.actbio.2013.02.016] [PMID: 23419553]

[509] Han F, Zhou F, Yang X, *et al.* A pilot study of conically graded chitosan–gelatin hydrogel/PLGA scaffold with dual-delivery of TGF-β1 and BMP-2 for regeneration of cartilage–bone interface. J Biomed Mater Res B Appl Biomater 2014.
[PMID: 25385571]

[510] Ben-David D, Kizhner T, Livne E, Srouji S. A tissue-like construct of human bone marrow MSCs composite scaffold support *in vivo* ectopic bone formation. J Tissue Eng Regen Med 2010; 4(1): 30-7.
[PMID: 19842114]

[511] Koç A, Elçin AE, Elçin YM. Ectopic osteogenic tissue formation by MC3T3-E1 cell-laden chitosan/hydroxyapatite composite scaffold. Artif Cells Nanomed Biotechnol 2015; 44(6): 1-8.
[http://dx.doi.org/10.3109/21691401.2015.1036998] [PMID: 25968048]

[512] Benya PD, Shaffer JD. Dedifferentiated chondrocytes reexpress the differentiated collagen phenotype when cultured in agarose gels. Cell 1982; 30(1): 215-24.
[http://dx.doi.org/10.1016/0092-8674(82)90027-7] [PMID: 7127471]

[513] Bohner M, Loosli Y, Baroud G, Lacroix D. Commentary: Deciphering the link between architecture and biological response of a bone graft substitute. Acta Biomater 2011; 7(2): 478-84.
[http://dx.doi.org/10.1016/j.actbio.2010.08.008] [PMID: 20709195]

[514] Jones AC, Arns CH, Hutmacher DW, Milthorpe BK, Sheppard AP, Knackstedt MA. The correlation of pore morphology, interconnectivity and physical properties of 3D ceramic scaffolds with bone ingrowth. Biomaterials 2009; 30(7): 1440-51.
[http://dx.doi.org/10.1016/j.biomaterials.2008.10.056] [PMID: 19091398]

[515] Karageorgiou V, Kaplan D. Porosity of 3D biomaterial scaffolds and osteogenesis. Biomaterials 2005; 26(27): 5474-91.
[http://dx.doi.org/10.1016/j.biomaterials.2005.02.002] [PMID: 15860204]

[516] Bohner M, Loosli Y, Baroud G, Lacroix D. Commentary: Deciphering the link between architecture and biological response of a bone graft substitute. Acta Biomater 2011; 7(2): 478-84.
[http://dx.doi.org/10.1016/j.actbio.2010.08.008] [PMID: 20709195]

[517] Guda T, Walker JA, Singleton B, *et al.* Hydroxyapatite scaffold pore architecture effects in large bone defects *in vivo.* J Biomater Appl 2014; 28(7): 1016-27.
[http://dx.doi.org/10.1177/0885328213491790] [PMID: 23771772]

[518] Zhang Q, Lu H, Kawazoe N, Chen G. Pore size effect of collagen scaffolds on cartilage regeneration. Acta Biomater 2014; 10(5): 2005-13.
[http://dx.doi.org/10.1016/j.actbio.2013.12.042] [PMID: 24384122]

[519] Aarvold A, Smith JO, Tayton ER, *et al.* The effect of porosity of a biphasic ceramic scaffold on human skeletal stem cell growth and differentiation *in vivo.* J Biomed Mater Res A 2013; 101(12): 3431-7. [http://dx.doi.org/10.1002/jbm.a.34646] [PMID: 23568640]

[520] Horst M, Milleret V, Nötzli S, *et al.* Increased porosity of electrospun hybrid scaffolds improved bladder tissue regeneration. J Biomed Mater Res A 2014; 102(7): 2116-24. [http://dx.doi.org/10.1002/jbm.a.34889] [PMID: 23893914]

[521] Despang F, Bernhardt A, Lode A, *et al.* Synthesis and physicochemical, *in vitro* and *in vivo* evaluation of an anisotropic, nanocrystalline hydroxyapatite bisque scaffold with parallel□aligned pores mimicking the microstructure of cortical bone. J Tissue Eng Regen Med 2013. [PMID: 23585334]

[522] Hosseinkhani H, Hosseinkhani M, Khademhosseini A, Kobayashi H. Bone regeneration through controlled release of bone morphogenetic protein-2 from 3-D tissue engineered nano-scaffold. J Control Release 2007; 117(3): 380-6. [http://dx.doi.org/10.1016/j.jconrel.2006.11.018] [PMID: 17207881]

[523] Gothard D, Smith EL, Kanczler JM, *et al.* Tissue engineered bone using select growth factors: A comprehensive review of animal studies and clinical translation studies in man. Eur Cell Mater 2014; 28: 166-207. [PMID: 25284140]

[524] Yamamoto M, Takahashi Y, Tabata Y. Controlled release by biodegradable hydrogels enhances the ectopic bone formation of bone morphogenetic protein. Biomaterials 2003; 24(24): 4375-83. [http://dx.doi.org/10.1016/S0142-9612(03)00337-5] [PMID: 12922150]

[525] Zhao J, Shinkai M, Takezawa T, Ohba S, Chung UI, Nagamune T. Bone regeneration using collagen type I vitrigel with bone morphogenetic protein-2. J Biosci Bioeng 2009; 107(3): 318-23. [http://dx.doi.org/10.1016/j.jbiosc.2008.10.007] [PMID: 19269600]

[526] Shin JH, Kim KH, Kim SH, *et al. Ex vivo* bone morphogenetic protein-2 gene delivery using gingival fibroblasts promotes bone regeneration in rats. J Clin Periodontol 2010; 37(3): 305-11. [http://dx.doi.org/10.1111/j.1600-051X.2009.01522.x] [PMID: 20041973]

[527] Jeong CG, Zhang H, Hollister SJ. Three-dimensional polycaprolactone scaffold-conjugated bone morphogenetic protein-2 promotes cartilage regeneration from primary chondrocytes *in vitro* and *in vivo* without accelerated endochondral ossification. J Biomed Mater Res A 2012; 100(8): 2088-96. [http://dx.doi.org/10.1002/jbm.a.33249] [PMID: 22615065]

[528] Park SY, Kim KH, Gwak EH, *et al. Ex vivo* bone morphogenetic protein 2 gene delivery using periodontal ligament stem cells for enhanced re-osseointegration in the regenerative treatment of peri-implantitis. J Biomed Mater Res A 2015; 103(1): 38-47. [http://dx.doi.org/10.1002/jbm.a.35145] [PMID: 24616330]

[529] Chen M, Song K, Rao N, Huang M, Huang Z, Cao Y. Roles of exogenously regulated bFGF expression in angiogenesis and bone regeneration in rat calvarial defects. Int J Mol Med 2011; 27(4): 545-53. [PMID: 21327326]

[530] Yang C, Liu Y, Li C, Zhang B. Repair of mandibular defects by bone marrow stromal cells expressing

the basic fibroblast growth factor transgene combined with multi-pore mineralized Bio-Oss. Mol Med Rep 2013; 7(1): 99-104.
[PMID: 23139139]

[531] Takafuji H, Suzuki T, Okubo Y, Fujimura K, Bessho K. Regeneration of articular cartilage defects in the temporomandibular joint of rabbits by fibroblast growth factor-2: a pilot study. Int J Oral Maxillofac Surg 2007; 36(10): 934-7.
[http://dx.doi.org/10.1016/j.ijom.2007.06.007] [PMID: 17826961]

[532] Hiraoka Y, Yamashiro H, Yasuda K, Kimura Y, Inamoto T, Tabata Y. *In situ* regeneration of adipose tissue in rat fat pad by combining a collagen scaffold with gelatin microspheres containing basic fibroblast growth factor. Tissue Eng 2006; 12(6): 1475-87.
[http://dx.doi.org/10.1089/ten.2006.12.1475] [PMID: 16846345]

[533] Soker S, Machado M, Atala A. Systems for therapeutic angiogenesis in tissue engineering. World J Urol 2000; 18(1): 10-8.
[http://dx.doi.org/10.1007/PL00007070] [PMID: 10766038]

[534] Liu Q, Huang Y, Lan Y, *et al.* Acceleration of skin regeneration in full-thickness burns by incorporation of bFGF-loaded alginate microspheres into a CMCS-PVA hydrogel. J Tissue Eng Regen Med 2015.
[http://dx.doi.org/10.1002/term.2057] [PMID: 26118827]

[535] Sekiya I, Vuoristo JT, Larson BL, Prockop DJ. *In vitro* cartilage formation by human adult stem cells from bone marrow stroma defines the sequence of cellular and molecular events during chondrogenesis. Proc Natl Acad Sci USA 2002; 99(7): 4397-402.
[http://dx.doi.org/10.1073/pnas.052716199] [PMID: 11917104]

[536] Ying B, Chen K, Hu J, *et al.* Effect of different doses of transforming growth factor-β□ on cartilage and subchondral bone in osteoarthritic temporomandibular joints. Br J Oral Maxillofac Surg 2013; 51(3): 241-6.
[http://dx.doi.org/10.1016/j.bjoms.2012.05.014] [PMID: 22763343]

[537] Choi B, Kim S, Fan J, *et al.* Covalently conjugated transforming growth factor-β1 in modular chitosan hydrogels for the effective treatment of articular cartilage defects. Biomater Sci 2015; 3(5): 742-52.
[http://dx.doi.org/10.1039/C4BM00431K] [PMID: 26222593]

[538] Herberg S, Kondrikova G, Periyasamy-Thandavan S, *et al.* Inkjet-based biopatterning of SDF-1β augments BMP-2-induced repair of critical size calvarial bone defects in mice. Bone 2014; 67: 95-103.
[http://dx.doi.org/10.1016/j.bone.2014.07.007] [PMID: 25016095]

[539] Tachi K, Takami M, Sato H, *et al.* Enhancement of bone morphogenetic protein-2-induced ectopic bone formation by transforming growth factor-β1. Tissue Eng Part A 2011; 17(5-6): 597-606.
[http://dx.doi.org/10.1089/ten.tea.2010.0094] [PMID: 20874259]

[540] Yamashima T, Yoshimura K, Morita O, Kobayashi K. [Histological study of bone regeneration using vascular endothelial growth factor on rat mandibular bone defect]. Nippon Hotetsu Shika Gakkai Zasshi 2005; 49(5): 726-35.
[http://dx.doi.org/10.2186/jjps.49.726] [PMID: 16237258]

[541] Wernike E, Montjovent MO, Liu Y, *et al.* VEGF incorporated into calcium phosphate ceramics promotes vascularisation and bone formation *in vivo.* Eur Cell Mater 2010; 19(3): 30-40.

[PMID: 20178096]

[542] Feng L, Wu H, e L, *et al.* Effects of vascular endothelial growth factor 165 on bone tissue engineering. PLoS One 2013; 8(12): e82945.
[http://dx.doi.org/10.1371/journal.pone.0082945] [PMID: 24376611]

[543] Marx RE, Carlson ER, Eichstaedt RM, Schimmele SR, Strauss JE, Georgeff KR. Platelet-rich plasma: Growth factor enhancement for bone grafts. Oral Surg Oral Med Oral Pathol Oral Radiol Endod 1998; 85(6): 638-46.
[http://dx.doi.org/10.1016/S1079-2104(98)90029-4] [PMID: 9638695]

[544] Chen TL, Lu HJ, Liu GQ, *et al.* Effect of autologous platelet-rich plasma in combination with bovine porous bone mineral and bio-guide membrane on bone regeneration in mandible bicortical bony defects. J Craniofac Surg 2014; 25(1): 215-23.
[http://dx.doi.org/10.1097/SCS.0000000000000420] [PMID: 24406581]

[545] Kawasumi M, Kitoh H, Siwicka KA, Ishiguro N. The effect of the platelet concentration in platelet-rich plasma gel on the regeneration of bone. J Bone Joint Surg Br 2008; 90(7): 966-72.
[http://dx.doi.org/10.1302/0301-620X.90B7.20235] [PMID: 18591611]

CHAPTER 5

Cell Sheets Engineering And Transplantation in MSCs Regeneration

Jing Wang[1,*], Renhuan Huang[2], Shengcai Qi[1], Rui Zhang[1], Liyan Lu[1], Fan Jian[2], Dan Wang[1], Yuanzhi Xu[1], Hua Wang[1], Haiyan Zhu[1] and **Xu Zhang[1]**

[1] *Department of Stomatology, Shanghai Tenth People's Hospital, Tongji University School of Medicine, Shanghai, China*

[2] *State Key Laboratory of Oral Diseases, West China School of Stomatology, Sichuan University, Chengdu, China*

Abstract: Transplantation of mesenchymal stem/stromal cells (MSCs) is an emerging treatment for various hard-to-treat diseases. In recent decades, cell sheet technology (CST) has been exploited as a novel promising scaffold-free approach for cell culturing and delivering in tissue regeneration. Results have demonstrated that MSCs facilitate cell viability and function in the CST co-culture system to improve transplantation efficiency. There are extensive interests in developing strategies to enhance the formation of cell sheets with MSC for downstream applications. Novel fabrication techniques have been emerging in recent years, including temperature-responsive system, electroactive responsive system, oxide surface assistant electrochemical system, light responsive system, PH responsive system, magnetite nanoparticles and magnetic force system. This chapter summarizes recent studies in cell sheet engineering and its combination with MSC for tissue regeneration.

Keywords: Cell sheet technology, Co-culture, Mesenchymal stem cell, Micro-patterning, Tissue engineering, Vascularization.

[*] **Corresponding author Jing Wang:** Department of Stomatology, Shanghai Tenth People's Hospital, Tongji University School of Medicine, No. 301, Middle Yanchang Road, Shanghai 200072, P.R. China; Tel: 86-21-66301726; Fax: 86-21-66301725; E-mail: dentist_tracy@hotmail.com.

Jing Wang & YunFeng Lin (Eds.)

1. INTRODUCTION

Recent years have witnessed the boosting development of regenerative medicine, which is targeted for functional restoration of damaged tissues with the transplantation of cells, tissues or even organs. In recent decades, substantial progress has been achieved in the realms of cell-based therapies and tissue engineering in regenerative medicine, though challenges still remain. For example, direct injection of single-cell suspensions, the most common method for cell transplantation therapy, cannot avoid huge cell loss in the host tissue site, and the island-like cell aggregation also discounts the healing [1, 2]. In addition, biodegradable scaffolds, one of the main components in tissue engineering, may lead to infiltration of pathological fibrous tissue or extracellular matrix (ECM) during scaffolds degradation [3, 4]. To overcome these issues, novel alternative approaches are needed.

Among all the constructive efforts, Okano and Yamada brought up the methodology named "cell sheet engineering" in the 1990s [5, 6], also referred as cell sheet technology (CST) in this book. CST is relied on the utilization of a temperature-responsive polymer, poly-N-isopropylacrylamide (PIPAAm), as the culture surfaces covalently grafted on conventional culture polystyrene dishes or glass substrates at nanometer level thickness. The hydrophilic/hydrophobic feature of the PIPAAm surface can be reversibly alternated in response to external temperature changes, which leads to thermally modulated cell adhesion and detachment. The spontaneous cell detachment facilitated by PIPAAm-grafted surfaces, enables a non-invasive harvest of cultured cells, in avoid of using trypsin, dispase and collagenase, which could cause cell damage and cell death during conventional cell culture for harvesting. All the cell surface proteins, external cell matrix and cell-cell/cell-ECM interactions are well preserved in the scaffold-free intact monolayer sheet [5, 6]. Furthermore, the collected monolayer sheet can either be directly transplanted to targeting sites straightway or accumulated to produce three-dimensional tissue-like structures with little cell migration in the transplanted site [7]. According to their sizes and shapes cell sheets can be classified into the following types (Table **1**).

There are other cell sheet materials, in addition to temperature-responsive polymers, in response to external stimuli to trigger cell detachment and promote transfer efficiency. So far the well investigated external stimuli that could influence the features of cell sheet materials include delicate mechanical force [8], pH [9], ionic concentration [10], magnetic force [11, 12], electrochemical polarization [13], and light [14], just to name a few. Temperature-responsive PIPAAm and its derived polymer membrane is the most frequently used in CST due to the safety and simplicity consideration [15 - 18].

Table 1. The advantages and disadvantages of the methods of cell sheet technologies.

Forms	**Structure/ cell types**	**Characteristics**	**Application**
Monolayered Cell Sheet	2D homotypic heterotypic	The basis of CST. Easy to fabricate. Either be directly transplanted to host sites or be further constructed. It is fragile, tends to shrink, wrinkle, clump and break when manipulating. Suitable for thin tissue transplantation without vascularization supply.	skin corneal epithelium bone periodontium urothelium islet
Multilayered Cell Sheet	3D homotypic hoterotypic	Most frequently applied and promising form of CST When MCSs are mounted, ECM deposited trans-sheets quickly. More stable structure, mimic to the real microenvironment with better cell interactions. Suitable for regenerating tissues or organs with high-density. Need vascularization supply.	bone, cartilage, periodontium kidney liver heart, blood vessels skin
Cell Sheet Fragment	up to1mm in diameter	Injectable Easier to transport to some narrow and deep locations with minimized damaging to the structure and the efficacy of cell transplantation is raised.	bone heart periodontium
Cell Sheet Pellet	3D microtissue	Formed by wrapping a cell sheet to a small global structure, which adds mechanical support and reduces instability. While the cellular bioactivity and nutrition supply in CSP for generating large constructs remain an issue.	dental-pulp

CST has been widely used along with various cell types for the regeneration of different types of tissues, including cardiac tissue, bone, tendon, cartilage, liver, corneal, bladder, pancreas, renal, skin, esophageal, blood vessel, and periodontium [4, 19 - 31]. Up to now, several clinical studies have shown curing effects of CST products for skin, cornea, esophagus, myocardium, cartilage, and peridontal regeneration [26, 32 - 38] (Fig. **1**). This part will be elaborately illustrated in the later section.

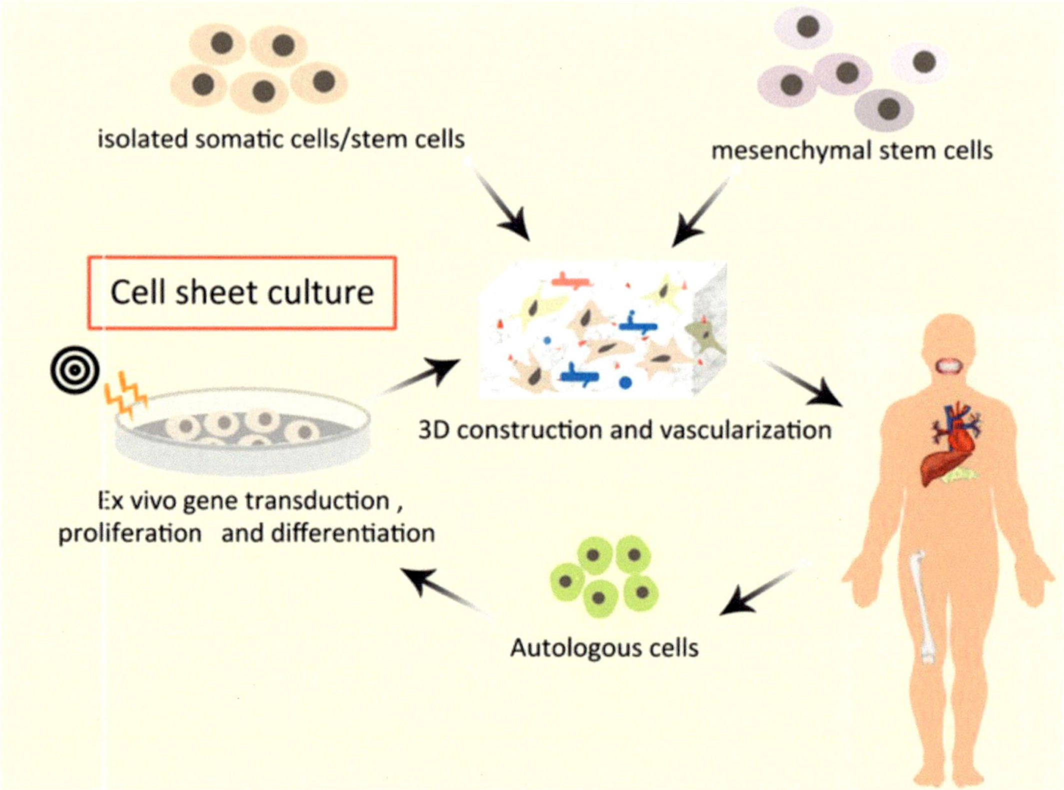

Fig. (1). The ideal strategy of cell-sheet-based therapy for tissue regeneration to treat heart, pancreas, liver, periodontal diseases and bone defects. Cell sheets are generated from isolated somatic cells or ES/iPS-derived cells, co-cultured with mesenchymal stem cells (MSC) using temperature-responsive culture dishes. To enhance specific protein expression, certain extra strategies may be applied, including ex vivo gene transduction, growth factors administration, *etc*. Single-layer sheets are stacked to form a 3D structure with vascularization. In the end, the mature CST products with satisfied function and shape are transplanted into the target sites of patients.

Moreover, a variety of stem cells have been introduced to cell sheet fabrication, and demonstrated obvious advantages over CST. Mesenchymal stem cells (MSCs),a group of self-renewal, multipotent stem/stromal cells derived from various adult tissues, are one of the promising seed cells in tissue engineering. MSCs can differentiate into different variety of cells, mainly osteoblasts, chondrocytes, adipocytes, epithelial cells, myocytes, and muscle cells [39]. The application of MSCs in CST has provided another abundant cell source. By co-cultering with somatic cells and/or administration of exogenous stimuli, stem cells can differentiate into specific cell linages with equivalent biofunctions and partially remain stemness [40]. Besides, MSCs are also a reservoir of multiple growth factors and cytokines, facilitating the cell bioactivity and angiogenesis, such as insulin-like growth factor-1 (IGF-1), vascular endothelial growth factor (VEGF), transforming growth factor-β1 (TGF-β1), and hepatocyte growth factor(HGF), *etc.* [41 - 44]. Thus, the transplantation of MSCs cell sheet not only restored cell/ECM in the defect sites, but also demonstrated evident improvements in systolic function, infarct wall attenuation, and fibrotic length after treating with MSCs sheets [45].

Moreover, MSCs are potentially immunosuppressive, and can be expanded easily *ex vivo* [30, 42]. MSCs are basically applied in two ways in CST: 1) as feeder cells to support the survival, proliferation, and differentiation of co-cultured somatic/stem/progenitor cells *in vitro*; or 2) as the resource for a specific cell type (Fig. **2**). MSCs sheet transplantation has demonstrated evident therapeutic effects in the animal models of bone defect, myocardial infarction as well as periodontal defect [46 - 49]. Vascularization is the key in constructing functional 3D tissue products as well as their survival after transplantation. In recent cases, the utilization of MSCs to generate prevascularization networks in engineered products prior to transplantation has drawn extensive interests and successful applications have been reported *in vitro* and in various animal models [23, 50 - 52]. Based on the innovative properties of CST with MSCs, CST is therefore suggested as one of the most promising technique with clinical potential in the near future.

Despite the therapeutic outcomes achieved *via* CST, certain challenges remain [53, 54]. Firstly, functional 3D cell sheet construction with satisfied stability after transplantation is required. Second, since tissue or organ transplantation requires a enormous number of high-quality cells of desired types, methods of large-scale and fast production of cell sheets is needed. The cell counts of the lab-prepared products are far from enough for the restoration of large defects. The emerging novel techniques and automated manufactures in CST have been established to address these issues, which will be discussed later.

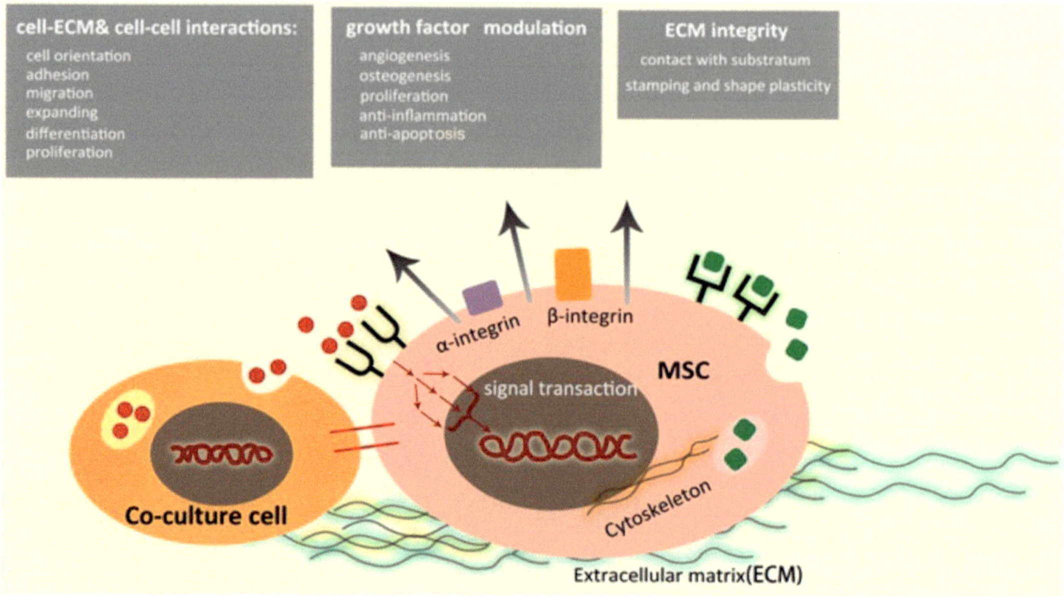

Fig. (2). Schematic plot of crosstalks between mesenchymal stem cell and the co-culture cells, and cell-ECM (illustrations are not to scale). Molecules or proteins in the ECM play an important role in the adhesion, migration and orientation of cells and transport the signals released by cells. With the involvement of specific receptors on the cell membrane, MSCs interact with the co-cultured cells or cells in the host tissues through secretion (autocrine or paracrine) of growth factors influencing cell proliferation, differentiation, angiogenesis, anti-apoptisis, and anti-inflammation. The integrity of cell-ECM is significant to mimic the natural-occurring tissues.

This chapter will provide a depiction of the latest strategies, applications and influence factors of CST focusing on the role and of MSCs, as well as the recent translational researches on CST.

2. FACTORS INFLUENCE THE TRANSPLANTING ABILITY OF MSCs IN CELL SHEETS

Procedures in the CST from cell culture, cell sheets stacking and 3D tissue-like structure transplantation contribute to the success of CST in regenerative medicine. Herein, we summarized several factors that may influence the transplantation ability of MSCs cell sheets in CST.

2.1. Cell Source and Donor-Specific Factors

In addition to MSCs derived from typical tissues such as bone marrow [31, 55 - 58], adipose tissues [45, 59, 60], teeth [61], periodontal tissues [62], new resources of MSCs including endometrial gland [63], ethmoid sinus mucosa [64], induced-pluripotent stem cells [65], showed equivalent differentiation capability with BMMSCs, implying more option considerations in the regeneration of specific tissues.

Although MSCs from different sources share similarities in immunophenotype, cell morphology, and differentiation characteristics *in vitro*, they differ in particular biological properties. Differences have been reported in the genomic methylation level, transcription behavior, and *in vivo* assessment among MSCs from white fat tissue, human bone marrow, umbilical cord, and skin [66]. BMSCs showed a superiority in chondrogenic and osteogenic differentiation potential than ASCs [66, 67], whereas other reports revealed similar efficiency in bone generation of the BMMSCs with that of human skeletal muscle-derived stem cells [68]. ASCs were reported to demonstrate higher angiogenic efficacy and adipogenic capacity than BMMSCs [69, 70]. In the regeneration of dentine, dental pulp derived MSCs (DPSCs) showed better dentinogenic differentiation ability than BMSC and ASCs [71]. Cell sheet of periodontal ligament cells (PDLCs) outweighed that of BMMSCs in periodontal regeneration [72]. These diverse outcomes may be related to different culture medium, micro-environment of transplantation sites and external stimuli used in different studies, and the underlying mechanisms remained to be elucidated [67].

MSCs from different tissue or area showed different features in regards to their expansion and differentiation capability as well as the immunology characteristics. For instance ASCs isolated from omental and subcutaneous adipose depots presented distinguished adipogenic differentiation capacity, though with resembling immunophenotypes [73]. Human amnion- and chorion-derived MSCs also showed diversities in soluble factor secretion and angiogenic/immuno-suppressive function [74].

On the other hand, donor-specific factors, including aging, smoking and underlying diseases of the donors, have also been considered influencing the differentiation potential of stem cells [75, 76]. It was reported that human PDL tissue derived from donors of different age could also provide functional PDLSCs, which could also be differentiated under an osteogenic/adipogenic micro-environment. Nevertheless, the proliferative and migratory capacity as well as the differentiation potency of PDLSCs decreased while age increasing. Results of another study [77] demonstrated that a sheet of PDLSCs extracted from elder donors had a weaker osteogenic capacity in contrast with those from younger donors. It was suggested that the age-related decrease in the cell matrix contents and osteogenic capacities of PDLSCs sheets should be given certain attention in CST research and clinical application.

These profound differences of MSCs from different source may provide implications for their clinical application. For instance, despite the advantages of BMMSC, they are restricted from extensive clinical use due to painful extraction procedure, low stem cell yield and weak multipotent ability after repeatedly passage. Certain techniques to improve the multi potency of ASCs or to apply induced-pluripotent cells may be considered as solution to these problems [65, 67].

2.2. Substrate Characteristics and Soluble Factors

Stem cells are sensitive to the biophysical and biochemical cues of the microenvironment and thus vast efforts in developing strategies have been made to enhance the formation of stem cell sheets for downstream applications [78]. Many investigations have suggested that substratum properties in the cell culture or transplantation site microenvironment including the mechanical, hydrophobicity and topographical parameters can influence cellular functions such as cell attachment, proliferation, migration, and differentiation by up-regulating several MSCs resident genes [16, 78 - 80].

Specifically, the layer thickness and external characteristics determine the hydrophilicity of the modified PIPAAm, which is crucial for effective thermal modulation on cell adhesion and detachment [81]. In early application of PIPAAm substratum, it possessed several technical challenges to achieve reversible detachment because of different cell affinity onto the PNIPAAm-grafted surface. It would take extra time for low-affinity cells to reach a confluent monolayer, and even hours for high-affinity cells to get fully detached, which may increase the risk of bacterial contamination, adhesion molecule loss, and possible cell damage from long-time exposure to low temperature [82]. In recent decade, much effort has been made to facilitate effective cell adhesion and rapid detachment, including alternating novel substrates, adding hydrophilic group [83], adjusting ingredients ratio [84], reaching fitted polymer thickness/density, providing appropriate external force (physical or magnetic force), *etc.* [11, 12, 45, 50, 85, 86]. To meet with various intrinsic cell adhesion and detachment properties, different types of PIPAAm coating methods have been developed, such as ultraviolet radiation, electron beam (EB) irradiation-induced polymerization for thin PIPAAm hydrogel layer CST [87 - 90], and living radical polymerization atom for a more precise control of PIPAAm layer thickness, polymer chain length and density [91 - 93]. Another method is to co-coat with hydrophobic segments, for instance poly(n-butyl methacrylate)-b-PIPAAm with PIPAAm copolymer. Novel polymer materials have also been utilized in cell sheet culture, revealing thermo-responsible features such as hydrogel, poly(N-isopropyl-acrylamide-co-butylacrylate) [84], and polyvinylidene difluoride membrane [94]. In order to fabricate soft tissues, grafting of pNIPAm on polydimethylsiloxane (PDMS) surfaces was introduced [95]. The fabrication part will be depicted in detail in the 5.4 section.

On the other hand, as one of the important aspects in tissue engineering, growth factors also have a pivotal role in MSCs sheet. Classical approach to promote vascularization is to decorate or supplement the culture medium with pro-angiogenic factors such as VEGF, and bFGF. In conventional CST culturing of somatic cells, exogenous growth factors and cytokines must be continuously administered to maintain the epithelial function during capillary formation. Previous studies indicated that MSCs promote vascularization not only by differentiating into endothelial cells but also by secreting angiogenic growth factors. It has been reported that MSCs secrete a variety of soluble growth factors, including VEGF, TGF-b, HGF, PGE2, IL-10, NO, indolamine2, 3-dioxygenase (IDO), heme oxygenase-1 (HO-1), and human leukocyte antigen-G(HLA-G).

These growth factors secreted by MSCs have been shown to modulate inflammation, inhibit apoptosis, maintain the stemness of MSCs and also increase angiogenesis [42, 44, 74, 96, 97]. The combination of ECs and fibroblasts CST transplants also stimulated the levels of HGF, placenta growth factor (PlGF) and matrix metalloproteinase-9 (MMP-9) in the targeted sites [43].

Although a wealth of literature has shown the effective facilitating impact of the paracrine function of MSCs, the growth factors secreted by MSCs may be insufficient to provide complete restoration of the desired functioning or massive fabrication of tissue engineering products. Additional strategies have therefore been proposed to enhance the MSCs paracrine capacity. Some researchers suggest that genetic modification of the MSCs sheet for certain growth factor overexpression may improve angiogenesis and augment the restoration function. Yeh and colleagues [98] prepared the ASCs sheet with overexpression of VEGF genes under transduction of hybrid baculovirus (BV), an emerging gene delivery vector, to enhance VEGF secretion in the treatment of myocardial infarction. The BV-ASC showed robust and prolonged overexpression and secretion of VEGF protein. After implanting the ASC sheet expressing VEGF to epicardial area of rabbit MI models, the infarct area was reduced and cardiac functions were almost recovered. Besides, it was also observed that an inactive myocardial wall attenuation, suppressed myocardium fibrosis and enhanced blood vessel formation.

Addition of extra growth factors has also been suggested. It was shown that platelet-rich fibrin (PRF) promoted cell proliferation and stimulated the expression of osteogenic factors in MSC sheets *in vitro* by upregulating related genes. Then the MSC+PRF combination produced more and denser bone, compared with the MSCs only, after being injected into a nude mice model [99]. Vitamin C was also used to develop BMSC sheet and PDLSC sheet which was able to increase specific ECM proteins and proliferate stem cells in CST [100 - 102]. Local application of simvastatin, Choukroun, and osthole in addition to MSC sheet promoted the proliferation and osteogenic differentiation of BMMSCs [99, 103, 104].

2.3. 3D-Culture and Pre-Vascularization

In 3D structure, cell-cell or cell-ECM interactions have been promoted greatly compared with that in the 2D structure, while dynamic culture surpasses a static one in aspects of cellular vitalities increase during the fabrication of thick engineered tissues [105]. By stacking multiple homotypic and heterotypic cell sheets, an effective biofabrication of functional 3D tissue or even organs can be achieved.

For the procedures of constructing 3D structure, Jun and colleagues [106] invented a novel way to transfer cell sheet *via* stamping it on temperature-sensitive hydrogels coated with cell-adhesive tubes on the surface. The hydrogels were synthesized from functionalized Tetronic® (a kind of copolymer), and the combination of fibronectin or cell-adhesive peptide (RGD)add cell adhesion properties on the hydrogel. The hydrogel complexions showed spontaneous thermally expansion in size within 10 min when the temperature reduced from 37°C to 4°C. Recently, Kim *et al.* [107] introduced a soft instrumented cell-culture platform equipped with ultrathin stretchable gold nanomembrane sensors and patterned graphene-nanoribbon (~150 and ~5 nm, respectively) cell aligners on a biocompatible PDMS substrate, whose modulus (~36.2 kPa) matches with that of target skeletal muscle tissues (45.9 kPa). This multifunctional platform could align the plated cells and *in situ* monitor cell physiological biomarkers during proliferation and differentiation. In addition, it yielded highly efficiency in delivering cell sheets *via* transferring printing mechanism *in vivo*. Utilization of such profound findings of stem cells in clinical application will provide opportunities for emerging novel strategies of tissue engineering and cell therapy technologies.

Vascularization is the key to success in fabricating cell-dense tissues. The survival of avascular tissues and thin tissues can be supported on merely simple diffusion of oxygen and nutrients without an additional vascular supply. The extent of simple diffusion is limited within approximately 100–200μm [108], which is insufficient for the survival of thick, cell-dense and actively-metabolic tissues, such as liver, heart, or kidney, since simple passive diffusion prevents the transportation of nutrients and may lead to a stack of metabolic wastes which could result in improper cell integration or cell death. In CST, the number of cell sheet layers in 3D structure is therefore limited, where neither core ischemia nor hypoxia. Hence, one of the main obstacles of current CST, as well as tissue engineering, is how to provide sufficient blood supply to grafted tissue substitutes in the early post-transplanted period.

One of the methods to resolve this problem could be recreation the microvascular networks within engineered tissues *in vitro* by fabricating transplantable 3D constructs before transplantation. An engineered full-thickness human skin with capillary-like networks facilitated an effective anastomosis *in vivo* [109]. Besides, in ischemic hearts, the transplantation of engineered vascularized myocardial tissues promoted angiogenesis in addition to strengthening the cardiac function [110]. Therefore, it is suggested that the induction of prevascularization in engineered tissue constructs fasten the integration with the host vascular network.

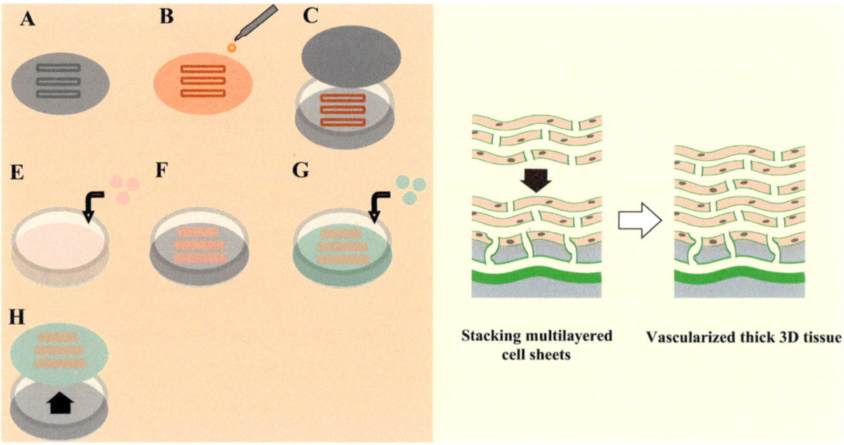

Fig. (3). The procedures of micro-printing a heterotypic cell sheet: **(A)** Fabrication and inspection of stratified polymer stamp on the substrate. **(B)** Administration of ECM onto the surface of stamp. **(C)** Microcontact printing. **(D)** Seeding the 1st cells. **(E)** Brief incubation and medium change. **(F)** Seeding 2nd cells. **(G)** After brief incubation, medium change, and cell culture, the harvest of cell sheet by reducing the temperature of temperature-responsive cell culture dish. (Adopted from the figure in [112]).

Previously, it has been reported that endothelial cells (ECs) play an essential role to fabricate vascularized 3D CST tissue. *Via* co-cultured three-dimensionally with other cell types, ECs spontaneously formed capillary-like networks *in vitro* using scaffold-based or scaffold-free 3D cell culture medium [17, 111, 112]. The coculture of ECs and fibroblasts or/and myoblast cells has been reported to promote vascular-like networks [113]. By successively seeding smooth muscle cells, fibroblasts and ECs to the cell sheets wrapped on a mandrel, newly formed and functional blood vessels were fabricated [114, 115]. In contrast with the scaffold-based technologies, CST-based 3D sandwich co-culture system has more advantages on controlling the cellular density and cell arrangement of inserted ECs or fibroblasts by the modification techniques of culture surfaces such as micro-patterning and micro-texture [116, 117]. With the aid of micro-patterning or micro-printing technique, co-culture of different cells was obtained in CST. The process is schematically plotted in Fig. (**3**).

MSCs have been proved to be a promising cell resource in the co-culture system. Epithelial progenitor cells have been co-cultured with MSCs in order to produce vascularized supply in bone regeneration. Researchers implanted cell sheet–EPCs complexes in nude mice, and the result revealed that the incorporation of EPCs could produce a vascular network and thus increase bone formation for CST-based bone engineering [118].

Moreover, a close relationship has been found between MSC and pericytes, cells on the inner side of blood vessels which impact over, contributing to vascularization, vessel stability and regulation of microvascular blood flow [119]. Considering the shared cell markers and functions, some researchers even suggest that pericytes are MSC-like progenitor cells [120]. Mendes *et al.* [121] fabricated co-cultured cell sheets of pericyte (CD146) cells and human umbilical vein endothelial cells (HUVECs) on an hBMSCs monolayer maintained in osteogenic medium for 7 days. The cell sheet expressed rich collagen and osteogenic capacity, indicating certain cross-talks between the two types of co-cultured cells. The proposed constructs also showed favorable osteogenic and angiogenic potential *in vivo*, assessed by the expression of osteocalcin and angiogenic markers overexpression. Besides, the diameter of human CD146+ blood vessels presented a higher average vessel diameter in the co-cultured CST condition.

In some cases, more than one kind of MSCs are co-cultured, among which one provides a mimic ECM micro-environment for the other. For instance, in the co-culture systems of HPDLSCs/DFCs and PPDLSCs/DFCs established by Liu *et al.*

[61], both HPDLSCs and PPDLSCs expressed prior levels of Klf4,Sox2, and Oct4,which are transcription factors related with self-renewal and pluripotency, after co-cultured with DFCs. This result implied that DFCs improved the comprehensive potency of HPDLSCs and PPDLSCs, by providing a favorable microenvironment as demonstrated by colony-formation and cell cycle analyses.

In the field of islet CST regeneration, the protective role on islet cells exerted by MSCs have been reported, through secreting HGF and TGF β1, and thus establishing a βcells supportive microenvironment [122]. The improvement of trophic effect in an MSC-islets sheet was observed compared to MSCs or islet cells alone, the underlying mechanism of which was proposed that MSC-sheet--generating ECM could promote the engraftment. Hirabaru *et al.* [30]applied a CST approach by combining islets with MSC sheets (islets + MSC sheet) to transplantation in severely immune-deficient mice with streptozotocin-induced diabetes. Restored islet function and viability was shown *in vitro*. A normal level of blood sugar was maintained for at least 84 days in the recipient mice post to transplantation, and neovascularization around the transplant was observed.

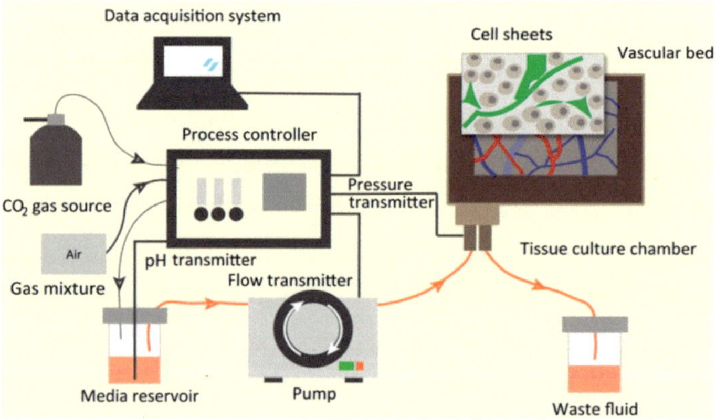

Fig. (4). The schematic graph of bioreactors for cardiac cell sheets. **(A)** The bioreactor consists of a delivery pump, custom-made tissue culture chamber, pH transmitter, flow transmitter, pressure transmitter, CO_2 gas source, process controller, and data acquisition system. The arrows of different colors indicate the direction of gas, culture medium and waste flows respectively. A multi-layered cell sheet is perfused in bioreactor with a vascular bed. **(B)** This vascular bed with collagen gel microchannels mimics the in vivo micro-environment. After the connection between newly formed microvessels and the collagen gel microchannels is formed, another triple-layered cardiac cell sheet is placed on the existing cell sheet. By repeating the sheet layering process, the subsequently added cell sheets are perfused with fresh medium through the newly-created vessels (right illustration). (Adopted from the figure in [17]).

Other alternative methods have been developed to facilitate the formation of microvascular networks within tissue engineered constructs, such as the delivery of specific genes or growth factors, along with the creation of bioreactors [98, 123 - 127]. Complex culturing system to fabricate clinical-level cell sheets and automated manufacture is a key for tissue-engineered therapeutic in the coming future [128, 129]. A schematic graph of the automated process is presented in Fig. (**4**).

2.4. Immune Responses to Transplanted Cell Sheet

Due to the injuries of host tissues often caused by transportation procedure, the immune responses remain one of the major causes for a transplant failure. Besides, under certain pathological conditions like periodontitis and heart infarction, the target sites are already infiltrated with inflammatory cells. Neutrophils and T cells involved in the inflammation process secrete TNF-α, IL-1B, IL-4, IL-6, and IL-18, and form an environment risking the survival of transplanted cells and engineered grafts.

The scaffold-free cell sheet in CST transplantation thus triggers less immune reaction compared with the conventional scaffold-based transplantation. MSCs are reported effective in attenuating immunoreactions in graft despite the diseases condition of hosts. Chondrocyte sheet has been fabricated to act as an immune-elusive carrier to protect islet cells from immune reactions after transplantation, by wrapping islet cells inside [20]. The role of MSC-CST in modulating chronic inflammation, has been reported recently, which may involve the activity of macrophages [30, 61, 122, 130 - 132]. In the treatment of heart failure, In the study of transplanting adipocyte cell sheets in a mouse MI model conducted by Imanishi and colleagues [133], abundant adiponectin was found secreted by adipocyte cell sheets. Adiponectin is a circulating protein with various cardiac protective effects, such asanti-inflammation, anti-fibrosis and cardiac hypertrophy. Three months after transplantation, adiponectin could be still detected in plasma. Attenuated level of TNF-αexpression in the margin area of the infarcted sites of the heart was also observed along with the improved infarct size, fibrotic area, and hypertrophy of cardiomyocyte.

2.5. Combination With Scaffold Materials

Both CST and scaffold-based methods have advantages and weaknesses. On the one hand, engineered tissues produce abundant ECM as endogenous scaffold in CST culturing. However, while maintaining the intact cell-ECM interaction, they are too fragile to be further fabricated into cell-dense tissues or organs with specific shape(such as tubular bone and vessel)due to the lack of enough support. On the other hand, as in the conventional scaffold-based methods, direct planting seed cells onto the scaffold results in unsatisfied cell expansion, proliferation, adhesion and utilization. So in recent years, the two complementary approaches are combined in tissue engineering application. Biocompatible and degradable scaffold materials, mostly nanomaterials, provide strong outer support to cell sheets, strengthen the resistance and mold the tissues into desired shape. Meanwhile the high-density and vital cells produced by CST improved the vitality and bioactivity of the engineered tissues [134]. For example, mXPCL films with coating modifications were used for developing stratified tissue from layered co-cultures [135].

Adding scaffold materials also provides a new way to design micro-patterning and construct 3D structure [136]. In the method described by Rayatpisheh *et al.* [115] NIPAm-grafted PDMS substrate was combined with an aligned electrospun scaffold of polycaprolactone (PCL) inducing alignment of the adherent cells. Upon cooling, the scaffold and its adherent cells were wrapped on a mandrel with the cell sheet detachment, which leads to the generation of tubular constructs with circumferentially aligned blood vessel cells and media.

The combination materials can range from a decellularized matrix to a mixture of tiny particles to plant the cells. In the study conducted by Mouthuy P.A *et al.* [137] electro spun membranes made by poly(lactic-co-glycolic acid) (PLGA) combined with collagen type I and hydroxyapatite (HA) were layered with hMSCs sheets. What's more, carbon nanotubes (CNTs), graphene, magnetic particles, polyethylenimine-alginate(PEI-al) nanocomposites, coral, surface-modified titanium and zirconia have also been utilized with cell sheet [42, 101, 107, 138 - 142].The combination of scaffold materials may indeed strengthen properties of CST *in vitro* to certain degrees. The inflammation caused by the degradation of scaffold materials, however, should also be considered in transplantation.

3. METHODS FOR FABRICATION OF CELL SHEETS

3.1. Fabricate Cell Sheet Using Temperature-Responsive System

PIPAAm is mainly used to fabricate cell sheets in temperature-responsive system. Because the lower critical solution temperature (LCST) of PIPAAm is 32°C, under which will change the characteristics of PIPAAm from hydrophobic to hydrophilic. After nanometer-scale thickness of the immobilizing PIPAAm covered on ordinary tissue culture polystyrene (TCPS) surfaces, the simple temperature changes from 37°C to 20°C will control the cell adhesion and detachment. On PIPAAm surfaces, the adherence, spreading and proliferation of various cell types are unaffected and similar to the normal TCPS at 37°C. Once the incubation temperature is reduced under the LCST, the conversion of the grafted PIPAAm will change from hydrophobic to hydrophilic, which leads to spontaneously detachment of all cultured cells (Fig. **5A**). Poly N-vinylcaprolactam (PNVCL) is another type of thermoresponsive polymer, which was successfully firstly synthesized by initiated chemical vapor deposition (iCVD), known to be nontoxic, water soluble and biocompatible. A sharp LCST transition of the iCVD PNVCL was observed at 31°C, which will almost do not affect the viability of cell sheets and differentiation potential, compared to 20°C. The thermoresponsive PNVCL surface enabled attachment and detachment of cells based on the thermally modulation, for a hydrophilic/hydrophobic alteration controlled by the induction temperature change [143]. Based on the temperature-responsive surfaces, a strategy of "cell sheet tissue engineering" was developed, which has been widely used in tissue engineering afterwards. Cultured cells sheets are harvested in intact state along with their deposited ECM (Fig. **5A** and **B**). The cell sheets are formed on the thermoresponsive surfaces, which allow cultured cells to be harvested without proteolytic enzymes such as trypsin or dispase, enzymes will result in many disadvantages. For example, cell damage, loss of differentiated phenotypes, and destroyed ECM. ECM can be transplanted to the tissue bed directly and could be used to create 3D structures in tissue repair therapy. Temperature-responsive system was the most popular methods to fabricate the cell sheets, which includes electron beam lithographic (EBL) system, plasm treated system, ultrathin responsive system, thermoresponsive surfaces system, spin-coated system, iCVD system and so on.

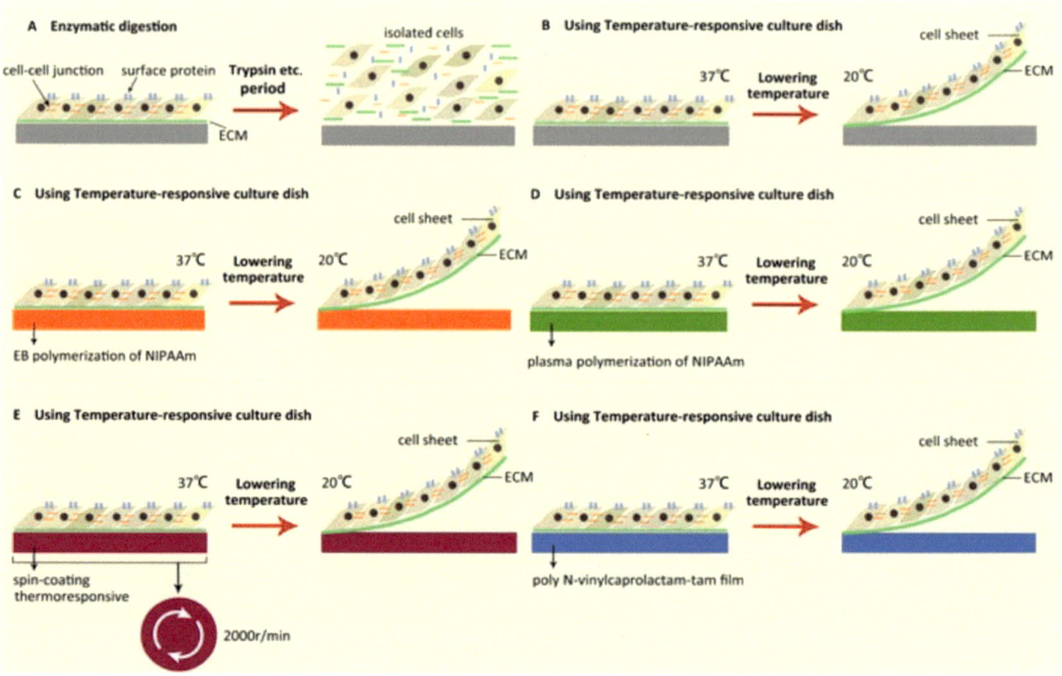

Fig. (5). Fabricate cell sheet using temperature-responsive system. **A:** confluent cells are harvested separately by enzyme digestion. **B:** a contiguous cell sheet by lowering the temperature-responsive culture dish. **C:** cell sheet formed on electron beam lithographic system. **D:** cell sheet detaches from plasma polymerization of NIPAAm. **E:** cell sheet formed on spin-coating thermoresponsive system. **F:** cell sheet detaches frompoly N-vinylcaprolactam-tam film. ECM, extracellular matrix; EB, electron beam.

3.1.1. Electron Beam Lithographic (EBL) System

An EBL system was developed to fabricate micrometer- and nanometer-patterned PIPAAm layers, which is a simple method for direct graft-polymerization without any photo-sensitive resists. Multiple processes were required in fabricating conventional nano-patterned grafted polymer layers. However, irradiation with the EBL system is the simplest way to fabricate nano-patterned polymer-grafted surfaces, which was able to control the patterning and graft-polymerization of the monomer solution simultaneously. The most employed method for producing thermoresponsive surfaces is PNIPAAm onto tissue culture polystyrene (TCPS) for cell sheet engineering [144, 145]. The compound N-isopropylacrylamide is used in EBL system. EBL exposure could make the compound N-isopropylacrylamide polymerized and grafted directly onto hydrophilic polyacrylamide-grafted glass surfaces (Fig. **5C**). The patterned area of the grafted PIPAAm can be controlled on a nanometer-scale through changing the irradiated area. This system could fabricate a minimal stripe pattern with a 200 nm line-width onto the surface. Cells adhered on the surfaces were oriented and elongated along the grafted PIPAAm at temperature above the LCST. The pattern interval of the grafted PIPAAm plays a key role in the magnitude of the spreading angle and elongation of cell adherence. Once culture temperature was lower than the LCST, the cell sheet was detached from the surface. This EBL polymerization of IPAAm on glass substrates permits the ability to harvest cell sheets with intact cellular morphologies and functions, which can be controlled by micrometer and nanometer scale surface features and easy fabrication of the surface patterning. This patterning system could be used in many aspects, such as functional cell sheet engineering, cell-based bioreactors, and the construction of 3D cell structures in tissue engineering.

The process will be briefly described as follows. A patterned thermoresponsive PIPAAm was prepared using an EB lithography system on PAAm grafted glass substrates. The IPAAm was firstly dissolved in 2-propanol, then used as the coating substrate to cover the PAAm-grafted glass substrate. This coated substrate was then grafted onto the PAAm grafted surface in a vacuum chamber and polymerized with EBL system. A computer-managed EB scanner is used to polymerize PIPAAm in a nanometer or micrometer sized formation. The IPAAm monomers were directly polymerized onto the PAAm grafted surfaces only in EB exposed areas. The unexposed IPAAm monomers and the conductive polymer layers were washed away with pure water after the EB exposure [146]. Thickness of the PIPAAm grafted layer plays a crucial role in thermoresponsive alteration in hydrophilic/hydrophobic property which will result in cell adhesion/detachment [144]. The grafted PIPAAm layers in the thickness ranged of 15–20 nm is best for thermoresponsive cell adhesion and detachment [147, 148]. The thickness over 30 nm grafted PNIPAAm layer could not be used, because cell will not adhere on it at any temperature [144, 145]. This patterned cell sheet technique has been reported applied in fabricating muscle cell sheets [145, 146], hepatic cell sheet

[149], keratin cell sheet [148], cornea with efficient shrinkage and relaxation in a specific direction.

3.1.2. Plasma Treated System

Surface immobilized PNIPAAm is currently used as a biosensor in a wide variety of tissue engineering. Plasma treated system generated the type of thermoresponsive NIPAAm surface through covalently attached to a solid substrate by the plasma induced polymerization [150, 151]. In the system, plasma polymerization can form and deposit an adhesion-promoting and functional plasma PNIPAAm layer on the outer surface, and the power is gradually decreased in the progress. Uncross linked IPAAm monomers and the conductive polymer layers were then removed with cold deionized water (Fig. **5D**). When coated substrates were placed in the temperature higher or lower than LCST, plasma-deposited NIPAAm polymers on the substrate showed a remarkable retention of the monomer integrity [152] and preserved the PNIPAAm phase transition [153], which was responsible for the detachment of cell sheets from the PNIPAAm surface. Plasma polymerization of NIPAAm fabricated thermally responsive coatings without specially prepared substrates and enabled the production of functionalized polymers in a simple one-step method [153].

The sum frequency generation (SFG) vibrational spectra was used to analyses the characteristic of this type of graft above the LCST, the result suggested that the hydrophobic isopropyl side groups were oriented towards the aqueous environment. However, the spectra of the graft obtained at room temperature showed a disordering of the isopropyl groups. It was suggested that plasma polymerization of NIPAAm could not participate in hydrogen bonding correctly when exposed to the aqueous environment below the LCST. Finally, plasma polymerization of NIPAAm did not show obvious effects for cell adhesion, proliferation and differentiation among different batches and thickness of plasma-deposited polymers [150, 151]. This means cell adhesion and proliferation seem not to be sensitive to the thickness of plasma-polymerized PNIPAAm substrates. Plasma treated system may be used to construct endothelial cell sheet and retina cell sheet.

3.1.3. Ultrathin Responsive System

Argon plasma made pre-oxidation on polydimethylsi-Loxane (PDMS) surface was found capable of stimulating the decomposition to produce free radicals under the ultraviolet line and trigger a polymerization reaction of NIPAAm. The ultrathin responsive system produced thermoresponsive surfaces involving the partial entrapment of a copolymer of NIPAAm and 4-(N-cinnamoylcarbamide) methylstyrene (CCMS) onto TCPS, which was entrapped in traditional TCPS and cross linked the copolymer by UV light irradiation to induce the cinnamoyl groups dimerization [154]. The grafted polymerization PNIPAAm thickness by UV light irradiation can be controlled by the plasma treatment duration and concentration of the monomer. Thickness around 30 nm of polymerization PNIPAAm is good for SMC adherence and detachment to form cell sheets. Cell adhesion and proliferation were found not sensitive to the thickness of the functional layer which is similarly to plasma-polymerized PNIPAAm [155].

Another approach to utilization of TCPS surface in the ultrathin responsive system was a photoreactive 4-azidoaniline groups mixed NIPAAm copolymer [156, 157], which a heavily cross linked layer was produced on the surface. This method has been employed to generate micropatterned surfaces on thermo responsive regions by photolithography, then the unexposed solvent coating is washed away [156, 157]. Using this method, cell sheets formation was observer after fibroblasts cultured on the micropatterned substrates for short periods. Afterwards, the cell sheets were detached from the grafted polymerization PNIPAAm regions [156, 157].

3.1.4. Cell Adhesion Protein Coated Thermoresponsive Surfaces System

PNIPAAm membrane is not grafted on a substrate, but deposited by combination with collagen on the solid surface, such as TCPS and glass. The mechanism of cell adhesion protein coated thermoresponsive surfaces system is that the collagen on the solid surface causes the cell to adhere on the PNIPAAm membrane, while the PNIPAAm is dissolved when temperature is lower than 20°C. This system mainly contained copolymers of PNIPAAm and N-tert-butyl-acrylamide (NtBAm), varying ratios were synthesized in this system. The copolymer

PNIPAAm films were coated with cell adhesion promoters, such as collagen, poly-L-lysine, and laminin, to increase cell adhesion and proliferation on the films. Cell viability on this coated copolymer films was similar controlled with standard tissue culture. Cell detached from the films depended on copolymer composition, which does not affect the ECM proteins [158, 159]. Cultivation of cell sheets on PNIPAAm copolymers films was relatively simple, viable and achievable by this cell adhesion protein coated thermoresponsive surface system.

3.1.5. Spin-Coated System

PNIPAAm was mixed in hydrochloric acid solution with tetraethyl orthosilicate (TOES). Coverslips cover by the mixture of PNIPAAm and TOES was rotated 2000 R/min for 60s to form formation of a thin layer of PNIPAAm [160]. PNIPAAm and alcohol was mixed to coat the rotated coverslip to form the PNIPAAm film, which can result in rapid detachment of various cell sheets. This spin and coated approach offers a simple and cheap alternative to cell surface design in tissue engineering [161]. Deposition of the thermoresponsive thin films using PNIPAAm blended with a small amount of 3-aminopropyltriethoxysilane (APTES) fabricated *via* the spin-coating technique on glass slides or silicon wafers, which enhanced the retention of PNIPAAm. When temperature decreases below LCST, PNIPAAm chain extended and caused the cell detachment from the film (Fig. **5E**). The thermoresponsive films created using this method offered a new way for cell attachment and detachment without additional usage of adhesive proteins and digestive enzymes, resulting in intact morphology and function during cell sheets harvest [162].

3.1.6. Initiated Chemical Vapor Deposition (iCVD) System

iCVD system is the method to use initiated vapor deposition chemical process to produce the temperature responsive poly N-vinylcaprolactam (PNVCL) film which was verified safe in cell sheet engineering (Fig. **5F**). A thermoresponsive PNVCL is known to be nontoxic, biocompatible and water soluble. The one-step iCVD method without complex purification steps, could fabricate a highly pure PNVCL homopolymer film. In the iCVD process, radical polymerization took place after N-vinylcaprolactam monomers introduction without the functional caprolactam ring damaged. A sharp LCST of the iCVD PNVCL is 31°C. Owing to above characteristics, the iCVD PNVCL film has been applied in NIH 3T3 mouse fibroblasts cell sheet fabrication. The advantage of this system is that cells cultured on such thermoresponsive surfaces could be recovered as an intact cell sheet by simply lowering the temperature below 31°C, better for the cell viability than 20°C [143].

3.2. Electroactive Responsive System

Electroactive responsive system is the second popular method in fabricating cell sheets for tissue engineering. Electroactive power in the system is used to make the cell adhesion and detachment to form cell sheets, including electrochemical self-assembled monolayer golden surface, polyelectrolyte surface, sulfenyl golden surface, and electrochemical modified micro-patterning surface.

3.2.1. Electrochemical Self-Assembled Monolayer Gold Surface

The monolayer self-assembled monolayer (SAM) is used to modulate the ligand release from the gold surface. The ligand can be released after the oxidation of the active molecules on the single cells attached surface after electric stimuli. The ligand contains polypeptide (Gly-Arg-Gly-Asp-Ser, GRGDS), in which peptide (Arg-Gly-Asp, RGD) plays mainly functions (Fig. **6A**). RGD, connected to alkyl monolayer by the electrical activity of hydroquinone, promotes the cell attachment (Fig. **6**). Cells cultured on peptide monolayer containing RGD, will be detachment from the gold surface when dihydroxybenzene is oxidized into benzoquinone after electrochemical stimulation [163].

3.2.2. Polyelectrolyte Surface

Polymerization electrolysis layer was attracted to the electrical opposite surface which is generated by electrostatic interaction. However, electrochemical reaction caused the remove of the polymerization electrolysis layer away from the polyelectrolyte surface (PLL-PEG). The electrochemical response of PLL-PEG was covered on conductive titanium and insulating silicon oxide regions, which is a biosensing platform for controlling the titanium surface functionalization, and maintaining a protein-resistant background on the silicon oxide regions. A silicon-based substrate containing PLL-PEG can be micropatterned by alternative arrangement of conductive titanium and insulating silicon oxide. Then it is subjected to positive potential for a short time. Without any polyelectrolyte loss from the silicon oxide regions, the PLL-PEG was removed from the titanium stripes by the potentiostatic action, which ended into the cell sheets fabrication.

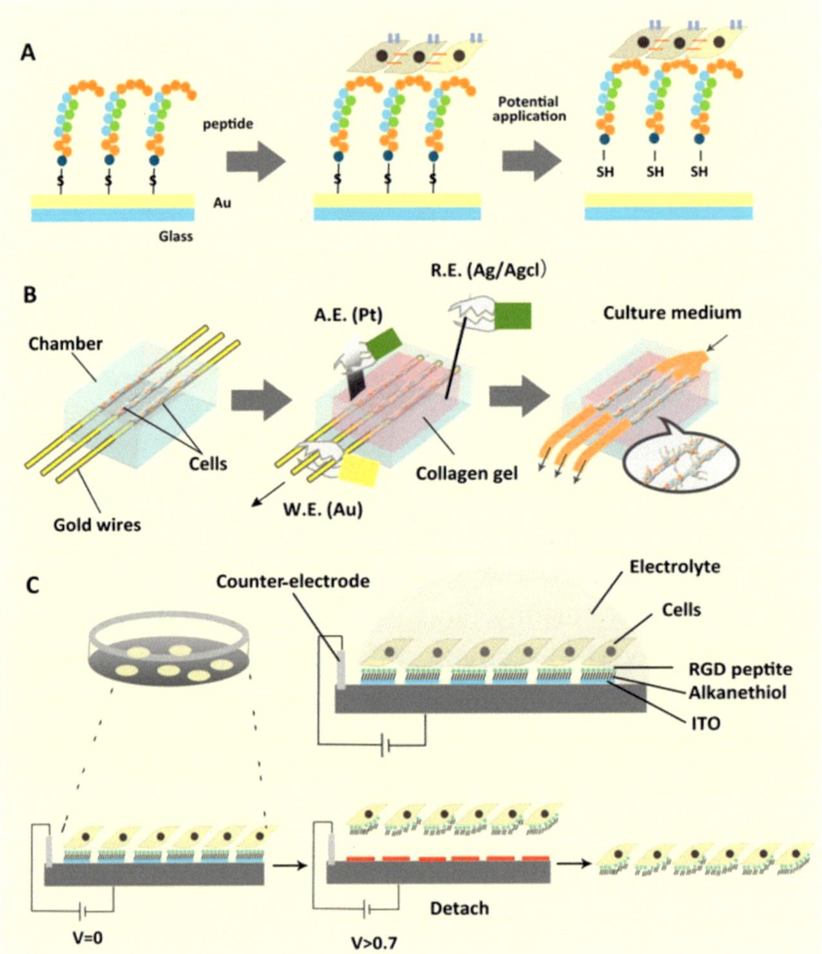

Fig. (6). Fabricated cell sheet using electroactive responsive system. **A and B:** Scheme for the fabrication of capillary-like structures in GeiMa by the electrical detachment of cells from a potential and extracting the rods. The oligopeptide CCRRGDWLC includes cysteine for bonding to gold, positively charged lysine and negatively charged glutamic acid, and the RGD sequence to promote cell adhesion. Cells attached to the surface are detached during the reductive desorption of the oligopeptide. **C:** Detachment of cell sheet in a spatially and temporally controlled manner. Cross sectional view of micro patterned ITO electrodes. The electrodes are modified with an alkanethiol SAM and RGD peptide.

3.2.3. Sulfenyl-Gold Surface

Gold–thiolate surface, precisely aligned within micrometer distances by an electrochemical approach, can be used in capillary-like structures fabrication. In this approach, a peptide containing cysteine residues at both end sides and a cell adhesion domain (Arg-Gly-Asp, RGD) in the center was designed. *Via* a gold–thiolate bond, cysteine with a thiol group can be adsorbed onto a gold surface. The cells adherence on the gold surface *via* the RGD to form cell sheets, and were detached by applying a negative electrical potential to cleave the gold–thiolate bond. This approach is applicable on a flat surface and various configurations. By applying this system, after transferred onto the sulfenyl-gold surface, human umbilical vein endothelial cells (HUVECs), can grow into the internal surface of the structure in the engineered blood vessel graft, thereby form vascular cell sheets *in vitro* [164].

3.2.4. Electrochemical Modified Micro-Patterning Surface

Electrochemical modified micro-patterning surface constructed the biointerfacial micro-patterned domains which can be switched electrochemically through the spatio-temporally controlled dissolution and adsorption of polyelectrolyte coatings. *Via* alkanethiol SAMs, cells cultured on the surface of indium tin oxide (ITO) can be attached to a gold electrode. Meanwhile, *via* reductive desorption of the SAM, cells can also be detached by applying a negative electrical potential in a very rapid manner. Furthermore, using photolithography can fabricate micropatterns of ITO electrodes, on which cells could be detached with single-cell resolution (Fig. **6B**). The potential of constructing one demand harvesting or elimination of one kind of cells from others under microscopy can be realized by this micro-patterning surface [165].

Previously, the creation of micropatterned cell co-cultures systems used special proteins or polyelectrolyte micro-patterns obtained by photolithography, which relied on the preferential attachment of one particular cell type (*i.e.* hepatocytes or astrocytes). Micro-patterning systems were then made in use of physical barriers, on which the attachment of the first cell population could be detached. Such micro-patterning system was suitable for any cell types. On two individual interlocking plates, to enable different cell populations organizes partially from different cell types, micromechanical strategy was developed, which provided the different cell types with at unable micro-spacing dynamic co-cultures. Nowadays, a new platform providing high versatility and simplicity enables the biointerface properties in a precise spatio-temporal manner.

A negative photoresist (SU-8) is spin coated on the ITO coated substrates to create a new micro-pattern by photolithography. SU-8can create mono and multilayers in the electrochemical dissolution of polyelectrolyte way by the underlying photoresist stencil, which mainly relied on polyelectrolytes shielded from the electrochemical treatment. This approach created a new method of micro-patterned cell co-culture [166]. With this technique, detachment of adhered cells will be achieved in the spatial controlling way. This electrochemical micro-patterning strategy might thus be a ideal tool to manage the distribution of different cell kinds in co-culture systems, which may create not only a 2D functional co-cultures system, but also a 3D tissue constructs to replicate the complex native tissues.

3.3. Light Responsive Cell Sheets Technology

The phenomenon of light-induced cell detachment is reported for light changing the hydrophobic property and hydrophily of metal oxides surface, especially ZnO and TiO_2.

3.3.1. Oxide Treated Surface Method

According to the biocompatibility and safety characterizations of TiO_2 nanodots, they showed a potential in cell culture of tissue engineering. Ultraviolet (UV) light 365nm (UV365) illumination was found able to change the hydrophobic property and hydrophily of TiO_2, resulting in detachment of the cells on a nano-TiO_2 coated quartz substrate after UV365 illumination (Fig. 7). This approach produced intact cell sheets demonstrating a promising way for rapid harvesting of cell sheets with good viability, which could therefore be used in the further culture processes and other applications in tissue engineering [167].

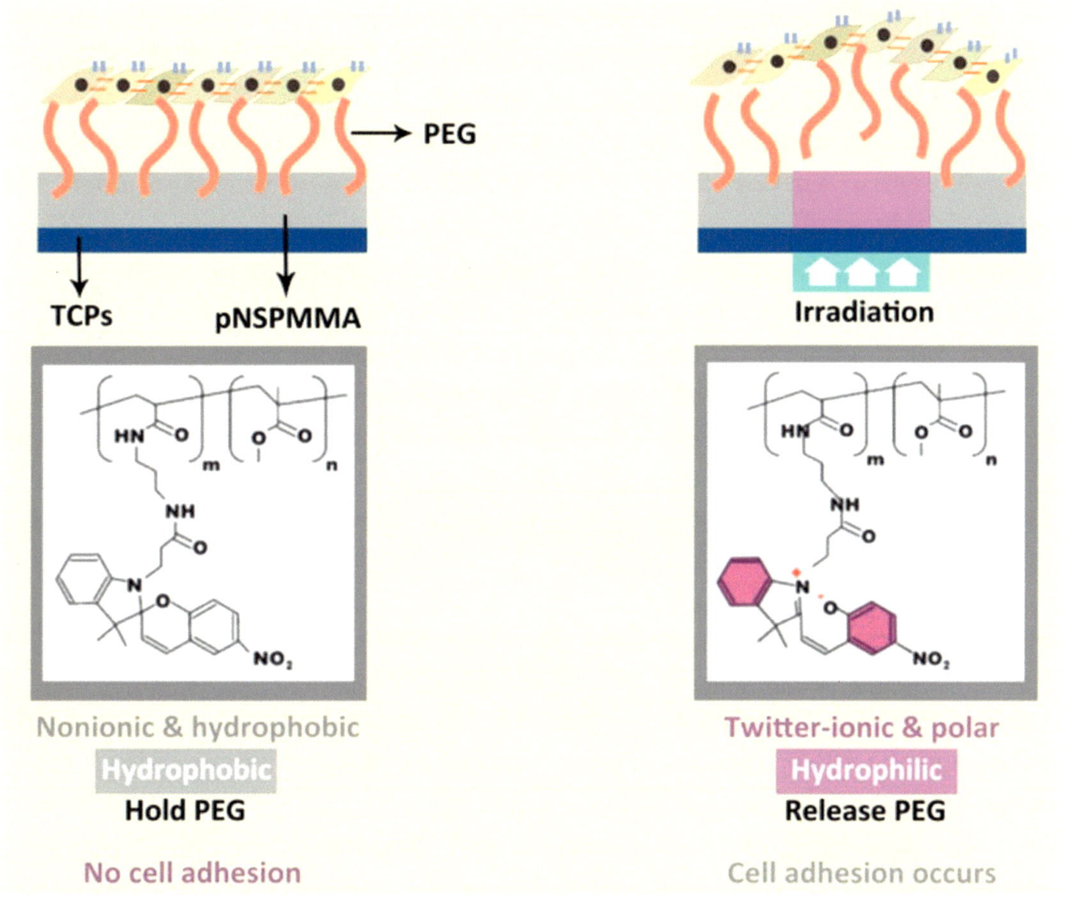

Fig. (7). Light responsive cell sheets technology. **A:** Polyethylene glycol (PEG) is used as a cell adhesion inhibitor in the photoresponsive culture surface (PRCS). PEG is immobilized via photoresponsive polymer, polynitrospyropyrane methylmethacryate (pNSpMMA) on tissue culture polystyrene dish. pNSpMMA is hydrophobic and keeps PEG molecules on surface so that cells are repelled from the surface before UV irradiation. **B:** Due to isomerization of the chromophore after UV irradiation, pNSpMMA changes hydrophobic to hydrophilic so that PEG is released from the surface to surrounding liquid. Spiropyran immediately returns to hydrophobic after irradiation and the irradiated surface allows cell adhesion.

3.3.2. Light Sensitive Molecule/Material Coatings Method

A photoresponsive culture surface (PRCS) will produce cell sheets for the property of photocontrol of cell adhesion. Because spiropyran chromophores as side chains was contained in a novel polymer material composed of poly(N-isopropylacrylamide) [167]. Spiropyran is a photoresponsive molecule, and UV irradiation will reversibly change the hydrophilic of nonionic spiropyran, which will back again by visible light irradiation (Fig. **7B**). This photosensitive surface mainly contain nitrobenzospiropyran and methyl methacrylate, poly(NSP-c--MMA). Compared to the unexposed surface, the water contact angles was decreased and diameter of a water drop was increased in UV irradiation of the photoresponsive surface of poly(NSP-co-MMA)-coated plates. Simple and patterned light irradiation will induce cell sheets detachment from poly(NSP-c--MMA)-coated glass plates, which was relied on adsorbing light-induced detachment of fibrinogen on poly(NSP-co-MMA) coated glass plates [168]. The cell adhesion on the PRCS could be promoted through previous UV irradiation, which would be reset by the visible light irradiation. UV irradiation, along a micropattern on the PRCS can fabricate a viability cells patterning with a line width of 200 micrometer line, which allowed cells seeded uniformly in advance [169]. The viability and differentiation potential of cell sheets were maintained sufficiently after UV light irradiation.

Benzocarbonyl, a photo switchable 4-[(4-aminophenyl)azo], is another photoresponsive molecule. A set of RGD cyclo(−RGD−) peptides with a benzocarbonyl central unit is synthesized as spacer between the acrylamide anchor and the RGD peptides. These peptides are used to cover PMMA (poly methyl methacrylate) surfaces. Benzocarbonyl can change the cell adhesion properties depending on the orientation and distance of the RGD peptides, achieved by controlled adhesion stimulation of UV irradiation with 366 or 450 nm light [170].

3.3.3. Light Sensitive Molecular Combined With Protein Coatings Method

The light sensitive surface is formed by the RGD peptides combination with the light sensitive molecule. Firstly, RGD peptides were immersed in a solution containing 4-dimethylaminopyridine and triethylamine. After reaction, the photosensitive linker and characterization of the photolabile surface layer was formed. UV spectroscopy can therefore detach the cell sheet from the surface layer [171]. With application of such photosensitive surface, human umbilical vein endothelial cells sheet could be fabricated. The design of the reasonable photosensitive crosslink is the most important step in this approach.

3.3.4. Nanoengineered Gold Coating Surface

Nanoengineered gold nanoparticle (AuNP) surfaces fabricate the cell sheets in a nonthermal manner, which resulted in detachment from the surface after triggered by laser irradiation. AuNP surface is a kind of nonthermal photochemical system. The mechanism of cell detachment was proposed for reactive oxygen species, which will be produced from illumination of gold nanoparticles under green laser light, and reactive oxygen species would cause the cell detach from the surface. Such detachment process also depended on the exposure time, laser power, AuNP patterning profile and cell viability [172]. Cell detachment from AuNPs was activated by laser light with corresponding wavelength. The amount of produced ROS within the periodic structures of AuNP stripes depended on lithographic patterning of nanoparticles. Hydrophilic glass slides were covered by uniform AuNP distribution, for which positively charged gold nanoparticles were prepared from an aqueous colloid suspension. AuNP patterning relied on three methods, including lithographic microcontact printing, laser beam profiling and selective gold nanoparticle deposition. Cells migrated from unirradiated areas on AuNP surface could reattach and recover onto the surface, which is suitable for wound healing and tissue engineering, because cells spatial organization were controlled in this way. The technique can be applied for both multiple cell detachment and single cell detachment.

3.3.5. PMB-PL Surface

Poly(2-methacryloyloxyethyl phosphorylcholine-co-n-butyl methacrylate) (PMB) bearing 4-[4-(1-hydroxyethyl)-2-methoxy-5-nitrophenoxy] butyric acid (PL) surface is a photoreactive and cytocompatible phospholipid polymer surface, which is a suitable platform to dynamic control cell adhesion and detachment. The PMB-PL was constructed with the amphiphilic and water-insoluble substance. The cell adhesion and detachment on the PMB-PL surface was controlled through the photoirradiation process. PMB-PL contained two parts: the MPC units and PL groups. The former provided cytocompatibility and the photoreactive, while the latter provided cell attachment points. The PL unit will split from PMB-PL modified surface after photoirradiation, then the cell will detach from the surface. The efficiency of cell detachment was obtained approximately half of the cell counts after the initial adherence. The proliferation and morphology of cells did not change in this approach compared to conventional enzymatic digestion. The functionalities and characteristics of cells detached from the PMB-PL surface were maintained the same as normal cells. On this PMB-PL surface, the human epidermoid carcinoma cancer cells sheets and primary murine embryonic fibroblast cell sheets were gained [173].

3.4. PH Responsive Material Coating Method

PH reaction system as a drug carrier is widely studied. Due to cell normal physiological functions, the pH range is limited from 6.8 to 7.4, which limited the research and application of the system, which mainly included ITO surfaces and chitosan surfaces.

3.4.1. Indium Tin Oxide Surface

Conductive indium tin oxide (ITO) electrodes were coated with deposition of cationic poly (allylamine hydrochloride) (PAH) and anionic poly (styrene sulfonate) (PSS) polyelectrolytes layer-by-layer, on which human placenta-derived mesenchymal stem cells sheets would be fast and viable formation This PAH/PSS substrates promoted cell adhesive and maintained proliferation property for all cell types, and stimuli responsive surfaces. The cells cultured in PAH/PSS substrates is suitable for interacting with proteins, which was from the medium or produced by the cells and would adsorbed on the polyelectrolyte films. When subjected to the PH lowering, a few hundreds of nanometer extend above the thin PAH/PSS films. Both local pH lowering induced by electrochemical and general decrease of the environmental pH at the biointerface could instigate that intact stem cell sheets rapidly detached *via* perturbation of the protein adlayer where cells adhere. The gained stem cell sheets remained viability and differentiation capacity. This novel approach provides an ideal and cytocompatible method for the fabrication of human stem cell sheets, which will be used in future clinical applications and tissue engineering [174].

3.4.2. Chitosan Surfaces

Chitosan, as the second-most abundant nature polysaccharide, has a similar structure to Glycosaminoglycans (GAGs), which are the main components of the ECM. As we known, GAGs modified Scaffolds could promote cell proliferation and differentiation. Compared with GAGs, chitosan keeps relatively lower cost, which results in exploring an amino group protection-deprotection strategy to imitate GAG structures is necessary. The strategy was explored for incorporating sulfonate or carboxyl groups into the chitosan structure [175]. The presence of chitosan modified by primary amine of glucosamine residue is a pH-responsive polycation and its pH-sensitive characteristic is around pH 7.4.Without additional enzymatic treatment and extensive washing steps, this pH-responsive chitosan is able to control cell detachment based on environmental pH changing, which is useful to recover cultured cells (Fig. **8**).

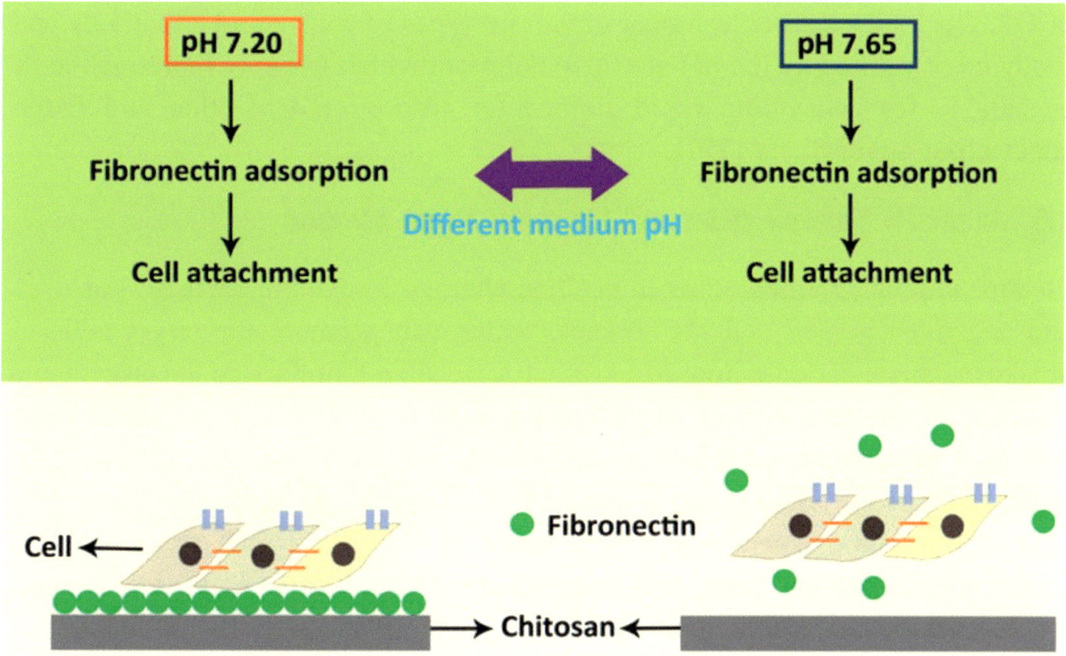

Fig. (8). PH responsive material coating methods on fabricating cell sheet. Schematic representation of the relationship among medium pH, fibronectin adsorption, and cell attachment on chitosan.

The other kind of the chitosan is membrane surface-sulfonated chitosan. The characteristic of this kind is reacted by sulfur-pyridine trioxide complex (SO_3/pyridine) in H_2O and N-sulfur-dimethyl form amide trioxide complex (SO3/DMF) in DMF. Once the surface-sulfonation was constructed in an aqueous alkaline medium, the surface-sulfonated chitosan significantly reduced the adhesion and activation of platelets. The underlying mechanism of this chitosan may be relied on the ionic interactions between the platelets membrane and the modified chitosan membrane [176].

The exhibited advantage of pH-sensitive chitosan is that the useful pH is not only suitable for the attachment and detachment of cultured cells, but also no damage to the cell. When the pH of culture medium is 7.65, over 90% of Hela cells would rapidly detach from chitosan surface. Meanwhile cells were attached and growth well on chitosan surface from pH 6.99 to 7.20. The medium pH could also regulate the fibronectin adsorbed and desorbed on chitosan. Most importantly and interestingly, the medium pH could be adjusted by simple altering environment pCO2. The attachment/detachment of various types of cells could effectively and easily be controlled by the pH-sensitive chitosan, which is easily reproducible. It is suitable for controlling rapid fibronectin adsorption/desorption and tissue engineering applications [177].

3.5. Magnetite Nanoparticles and Magnetic Force Method

Electrostatic interactions between positive charged Magnetite Cationic lipsomes and negative charged cell membranes, which can promote the target cells to uptake magnetic cationic liposomes (Fig. **9A**) leading to influence the growth and direction of the cells outside the magnetic field. In this system, magnetic force was used to construct and harvest cell sheets without enzymatic treatment. In order to improve adsorption, a positive surface charge was combined with magnetite cationic liposomes, which would be combined with target cells at certain concentration. Then, the magnetically labeled cells were seeded onto ultralow-attachment plates, under which a magnet was placed. After cultured for a short time, a magnetically labeled cell sheet was formed. If a cylindrical magnet was rolled onto the cell sheet, the tube-like structure could be formed. Then removing away the magnet, a tubular structure was left behind, which is suitable to create urinary tissue and vascular tissue (Fig. **9B**). Meanwhile, to fabricate multi-layered cell sheets, a covalently bound hydrophilic and neutrally charged hydrogel layer was used for the surface of the cell sheet. The multi-layered cell sheet could be constructed in low-calcium medium after culture with a magnet under the plate. Once removing away the magnet, the cell sheets will form and detach from the bottom of the plates. This approach has been termed as "magnetic force-based tissue engineering" (Mag-TE), for a magnet is used to harvest the sheets, which is believed as a promising approach for tissue engineering [176], such as skeletal muscle tissue, keratinocyte sheets, and bone tissue engineering.

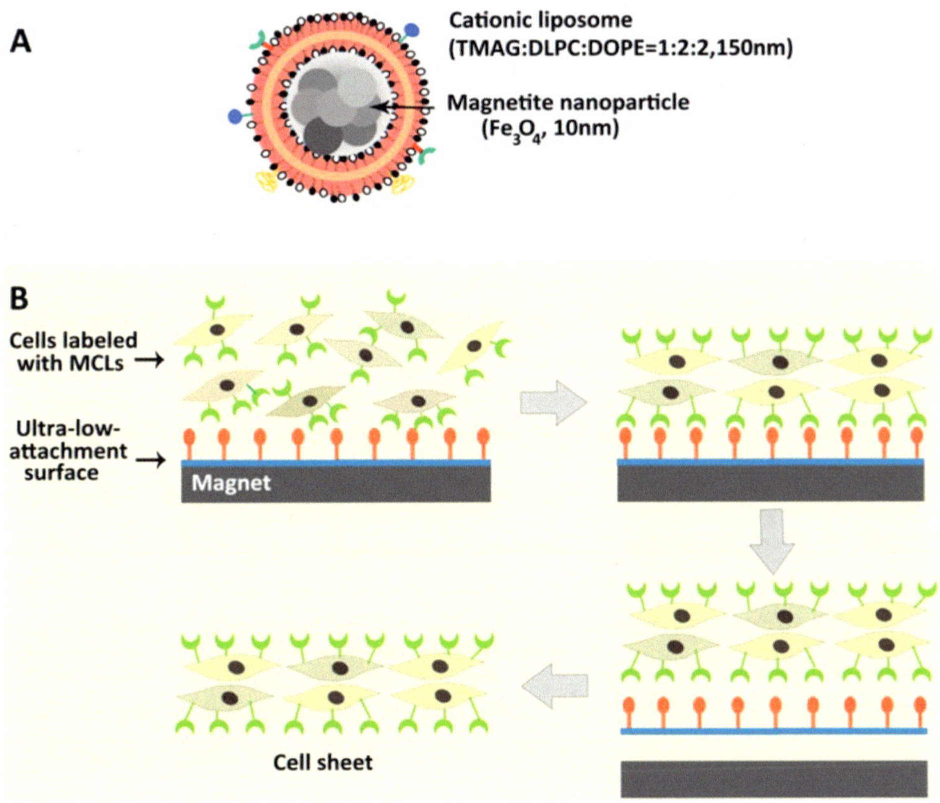

Fig. (9). Magnetite nanoparticles and magnetic force method to fabricate the cell sheet. **A:** Illustration of magnetite cationic liposome (MCL). **B:** Construction of cell sheet by Magnetic force. Cell labeled with MCLs are seeded onto an ultraslow-attachment plate, and a cylindrical magnet was then placed under the plate. Cells attracted to the dish by the magnetic force were cultured to construct a cell sheet. When the magnet was removed, the cell sheets were detached from the culture surface, and could be harvested without enzymatic treatment.

3.6. Others

3.6.1. Varies in Culture Medium

The enhancement of the stem cells proliferation caused by lower oxygen concentration boosted the development of oxygen concentration adjustment as a novel method to fabricate cell sheets. The oxygen concentration was reported able to affect the stratification of corneal epithelial cell sheets. Researchers cultured limbal epithelial cells under hypoxia, then normaxia during conflunce stage. The results showed that the cell sheets can be formed more quickly and efficiently [143].

The novel culture medium using signaling inhibitors was found effective for fabricating epidermal cell sheets. Activation of Notch-signaling in the suprabasal cell layer will promoted stratification and differentiation of basal cells. When Notch-signaling was activated, r-secretase left the Notch intra-cellular domain (NICD), which then translocated into the nucleus and bound with DNA to up regulate the target genes. Therefore, fabrication of epidermal cell sheets could be accelerated by culturing normal human epidermal keratinocytes with or without r-secretase inhibitors in the appropriate time.

3.6.2. Automatic Fabrication of Cell Sheets Using Manipulator Technique

Automated manufacturing of tissue-engineered products is common and economical. A recently-developed cell sheet manipulator technique using cell-adherent hydro gels has now made it easier to fabricate 3D constructs from cell sheets. The 3D tissue-engineered constructs can be fabricated *via* the cell sheet manipulator technique by an automatic apparatus. The manipulator technique can fabricate the cell sheets from temperature-responsive culture dishes and stack into a multilayered cell sheet (Fig. **10**). By optimizing the cell seeding and stacking conditions, multilayered cell sheets would be eventually reproducibly fabricated by the manipulator technique. A secure, cost-effective manufacturing system can be created to fabricate tissue-engineered products from cell sheets *via* combining the cell sheet manipulator technique and industrial robot technology in future.

Cell sheet compose of single cells is mainly used to repair the simple tissue defect, such as periodontal membrane [178], the skin [148], the cornea [179] and so on. The single cell sheet might be superimposed together to repair the single solid structure, such as heart [180]. The relatively complex structure might be repaired by superimposing different sources cell sheets, such as liver lobule [181].

Most of the natural tissues are consisted of many kinds of cells, which can be shaped by the cell sheet mixed with a variety of cells into a complex three-dimensional structure. We introduced several different systems to fabricate cell sheets, the advantages and disadvantages are shown in Table **1**. Each system should be chosen according to the necessary of tissue engineering.

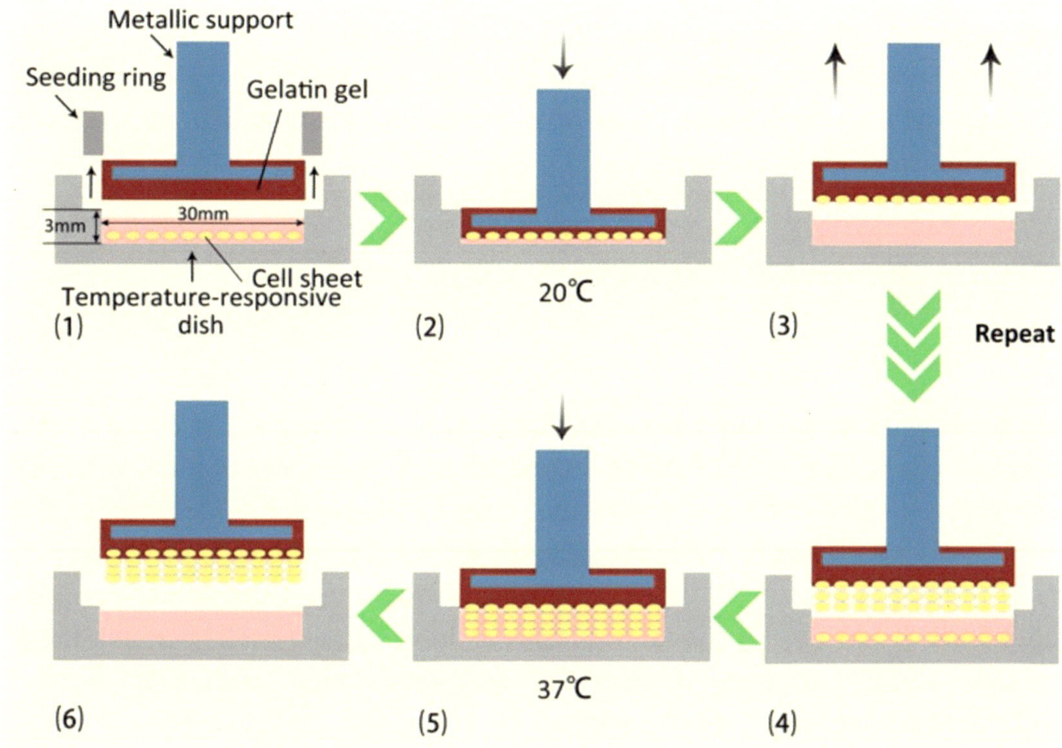

Fig. (10). Automatic cell sheet stacking process. (1-3) A cell sheet manipulator is placed on the cell sheet and held at 20 °C for a given duration. The first cell sheet is harvested on the bottom surface of the gel. (4-6) For many times, the manipulator is placed on the other cell sheet is harvested on the bottom surface of the gel. For the last dish, the manipulator is also placed on the last cell sheet.

3.7. The Future and Development of The Cell Sheet Technologies

Compared with the traditional tissue engineering method, the cell sheet engineering product is made of cells only, no need of trypsin treatment, which avoids the adverse effects of support and trypsin. At present, reactive systems are expensive, equipment and procedures are complex. All of those affected the clinical application in the future. In addition, the most widely used temperature responsive culture dishes are currently used. However, fabricating a single layer of cell membrane required more than 30 min, and the length of cooling time will seriously influent cell function. Therefore, a simple and convenient reaction system is necessary to be developed. The formation time of the cell membrane is the main issued need to be addressed at present. The further development of the reactive system will be helpful to satisfy the large number of cell sheets and the reconstruction of complex tissue in clinical application. According to the automatic application, automatic fabrication of cell sheets using manipulator technique will produce many kinds of common and economical cell sheets in tissue engineering in the future.

4. ANIMAL MODELS AND CLINICAL TRIALS OF CELLS SHEETS APPLICATION IN REGENERATION MEDICINE

4.1. Introduction

Regenerative medicine has been expected to have an important effect in curing severe diseases. The field of tissue engineering, as an important method in this aspect, has yielded several successes not only in the animal model studies, but also in early clinical trials of regenerative medicine. Cell sheet technology (CST) has developed fast in recent years as a new method in regenerative medicine. By using this technology, functional tissue sheets have been created to treat a lot of diseases including esophageal cancer, corneal dysfunction, cardiac failure, and tracheal resection (Fig. **11**). Moreover, by using the CST to generate three-dimensional tissues, the thick vascularized tissues, even organ-like tissues have been successfully harvested *in vitro* and some have been applied preclinically. This section is going to discuss the animal model and the application of CST in the regenerative medicine.

4.2. Animal Models

The key method of regenerative medicine is the replacement of the injured or missing tissue with suitable amount of cells or tissues. CST is used to fabricate various tissues. The transplantation of cell sheets showed good therapeutic effects in different kinds of damaged or defect animal models. Till recently, thin-layered tissues such as the cornea, periodontal tissue, esophagus, and articular cartilage, were fully regenerated and harvested *in vitro*. Transplanted cell sheets can replace the injured tissue, but also compensate the impaired function after implantation. This part will discuss the current progress of CST in curing the defect animal models and its underlying mechanisms.

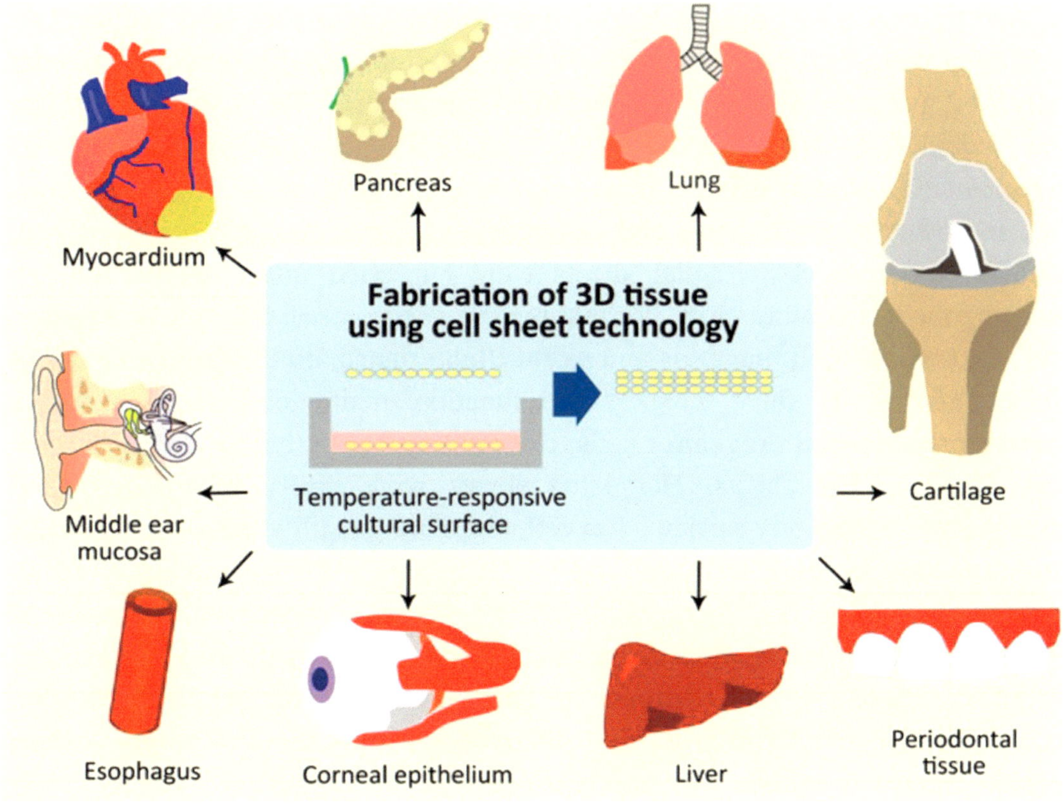

Fig. (11). Fabrication of various tissues by cell sheet technology(CST).

4.2.1. Corneal Epithelial

Three layers composing the cornea: the epithelium, stroma, and endothelium, play important roles in maintaining normal structure of cornea that is critical for vision. Limbal stem-cell deficiency from disease (Stevens–Johnson syndrome or ocular pemphigoid) or severe trauma (thermal or chemical burns) causes vision loss and corneal opacification. Under normal conditions, the conventional treatment is corneal transplantation from donors' cornea tissue. However, the shortage of donor eyes and the high risk of graft rejection limited the treatment of corneal pacification.

Attempts have been made to manufacture corneal epithelial graft constructs. Nishida [182] and Hayashid [183] reported that they have transplanted autologous tissue-engineered epithelial cell sheets manufactured from harvested oral mucosal epithelium in a rabbit model. Rabbit oral mucosal epithelial cells cultured on temperature-responsive culture surfaces with mitomycin-C-treated 3T3 feeder cells for 2 weeks produced confluent epithelial cell sheets. The putative progenitor cell population was assessed by colony-forming assays. Autologous transplantation of these cell sheets was performed for surgically manipulated eyes. Ocular surface reconstruction and cell phenotypic modulation were examined. Multilayered corneal epithelial sheets were harvested intact simply through reducing the temperature only. Without the use of proteases, the cellular function and features, cell-cell junctions and extracellular matrix on the basal side of the sheet, important to sheet integrity and function, maintained intact. A viable population of corneal progenitor cells, close in number to that originally seeded, was found in the sheets. Harvested sheets were easily manipulated and transplanted without any carriers. The cell sheets were readily adhesive to corneal stroma so that suturing was not required. Autologous transplantation to rabbit corneal surfaces successfully reestablished the corneal surfaces with restoration of transparency. Four weeks after transplantation, the epithelial stratification showed a resemblance to that in the corneal epithelium, although the keratin expression data retained characteristics of the oral mucosal epithelium. Corneal surface reestablishment in rabbits was highly successful.

These studies show that CST enables fabrication of viable, transplantable, tissue-engineered epithelial cell sheets that maintained putative progenitor cells from autologous oral mucosal epithelial cells. Promising clinical applications for autologous tissue-engineered epithelial cell sheets for the ocular surface reconstruction were indicated.

4.2.2. Retinal Pigment Epithelium (RPE)

Retinal pigment epithelium (RPE) plays a key role in many retinal diseases. The degeneration of RPE could cause retinitis pigmentosa as well as age-related macular degeneration, which finally results in the loss of vision. However, RPE cells do not proliferate when developed *in vivo*. Therefore, when damage occurs, mature RPE cells can grow only by cell enlargement. Regenerative RPE-based therapies, including allotransplantation, translocation or transplantation of autologous RPE cells, and even the use of alternative cell sources such as iris pigment epithelium, bone marrow, or ES cells, have been evaluated as method to restoring RPE function. Recently, researchers tried to use CST to create PRE cell sheets for transplantation.

In 2009, Yaji *et al.* [184] successfully transplanted carrier-free RPE cell sheets in a rabbit model. The RPE cells were isolated from pigmented rabbit eyes and cultured on temperature-responsive culture surfaces. Confluent cells, which were set as a monolayer with a cobblestone cell shape like native RPE, can be non-invasively harvested as a single contiguous cell sheet, which could maintains all cell structural as well as functional intercellular junctions. RPE cell sheets were transplanted into the subretinal space of albino rabbits by using vitrectomy. Seven days after transplantation, historical analysis and electron microscopy scan were conducted to evaluate the morphology and function of the rabbit eyes. Comparing with the injection of cell suspensions, this method showed that the RPE cell sheets attached to the host subretinal space more effectively. In most area the cell sheets maintained a monolayer structure. In some regions, however, they were found to be slightly folded or wrinkled. It demonstrated that tissue-engineered RPE cell sheets can be effectively transplanted beneath the neural retina. Furthermore, in order to confirm the clinical applicability of RPE cell sheet transplantation, electrophysiological examination and functional evaluations should be performed in the future.

4.2.3. Middle Ear Mucosal

For treatment of chronic otitis media including cholesteatoma and adhesive otitis media, the tympanoplasty with mastoidectomy is common used. During which, the normal middle ear mucosa should be preserved as well as possible. Otherwise, the function of gas exchange, sound conduction will be affected. And it would also causes adhesion and retraction of the tympanic membrane and the relapse of cholesteatoma. In order to prevent that, the use of oral mucosa, silastic sheeting, and collagen sponge has been reported. However, they are hardly feasible.

Kazuhisa Yamamoto *et al.* [185] developed a new approach in a rabbit model to use the nasal mucosal epithelial cell sheets for the middle ear mucosa repair after surgery, and evaluated the healing results by histology and computed tomography. The gas exchange function was also evaluated by measuring the maximum middle ear total pressure. They found that compared with the mucosa-eliminated control group, the narrowing of pneumatic space and the bone hyperplasia in the middle ear cavity were suppressed. The gas exchange function of the cell sheet-transplanted group was also found to be better. These results implied that the nasal mucosal epithelial cell sheet might be an effective graft material after middle ear surgery. The similar result was also found in Yaguchi Y's research [186]. They also used the CST to repair the middle ear mucosa, while the seeding cells were derived from the middle ear bulla. These studies indicated the CST would become a novel therapy in the future hopefully.

4.2.4. Periodontal Tissue

The method for regeneration of periodontium includes grafting, root surface conditioning, application of growth factors and guided tissue regeneration, among which the guided tissue regeneration may be the most promising. Although this method can regenerate periodontal tissue, the inadequate cells may affect the results of this method. The CST just can harvest cells from the temperature-responsive cell culture dish surface invasively and harmlessly. It can preserve the extracellular matrices, *e.g.* the type I collagen and fibronectin.

Many animal experiments have been made to testify the application of the CST in the regeneration of periodontal tissues [187 - 193] Y Tsumanuma *et al.* [191] used a canine one-wall defect model to compare the effect of different stem cell sheets for the regeneration of parodontium. The periodontal ligament cells (PDLCs), iliac bone marrow mesenchymal stromal cells (BMMSCs) and alveolar periosteal cells (APCs) were obtained from each dog. Three-layered cell sheets of different cell source supported by the woven polyglycolic acid were transplanted to the denuded root surface autologously. After eight weeks, they found more regeneration of both the well-oriented PDL fibers and new cementum in the

PDLCs group. The nerve filament was only observed in the PDLCs group. The amount of alveolar bone regeneration was observed the highest in the PDLCs group. Y Bai *et al.* [190] used the dental follicle cell (DFCs) sheets co-cultured with Hertwig's epithelial root sheath cells to restore the rat periodontal tissue. As we know, the DFCs contain the precursor cells of the periodontium. They used co-culture of DFCs with alveolar mucosa epithelial cells as experimental group and DFCs only as controls. After five weeks ,they found DFCs sheets exposed to HERS cells previously led to the formation of cementum-like tissues, as well as the periodontal ligament-like tissues. But in control groups, only fibrous tissues was produced. Thus, different tissue-derived stem cell sheets would have different potentials for the regeneration of periodontal tissue. And the inducer is also very important.

The application of PDLCs sheet technology in the regeneration of periodontal tissue has been tested in the periodontitis and teeth replantation. The canine model, miniature swine model and rat model have been made to testify the capability of regenerating periodontal tissues. In those studies, both new bone and cementum connected by the well-oriented collagen fiber were found. Y Zhou *et al.* [192] applied a new periodontal ligament cell-sheet delivery system to replanted teeth to promote periodontal healing in a canine defect model. The found, compared with the control group (no CST), the PDL cell-sheets group showed significant higher occurrence of alveolar bone healing (88.4%) than control group (5.3%). And the regenerated tissues showed high collagen type III, type I and fibronectin expression.

4.2.5. Tooth

Tooth regeneration has always been thought as an ideal approach for the restoration of missing tooth. The traditional methods involved harvest of seeding cells by trypsin digestion which results in alteration of their functions and properties by degrading both the cell-to-cell junction proteins and ECM. The CST just has overcome the previous shortcomings.

Yang *et al.* [194] proposed a strategy utilizing human DFCs sheets with treated dentin matrix (TDM) as a biological scaffold for root regeneration. The DFC sheets combined with TDM were implanted subcutaneously into the dorsum of mice. They used the scanning electron microscopy, immunofluorescence microscopy, transmission electron microscopy, immunohistochemistry and quantitative real-time PCR to study the effect of TDM on DFCs sheets. The investigators found a whirlpool-like alignment of the DFCs in multiple layers that were positive for integrinb1, COLI, fibronectin and alkaline phosphatase (ALP), suggesting the formation of ECM. DFCs sheets highly expressed DMP-1 and bone sialoprotein (BSP) under the effect of TDM, indicating their potential for osteogenesis and odontogenesis. Importantly, *in vivo*, TDM could induce and support DFCs sheets to develop new dentin-pulp like tissues and cementum-periodontal complexes that were positive for markers such as COLI, VIII factors and Nestin, cementum attachment protein (CAP) and DSP, implying successful root regeneration. Therefore, the TDM should be an important inducer for DFCs sheets to form tooth root. Thus, DFCSs combined with TDM may be a better method for the construction of tooth root.

4.2.6. Calvarial Defect

The large bony defects caused by tumors, trauma, or inflammation remains a significant clinical problem. Tissue engineering is a promising technique for bony regeneration. However, the traditional method is not conducive for seed cells adhesive to the scaffold and the secretion of ECM, so not favor for forming bone tissue structure. D Ma *et al.* [195] firstly used autologous bone marrow stromal cell sheet to reconstruct the rabbit's critical-size calvarial defects. They divided 40 adult New Zealand White rabbits randomly into 4 groups of 10 animals each:(1) empty, (2) demineralized bone matrix (DBM) alone, (3) DBM/cell suspension, and (4) DBM/cell sheet. They used radiographic, histomorphometric, and histologic analyses to evaluate the new bone formation. They found the defect treated with DBM/cell sheet showed the best bone formation than other 3 groups.

Another problem affecting the repair size of the bone defect might be the lack of sufficient blood supply. Some research workers [196] studied the coculture of endothelial progenitor cells with BMMSCs for bone tissue engineering. They found the direct contact of the two kind cells could form a prevascular network during coculture and increase osteoblastic activity. CD34+ cells that contain intensive endothelial progenitor and haematopoietic stem cells are capable of inducing neovascularisation. They have also been reported to differentiate into osteoblasts *in vitro*. So, the CD34+ cells have therapeutic potential for both vasculogenesis and osteogenesis. Guanghui Li *etc.* co-cultured the peripheral blood CD34+ cell (PB-CD34+ cells) and BMMSCs sheets to investigate the formation of bone in calvarial critical-size defects in rabbits. They found co-cultured cells showed better osteogenic differentiation than MSCs alone *in vitro*. And the co-cultured cell composites promoted bony regeneration more efficiently.

4.2.7. Jaw Defect

C Yao *et al.* [197] used the CST to construct tissue engineering bone to repair the mandibular defect in the dog. They induced the BMMSCs to osteoblasts, then fabricated cell sheets. They wrapped the poly lactic-co-glycolic acid (PLGA) with cell sheets and then implanted the cell sheets into the mandibular defect in the left side. The other side transplanted into PLGA only was used as control side. They found the optical density of the fresh bone in the experimental side was higher than that of the control side. Plenty of lamellar bones and Haversian system with red marrow formed in experimental side, but not in the control side.

4.2.8. Esophageal Mucosa

Endoscopic submucosal dissection (ESD), as a minimally invasive therapy for early esophageal cancer and Barrett's esophagus, has gained success in resecting superficial cancer en bloc and offering an accurate histological assessment for diagnosis [198, 199]. However, when ESD is applied to cure widespread superficial neoplasms, which often leads to loss of esophageal mucosa, severe postoperative esophageal stricture is inevitably occurred. The stricture may cause dysphagia, lack of nutrition and decrease in quality of life. Regenerative medicine and tissue engineering approaches have been put into attempt to prevent postoperative stricture by reconstructing normal tissues structurally and functionally. By using early re-epithelialization after ESD has been improved greatly.

CST, as an effective method to create epithelium layers, provides a solution to prevent severe esophageal constriction after ESD. The first animal models is set to transplant the fabricated autologous oral mucosal epithelial cell sheets in a canine model by Ohki *et al.* [200] They chose oral mucosal epithelial cells as seed cells. These cells, collected from beagle dogs, were cultured and harvested on temperature-responsive dishes. After ESD (5 cm in length, 180°in range), cell sheets were transplanted by endoscopy. The transplanted cell sheets were able to adhere to the underlying muscle layers and survive in the ulcer sites, which provide an intact, stratified epithelium. Four weeks post-op, complete wound healing, with no observable stenosis, was observed in the animals By contrast, in the control group, those animals only received ESD, were found to suffer from noticeable fibrin mesh and host inflammation, which lead to the esophageal stricture. The result showed the effectiveness of a combined endoscopic approach for the potential treatment of oesophageal cancers. By using the CST, the transplantation of the cultured layers can effectively improve wound healing and may prevent postoperative oesophageal stenosis.

Although oral mucosal tissue as a cell source has several advantages, the amount of resectable autologous tissue is limited. In another animal study, autologous skin was chosen to be an alternative cell source for regenerative therapy. In this study, Kanai *et al.* [201] selected eight pigs and divided them into two groups. All the pigs underwent circumferential esophageal ESD. In group I, fabricated autologous epidermal cell sheets were immediately transplanted to the central ESD sites and early re-epithelialization and mild fibrosis in the muscularis were observed by necropsy and histological assessment at one and two week post-ESD. However, in group II, only the circumferential ESD was performed. As a result, all pigs showed severe esophageal constriction after two weeks. It showed that fabricated autologous skin epidermal cell sheets, as an alternative choice, would be effective in preventing severe esophageal constriction after circumferential ESD.

4.2.9. Tracheal Replacement

Traditional tracheal reconstruction generally involves surgical methods such as resection and anastomosis of the peripheral regions. However, in cases where large portions of the host trachea are removed, it is quite difficult to reconstruct. Other tissues like tracheal allografts or autologous portions of gastrointestinal segments: are often used to replace the loss trachea. Tissue engineering and biomaterials approaches are alternative choices to solve the problem. In order to successfully engineer a bioartificial tracheal replacement, the regeneration of a functional epithelial lining is the crucial step. CST is a good way to provide enough epithelium layers to cover the lumen.

Kanzakiet *et al.* [202] set a rabbit model to investigate the transplantation and prognosis of tracheal epithelial cells which were harvested by CST. In their study, rabbit tracheal epithelial cells were cultured on temperature-responsive culture dishes and then noninvasively harvested as intact sheets. The tracheal epithelial cell sheets were transplanted to the luminal surface of the Dacron, which had been subcutaneously implanted four weeks in advance and allowed vessel infiltration and tissue migration. Four weeks post-op, it was observed that the tracheal grafts were covered by a mature, pseudo stratified columnar epithelium. In contrast, control constructs without cell sheet transplantation demonstrated only a thin, immature epithelium at the center of the replacement graft.

Therefore these results demonstrated that tracheal epithelial cell sheets can create an epithelial lining on the luminal surface. The combination of cell sheet transplantation and the implantation of traditional biomaterials suggest new way to apply CST to complex tissue reconstruction.

4.2.10. Thyroid Gland

Hypothyroidism is a common endocrine disorder with insufficient secretion of thyroid hormone from the thyroid gland. It is a normal complication of thyroidectomy and can cause a number of symptoms, such as tiredness, weight gain and poor ability to tolerate cold. The continuous oral hormone administration is necessary and indispensable to those patients who suffer from hypothyroidism.

Researchers had made their attempt to create thyroid tissue. In 2009, Arauchi *et al.* [203] used CST to reconstruct the three-dimensional thyroid tissue and transplanted into a hypothyroidism rat model to modulate the hormonal status. The cell suspension isolated from rat thyroid gland was cultured on temperature-responsive culture dishes. By reducing temperature, confluent cells were harvested a cell sheet without any enzymatic treatment. To create models, the 8-week-old Lewis rats underwent total thyroidectomy. And one week post-op,

Serum levels of free triiodothyronine (fT3) and free thyroxine (fT4) were found decreasing significantly. Then, thyroid cell sheet were transplanted into the rat model. One week after the cell sheet transplantation, thyroid function began to restore, and such improvement maintained for four weeks. Moreover, in these engineered thyroid tissues, follicle epithelial cells were observed by morphological and immunohistochemical analysis. The presence of functional microvessels was also detected. The results indicated that transplanted cell sheets in a hypothyroidism model can reorganize histologically to resemble a typical thyroid gland and restore function *in vivo*, which confirmed engineered thyroid tissue created by CST, can repair hypothyroidism rat models. And, therefore, cell sheet transplantation of endocrine organs may be an alternative choice for the therapy of hormonal deficiency. Further studies should be implemented to achieve more convincible results.

4.2.11. Articular Cartilage

Articular cartilage has a minimal ability for self-repair due to the lack of blood supply and low cellular density. Therefore, severe cartilage damage rarely resolves spontaneously. Generally, cartilage damage is treated by chondrocyte transplantation, microfracture or mosaicplasty. With the development of tissue engineering, research on techniques to repair articular cartilage damage using transplanted cells has been prompted. Although the clinical results of autologous chondrocyte implantation, problems with cell handling and scaffold imperfections still remain to be solved. CST is potentially able to overcome these obstacles.

Several research groups have taken steps in creating and transplanting layered chondrocyte sheets to repair the cartilage damage. Kaneshiro *et al.* [204] first tried to find a way to treat cartilage damage by using CST and successfully gained the chondrocyte sheets *in vitro*. Sato *et al.* [205] introduced two different methods for treating the cartilage defects without scaffold. One is using chondrocyte sheets to treat partial thickness defects of the articular cartilage. The other is using a chondocyte-plate for the core of an articular cartilageous defect. Their experiment showed that the layered chondrocyte sheets and chondrocyte-plates could maintain the cartilageous phenotype. Ebihara *et al.* [206] found that safranin-O staining may play a role in the repairing in a minipig model of full-thickness defects. They created such models to perform an experiment on the repair and regeneration of articular cartilage. In animals transplanted with layered chondrocyte sheets, good safranin-O staining and integration with surrounding tissues was achieved. However, poor safranin-O staining was identified in some of the animals, and the subchondral bone was poorly repaired in these animals. Kokubo's [207] experiment indicated that synoviocyte co-cultures could shorten the required culture period by providing an optimal environment for the preparation of chondrocyte sheets for tissue transplantation

For further application, a clinical study using cell sheet engineering for cartilage regeneration is still in early stage at Tokai University School of Medicine.

4.2.12. Lung Surface (Sealing of Lung Air Leak)

Intraoperative air leak is an unavoidable complication during pulmonary surgeries, and thus, the prevention of air leaks is a critical aim for patient management post-op. In order to solve this difficulty, fibrin glue and various other biological sealants have been applied. But the early clinical trials did not show significant improvement. As an alternative tissue engineering approach, a novel lung air leak sealant was recently developed using CST.

In 2007, Kanzaki *et al.* [208] used the rat and rabbit animal models to testify whether the CST is a better method to seal the air leak. The experiment was composed of two kinds of animals. In the rat group, lung and skin fibroblasts were chosen to seed on the temperature-responsive culture surfaces to create transplantable sheets. 26 male F-344 athymic rats underwent the transplantation of bilayer cell sheets of either lung or skin fibroblasts. Animals received lung punctures without cell sheet transplantation were set as controls. After 4 weeks, further analysis was done to evaluate the sealing improvement. While in the rabbit group, only skin fibroblasts were collected to harvest cell sheets in the same way. Eight New Zealand white rabbits transplanted autologous skin fibroblast sheets dishes onto the air leak site. Two or 4 weeks after transplantation, partial lung tissue including the transplanted sheets was resected for histological studies. The results demonstrated that mesothelial cells can migrate to cover the transplanted cells sheets, which confirmed excellent biocompatibility and integration with the host tissues. This study showed a bright prospect to reduce surgical complications during high-risk pulmonary resections.

Later in a year, the same investigators performed another study in a porcine model to examine CST for the closure of pleural defects. In this study, skin biopsies were obtained from juvenile swine and tissue sheets composed of dermal fibroblasts were created using *ex vivo* culture on temperature-responsive dishes. Tissue engineered autologous dermal fibroblast sheets were transplanted directly to the defects without the use of sutures or additional adhesive agents. Different from the conventional biological sealants, dermal fibroblast sheets can seal lung air leaks immediately and permanently without suture. The stable closure of the air leaks against airway pressure was confirmed by a water seal test. Dermal fibroblast sheets were sufficiently responsive to lung respiration movement and during a four-week follow-up period after transplantation, no recurrence of the air leak was observed. This experiment revealed that implanted fibroblast sheets can be a good sealant for lung air leaks. Experiments using large animal models have already been completed, and a clinical trial is currently being prepared [209].

4.2.13. Myocardium

After birth, the cardiomyocytes will lose the proliferative ability [210]. So, the heart is an important target organ for regenerative medicine.

Many reasons may induce the loss of cardiomyocytes, *e.g.* some injuries, myocarditis, myocardial infarction, genetic disorders, and continuous hemodynamic overload due to the hypertension and valvular heart diseases. The loss of some cardiomyocytes will make the remaining intact cardiomyocytes exposed to the pressure and volume overload, which will leads to left ventricular hypertrophy (LVH). However, persistent LVH will induce the up-regulation ofp53 and subsequent down-regulation of VEGF in the myocardium. This will fail to compensate the insufficient microvascular network, and lead to persistent apoptosis of cardiomyocytes [211]. So, this is a vicious cycles. How to break down this vicious cycle might be a promising strategy for treat heart failure.

Some groups have reported that cell sheet transplantation of adipose stem cells, Sca-1(+) cardiac progenitor cells, or neonatal rat cardiomyocytes to the infarcted heart could improve the cardiac function [212 - 214].Nowadays, many animal experiments have been carried out with many types of cells applied to the cell sheet fabrication which is used for the transplantation to the heart, including skeletal myoblasts (Rat, Hamster, Canine and Porcine models) [215 - 220], adipocytes(Mouse model) [221], cardiomyocytes (rat model) [212, 222], mesenchymal stem cell/stromal cells from adipose tissue (Rat and Mouse models) [214, 223], cardiomyocytes and endothelial cells (Rat model) [224], dermal fibroblasts with endothelial progenitor cells(Rat model) [225], human-induced pluripotent stem (IPS) cell-derived cardiomyocytes (Porcine model), primate embryonic stem (ES) cell-derived cardiac progenitor cells with adipose tissue-derived stromal cells (Monkey model) [226], and embryonic stem cell-derived cardiomyocytes with endothelial cells and mural cells (Mouse model) [227], menstrual blood (Rat model) [228, 229].

Skeletal myoblasts are the chief cell sources. The cell sheet transplantation of skeletal myoblasts(autologous or allogeneic cells), improved the cardiac function of many kinds of heart diseases [215 - 218]. The Skeletal myoblasts can express various growth factors abundantly, such as vascular endothelial growth factor (VEGF), hepatocyte growth factor (HGF), angiogenin, angiopoietins, stromal cell-derived factor-1, platelet-derived growth factor-BB, and placental growth factor. It can also express some proteases, for example the matrix metalloproteinases-2, -9 and -10 [230], especially under hypoxic conditions, which were beneficial for the angiogenesis [231, 232].

There were also many other method to enhance the function and strengthen the efficacy of transplantation in recent years for skeletal myoblast cell sheet engineering. One example is the co-cultured cell sheets with mesenchymal stem cells. It can enhance the secretion of several cytokines and growth factors derived from skeletal myoblasts significantly, including HGF, VEGF, and leptin. This is good for angiogenesis and improvement of cardiac function in the infarcted heart compared with sole cell sheets of mesenchymal stem cells or skeletal myoblasts [233]. In another method, myoblast cell sheets were covered with an omental flap, which enhanced the angiogenic growth factor expression and improved the cardiac function. These methods might be due to the modification of the vascularization of the cell sheets soon after transplantation [234].

Another injury to heart disease is the immune responses, which is provoked by tumor necrosis factor-α(TNF-α), interleukin (IL)-1B, IL-4, IL-6, and IL-18 [235, 236]. Although the antibodies of TNF-α[237, 238] and non-steroidal anti-inflammatory drugs [239] have already been applied for the ischemic heart diseases or heart failure, no benefits was found for the cardiac function. Meanwhile, some types of cells, such as mesenchymal stem cells, have been reported to attenuate the immunoreactions [240]. So, the cell sheet transplantation might be effective for attenuating inflammation in heart disease. Imanishi and colleagues have reported that adipocyte cell sheets from mice abundantly secrete adiponectin have multiple cardiac protective functions, including anti-inflammation, reducing fibrosis and cardiac hypertrophy [221]. 3 months after transplantation of adipocyte cell sheets, adiponectin can be detected in plasma. The TNF-αexpression and hypertrophy of cardiomyocyte are attenuated and mortality is improved.

So, cell sheet engineering contributes to the recovery of injured hearts not only through fabricating thickened and functional human cardiac tissue but also attenuating the injury from the immune responses. However, small numbers of remaining IPS cells may be engrafted in some cases, which might increase the risk of tumor formation. So, eliminating any remaining undifferentiated cells will make the cell sheet-based cardiac tissue regeneration more secure.

4.2.14. Hepatic Tissue

The liver plays an important role in mediating many physiological functions, *e.g.* protein synthesis, metabolism and detoxification.

For some severe patients with liver-based genetic and metabolic disorders such as glycogen storage disease, ornithine transcarbamylase deficiency, Crigler–Najjar syndrome and coagulation factor VII deficiency, orthotropic liver transplantation (OLT) is still the sole critical treatment. However, there are a serious issue of a shortage of liver organs from brain death and living donors [241]. Therefore, many studies have been made for development of alternative methods to OLT using hepatocytes.

Hepatocytes transplantation has been considered as a treatment for severe hepatic diseases. However, due to the insufficient mass transplantation, the genetic and metabolic hepatic diseases just can be partially corrected in many cases. In a rat model, a large fraction (>70%) of transplanted hepatocytes are rapidly rejected by the immune system through the Kupffer cells [242, 243]. So, for hepatocyte transplantation, a more effective therapy should be needed.

K. Ohashi [244, 245] used hepatocyte sheet to construct the functional liver tissue *in vitro* and transplanted it to ectopic site of mice. They found small tissues engrafted stably in the subcutaneous space of mice. Because the oxygen-demanding characteristic of the hepatocytes, it is very important to guarantee the vascularization for the survival of the transplantation of hepatocyte. They also stacked multiple hepatocyte sheets between two layers of basic fibroblast growth factor-induced vascular networks to create three-dimensional hepatic tissues in the subcutaneous space of mice. However, the long-term hepatic function and viability, the increase in the size of engineered hepatic tissues, and further functionalization should be studied further.

Besides the hepatocyte-based cell sheet, another regeneration method is functional CST based on gene transfer which can replace one of the functions in the liver. In a hemophilia A mice model, K. Tatsumi [246] used the blood outgrowth endothelial cells (BOECs) sheets to create 3–5-fold higher secretion of FVIII in subcutaneous sites compared with the Engelbreth–Holm–Swarm sarcoma gel method [247].

Although the mass and function of the transplanted hepatic tissues is difficult to maintain for a long time, the combination of gene transfer techniques and cell sheet engineering will developed new therapeutics for partially supporting the function of native liver with metabolic disorders.

4.2.15. Pancreatic Islets

The patients with Type 1 Diabetes Mellitus usually suffer from severe hypoglycemia which manifests unconsciousness, coma, or seizures by insulin injections. The pancreatic transplantation might be the proper treatment that restores euglycemia while the hypoglycemia could be avoid [248]. The islet transplantation was less invasive and demanded less technology than pancreatic organ transplantation [249]. However, the rapid loss of the transplanted islets induced by the instant blood-mediated inflammatory reaction (IBMIR) limited its application.

In order to improve the life quality of the patients with type 1 DM, the treatment should be more safe, readily available, and less-invasive. By comparing different transplantation site, the subcutaneous space is an excellent candidate site for transplantation of engineered islets, due to its easy conduction of transplantation *via* a less invasive approach under regional anesthesia and the easily repeatable implantation procedures. H. Shimizu [250] used a rat model to regenerate the islet tissues by CST. They found the dispersed rat islet cell can spread as a monolayer on Laminin-332 (Laminin 5)-coated dishes, and secreted insulin according to glucose levels. Because Laminin-332 can regulate the cellular attachment ion and spreading, and promotes cell survival in epithelial cells, their transplanted cell sheets are stable in subcutaneous spaces.

Besides, an increasing research on ES and IPS cells were expected to solve the problem of lacking donor cell. However, β-cell-like cells or hepatocytes differentiated from ES and IPS cells have not reproduced sufficient function at present.

4.2.16. Renal Tissue

The kidney participates in whole-body homeostasis by excretion of wastes and reabsorption of vital nutrients. It also regulated acid-base balance, electrolyte concentrations, extracellular fluid volume, and blood pressure. Sometimes it may be regarded as endocrine organs because it secreting hormones such as erythropoietin (EPO) and 1,25-Dihydroxyvitamin D3 (active vitamin D3). The decrease of 1,25-Dihydroxyvitamin D3 results in lower absorption of calcium from the intestine, causing hypocalcemia. And defective erythropoietin (EPO) production from the kidney not only causes anemia but also accelerates renal failure. Patients with decreasing 1,25-dihydroxyvitamin D3, EPO, and hormone secretion require repeated drug administration as well as specific blood condition monitoring. These insoluble problems with hemodialysis are severe issues for end-stage renal disease(ESRD) patients.

In 2013, a new approach by applying cell transplantation technology to adjust hormone secretion failure in the kidney has been developed in Sekiya's lab in Tokyo [251]. The renal cells were primarily cultured from porcine kidney. After refining the cells from kidney cortex, frozen renal cells were thawed and recultured on laminin-coated culture dishes with RGM-2. The renal cell-sheets were harvested by reducing the culture temperature ten days later. When they were harvested, the renal cell sheets shrink and became round because of their cytoskeleton. These sheets contained different kinds of renal cells that showed cyst-like formation, which was observed by the histological analysis and electron microscopy scanning. The synthesis of 1,25-dihydroxyvitamin D3 was determined, which demonstrated that these cultured renal cell sheets intro do have the hormone secreting function.

Furthermore, the researchers transplanted these renal cell sheets in a nude rat, and porcine EPO production in the plasma of the host rat or cultured medium was testified and measured. It showed that EPO production of renal cell sheets was both observed *in vitro* and *in vivo*. These results indicated that the renal tissues with function could be regenerated by CST. This study introduced a hormone-supplying treatment for renal dysfunctional diseases using engineered renal tissues. Moreover, since the renal cell sheets developed renal tubular-like structures *in vivo*, it holds promise for fabricating artificially engineered true renal tissue in the future.

4.2.17. Urothelial Tissue

Bladder augmentation, also called augmentation cystoplasty, is a surgical procedure used in those who lack adequate bladder capacity or detrusor compliance. Traditionally, gastrointestinal flaps were transplanted to provide a safe functional reservoir that allows for urinary continence and prevention of upper tract deterioration. However, the transplanted flap may induce severe complications such as lithiasis, urinary tract infection, and electrolyte imbalance. With the development of CST, contiguous urothelial cell sheets cultured *in vitro* should enable us to avoid these complications.

In 2003, Shiroyanagi *et al.* [252] first successfully cultivated the transplantable urothelial cell sheets *in vitro*. The seeded cells were collected from Canine urothelium. The transplantable urothelial cells sheets were obtained by utilizing a temperature-responsive cell culture method, and then examined by immunostaining and electron microscopy, revealing well-developed microridge, microvilli, and cell junction complexes. Conversely, these same cell features were destroyed by dispase treatment. This study proved that CST would be efficient in urinary tract tissue engineering in the near future.

And later in another experiment performed by the same investigators in 2004 [253], a dog model with demucosalized gastric flaps has been set to evaluate urothelium regeneration by CST. Urothelial cells, obtained from eight beagle dogs by partial cystectomy, were cultured for 3 weeks on temperature-responsive dishes to generate contiguous urothelial cell sheets. Then the cell sheets were autografted onto surgically demucosalized gastric flaps. After another 3 weeks after autografting, cell and tissue characteristics were compared between these urothelial cell sheet-grafted gastric flaps and native urothelium. Electron microscopy showed that five of the eight urothelial cell sheet-grafted flaps presented viable urothelial regeneration. Urothelial cell sheets were found to attached spontaneously to demucosalized tissue surfaces completely, with no suture or fixing, and developed into a stratified viable epithelium very similar to native urothelium. On the other three of the eight, flaps where there were severe haematomas, grafted cell sheets were not adherent and there was no urothelial regeneration. The study indicated this novel intact cell-sheet grafting method could rapidly produce native-like epithelium *in vivo*, and should be useful in urinary tract tissue engineering and surgical reconstruction.

In summary, the CST has some advantages, compared with the traditional Tissue engineering. It does not need to do trypsin digestion to seed cells. The extracellular matrix could be retained in the process of cultivation, which makes the organizational structure more similar to normal tissues, so as to realize the efficient regeneration of the structure and function of the damaged tissues or organs. The CST has a wide application in prospects of regeneration medicine. Many animal experiments have been made to apply this technology to tissue regeneration and obtained good results. However, there are still some problems needing to be resolved. For example, the malnutrition of the middle layers with the increase of the thickness of the cell sheets and poor mechanical strength of it. Another important factor that affects the consequence of CST is the cell source. For different tissue regenerations, an optimal seed cell is necessary.

Anyhow, the CST delivers a new method for tissue regeneration. By co-culture techniques and integrated with the scaffold, the cell sheet has the capability of being fabricated into three dimensional tissues for repair of large tissue defects. With the development of the CST, we can look forward to a comprehensive application of it.

4.3. Clinical Trails

Clinical trials using CST have already started in some tissues. Functional three-dimensional (3D) tissue can be easily fabricated by layering cell sheets without the use of scaffolds. CST has been applied in regenerative medicine for several tissues, and a number of clinical trials have already started. In addition, to fabricate more complex and functional 3D tissue, cell micro-patterning technology can be combined with cell sheet technology, and this interdisciplinary technology could produce interesting results. In this part, recent advances in temperature-responsive culture surfaces, cell sheet and cell micro-patterning technologies are summarized and discussed. In addition, the application of these technologies to regenerative medicine and the reconstruction of various tissues including heterogeneous tissues are also discussed.

4.3.1. Corneal Reconstruction

Damage to outermost epithelium of the eye balls from trauma or infection may cause the complete loss of corneal epithelial stem cells, and resulting in corneal opacification with severe visual loss. Under these circumstances, transplantation of alternative stem cell sources is needed for the restoration of a healthy cornea. A novel therapy using autologous epithelial cell sheets was introduced to improve the visual acuity of patients with unilateral or bilateral diseases through surrogating the damaged tissues with implanted cell sheets.

In unilateral disease patients, autologous epithelial cells can be isolated from the healthy contralateral eye with a simple biopsy. In patients with complete bilateral corneal epithelial stem cell deficiencies, an alternative autologous therapy extracting the patient's own oral mucosa as the source of epithelial stem cells was constructed.

Using a temperature-responsive culture dish, these autologous epithelial cells can be recovered as a monolithic epithelial cell sheet with both cell–cell junctions and deposited Extracellular matrix (ECM) on the basal surface. The ECM allows the cell sheet to attach rapidly and stabilize onto the host eye without any sutures. The treated corneal surfaces were totally protected from fluorescein penetration immediately after transplantation by the implanted cell sheet because they contained intact cell–cell junctions. The implanted autologous oral mucosal epithelial cell sheet demonstrated a similar histological structure to corneal epithelium and an adequate modulation in keratin expression profiles towards a corneal phenotype. Clinical results have demonstrated good results for the long-term maintenance of healthy ocular surfaces with significant recoveries in all vision lost patients.

A first clinical trial [254] for corneal reconstruction for four eyes of four patients with autologous mucosal epithelial cell sheets has resulted in the long-term maintenance of healthy ocular surfaces with the significant restorations of the vision of in all patients with corneal limbal epithelial stem cell deficiency (LSCD). The clinical study for 26 eyes of 25 patients in France with autologous oral mucosal epithelial cell sheets demonstrated that the cell sheets have been well-tolerated and a safe engineered tissue-product, and confirmed its efficacy on the regeneration of the ocular surface of the patients.

These work showed that CST is a well-tolerate and safe for the treatment of human LSCD. CST reestablished the patients' ocular surface (new epithelium and less vascularization or less vessel activity), leading to better visual acuity in cases of healthy stroma. CST should also be helpful in patients with total bilateral LSCD, whose corneal stroma is severely damaged and cannot receive a secondary graft (donor cornea) for the presence of neovessels, which may hinder its success. A decrease in the rejection risk for a secondary donor cornea graft was expected in recipients of a CST transplantation after one year. Further assessment with long-time follow-up has been still under evaluation.

4.3.2. Oral Mucosa

In 2006, Li *et al.* [255] prepared and investigated the multi-layer tissue engineered human oral mucosa growth after transplanted into mucosa defects patients. The seed cells were obtained from 3-month-old children excess oral mucosa with cleft lip surgery, isolated fibroblasts and epithelial cells were seeded in polylactic acid/glycolic acid (PLGA) collagen composite film culture, and then moved onto gas-liquid interface for composition to build stratified tissue engineered oral mucosa, which were transplanted into 7 benign oral tumor excised patients with left mucosa defects in the range of 1.8 cm*1.6 cm to 3.2 cm*2.6 cm. 1 case volunteers in transplant recipients 18 and 30 d, respectively, transplanted parts of the mucosa histological observation. The results showed the prepared tissue engineering oral mucosa grew well with the underlying cortex and epithelial layer structure by the PLGA film. The epithelial layer had 5-6 layers of cells straining was positive. The underlying cortex had 3-7 layers with negative staining. 7 patients grafted after 10 days, tissue engineering oral mucosa and the wound grew color than normal mucosa; 18 days later, transplanted area and normal mucosa still could be distinguished; 30 days later, there were indistinguishable boundaries and the wound reached I healing. Histological observation showed 18days wound on the skin and the below granulation tissue grew well, capillary proliferation, epithelial spikes forming, fibroblasts were oval; 30 days wound got more obvious collagenization and the transplanted area structure were similar to the surrounding normal tissue. In conclusion, the tissue engineering oral mucosa allograft repair, mucosal growth is good, but the source of the epithelial tissue healing needs further study.

As an excellent cell delivery system, cell patch tissue engineering technology is a traditional useful complement for tissue engineering and regenerative medicine development to provide new ideas and broad prospects.

4.3.3. Esophageal Mucosa

It has been testified in animal models that CST transplantation combined with ESD could enhance wound healing and provide a functional barrier to the esophageal lining. Thus, this novel method could reduce the postoperative complications and gain a better result of the surgery.

In 2012, a clinical study [256] on regenerating esophageal mucosa in nine patients who were removed superficial oesophageal neoplasm by ESD was performed by Okhi *et al.* The size of resected mucosa ranged from half to fully circumference. Autologous oral mucosal epithelial cell sheets, harvested from patients'oral mucosa tissues, were used. The cell-sheet size was approximately 6 mm in diameter. The cell sheets were successfully transplanted to ulcer surfaces without suturing by using an endoscope. After transplantation, within a median time of 3.5 weeks, complete re-epithelialization were observed. Except one patient who has a full circumferential ulceration expanding to the esophagogastric junction, all the other patients experienced no dysphagia, stricture, or other complications post-op.The operative procedure show the potential to prevent stricture formation after ESD and improve the quality of life of patients.

Furthermore, since December 2012, investigative clinical studies to examine the protocol as standardized regenerative medicine for esophagus diseases such as Barrettg by using an endowing performed [257].

4.3.4. Myocardium

Using autologous myoblast cell sheets to cure severe heart disease have also been started. In a clinical study, conducted by Sawa *et al.* [258], a patient suffered from the dilated cardiomyopathy, received the cell sheet therapy. About 10 grams of the patient's own skeletal muscle was obtained for the fabrication of the myoblast cell sheets. A total of 20 myoblast sheets were fabricated for transplantation onto five areas with a diameter of 40 mm and circular quadruple-layered. The patient got a good clinical condition including (1) the improvement of cardiac function, (2) no indications of arrhythmia, (3) the discontinuance of a left ventricular assisting device, and (4) no requirement of cardiac-transplantation. The patient was discharged from the hospital without using the device.

Cell sheet engineering(CST) has been used for the reconstruction of various tissues and effectively reconstructed some thicker vascularized tissues *in vitro*, which showed an enormous and wide-ranging potentials to cure patients with severe diseases and physically function. Clinical trials using CST includes repair of cornea epithelium, myocardium, esophagus, periodontal, and cartilage tissues. While, all of which are derived from autologous cells.

The profound success gained in corneal reconstruction demonstrated CST can offer a long-term maintenance of a healthy ocular surface with restoration of previously lost visual acuity. For curing severe heart disease, the patient got a remarkable improvement in clinical condition by transplanted the autologous myoblast cell sheets and avoid cardiac transplantation. To prevent ESD complication, CST played an important role in post-op re-epithelialization, which enhance wound healing and avoid dysphagia and stricture. Other clinical studies have also begun in the therapy of periodontal disease and cartilage regeneration. However, the longtime follow-up is still under evaluation. We believe that in near future, it will be more meaningful in restoring the health of patients worldwide.

CONFLICT OF INTEREST

The authors confirm that they have no conflict of interest to declare for this publication.

ACKNOWLEDGEMENTS

We would like to express our deepest gratitude to the colleagues in Department of Stomatology, Shanghai Tenth People gratitude to the relationships with other people or organ. This work is funded by Natural Science Foundation of Shanghai (15ZR1432700), National Natural Science Foundation of China (81201410, 81500806,81500808) and Fundamental Research Funds for the Central Universities (2014KJ090, 1501219075).

REFERENCES

[1] Hagège AA, Marolleau JP, Vilquin JT, *et al.* Skeletal myoblast transplantation in ischemic heart failure: long-term follow-up of the first phase I cohort of patients. Circulation 2006; 114(1) (Suppl.): I108-13.
[PMID: 16820558]

[2] Shimizu T, Sekine H, Yang J, *et al.* Polysurgery of cell sheet grafts overcomes diffusion limits to produce thick, vascularized myocardial tissues. FASEB J 2006; 20(6): 708-10.
[PMID: 16439619]

[3] Zhou Y, Chen F, Ho ST, Woodruff MA, Lim TM, Hutmacher DW. Combined marrow stromal cell-sheet techniques and high-strength biodegradable composite scaffolds for engineered functional bone

grafts. Biomaterials 2007; 28(5): 814-24.
[http://dx.doi.org/10.1016/j.biomaterials.2006.09.032] [PMID: 17045643]

[4] Masuda S, Shimizu T. Three-dimensional cardiac tissue fabrication based on cell sheet technology. Adv Drug Deliv Rev 2015; 96: 103-9.
[http://dx.doi.org/10.1016/j.addr.2015.05.002] [PMID: 25980939]

[5] Shimizu T, Yamato M, Kikuchi A, Okano T. Cell sheet engineering for myocardial tissue reconstruction. Biomaterials 2003; 24(13): 2309-16.
[http://dx.doi.org/10.1016/S0142-9612(03)00110-8] [PMID: 12699668]

[6] Yamato M, Okano T. Cell sheet engineering. Mater Today 2004; 7(5): 42-7.
[http://dx.doi.org/10.1016/S1369-7021(04)00234-2]

[7] Takaku Y, Murai K, Ukai T, *et al.* *In vivo* cell tracking by bioluminescence imaging after transplantation of bioengineered cell sheets to the knee joint. Biomaterials 2014; 35(7): 2199-206.
[http://dx.doi.org/10.1016/j.biomaterials.2013.11.071] [PMID: 24360579]

[8] Tadakuma K, Tanaka N, Haraguchi Y, *et al.* A device for the rapid transfer/transplantation of living cell sheets with the absence of cell damage. Biomaterials 2013; 34(36): 9018-25.
[http://dx.doi.org/10.1016/j.biomaterials.2013.08.006] [PMID: 23972478]

[9] Guillaume-Gentil O, Semenov OV, Zisch AH, Zimmermann R, Vörös J, Ehrbar M. pH-controlled recovery of placenta-derived mesenchymal stem cell sheets. Biomaterials 2011; 32(19): 4376-84.
[http://dx.doi.org/10.1016/j.biomaterials.2011.02.058] [PMID: 21458856]

[10] Zahn R, Thomasson E, Guillaume-Gentil O, Vörös J, Zambelli T. Ion-induced cell sheet detachment from standard cell culture surfaces coated with polyelectrolytes. Biomaterials 2012; 33(12): 3421-7.
[http://dx.doi.org/10.1016/j.biomaterials.2012.01.019] [PMID: 22300744]

[11] Ito A, Ino K, Kobayashi T, Honda H. The effect of RGD peptide-conjugated magnetite cationic liposomes on cell growth and cell sheet harvesting. Biomaterials 2005; 26(31): 6185-93.
[http://dx.doi.org/10.1016/j.biomaterials.2005.03.039] [PMID: 15899515]

[12] Shimizu K, Ito A, Yoshida T, Yamada Y, Ueda M, Honda H. Bone tissue engineering with human mesenchymal stem cell sheets constructed using magnetite nanoparticles and magnetic force. J Biomed Mater Res B Appl Biomater 2007; 82(2): 471-80.
[http://dx.doi.org/10.1002/jbm.b.30752] [PMID: 17279563]

[13] Kim SJ, Cho HR, Cho KW, *et al.* Multifunctional cell-culture platform for aligned cell sheet monitoring, transfer printing, and therapy. ACS Nano 2015; 9(3): 2677-88.
[http://dx.doi.org/10.1021/nn5064634] [PMID: 25687418]

[14] Hong Y, Yu M, Weng W, Cheng K, Wang H, Lin J. Light-induced cell detachment for cell sheet technology. Biomaterials 2013; 34(1): 11-8.
[http://dx.doi.org/10.1016/j.biomaterials.2012.09.043] [PMID: 23069710]

[15] Yang J, Yamato M, Shimizu T, *et al.* Reconstruction of functional tissues with cell sheet engineering. Biomaterials 2007; 28(34): 5033-43.
[http://dx.doi.org/10.1016/j.biomaterials.2007.07.052] [PMID: 17761277]

[16] da Silva RM, Mano JF, Reis RL. Smart thermoresponsive coatings and surfaces for tissue engineering: switching cell-material boundaries. Trends Biotechnol 2007; 25(12): 577-83.

[http://dx.doi.org/10.1016/j.tibtech.2007.08.014] [PMID: 17997178]

[17] Egami M, Haraguchi Y, Shimizu T, Yamato M, Okano T. Latest status of the clinical and industrial applications of cell sheet engineering and regenerative medicine. Arch Pharm Res 2014; 37(1): 96-106.
[http://dx.doi.org/10.1007/s12272-013-0299-8] [PMID: 24293063]

[18] Yamato M, Utsumi M, Kushida A, Konno C, Kikuchi A, Okano T. Thermo-responsive culture dishes allow the intact harvest of multilayered keratinocyte sheets without dispase by reducing temperature. Tissue Eng 2001; 7(4): 473-80.
[http://dx.doi.org/10.1089/10763270152436517] [PMID: 11506735]

[19] Chen F, Zhou Y, Barnabas ST, Woodruff MA, Hutmacher DW. Engineering tubular bone constructs. J Biomech 2007; 40 (Suppl. 1): S73-9.
[http://dx.doi.org/10.1016/j.jbiomech.2007.02.017] [PMID: 17553509]

[20] Lee JI, Nishimura R, Sakai H, Sasaki N, Kenmochi T. A newly developed immunoisolated bioartificial pancreas with cell sheet engineering. Cell Transplant 2008; 17(1-2): 51-9.
[http://dx.doi.org/10.3727/000000008783907035] [PMID: 18468235]

[21] Takagi S, Ohno M, Ohashi K, Utoh R, Tatsumi K, Okano T. Cell shape regulation based on hepatocyte sheet engineering technologies. Cell Transplant 2012; 21(2-3): 411-20.
[http://dx.doi.org/10.3727/096368911X605312] [PMID: 22793048]

[22] Liu Y, Ming L, Luo H, *et al.* Integration of a calcined bovine bone and BMSC-sheet 3D scaffold and the promotion of bone regeneration in large defects. Biomaterials 2013; 34(38): 9998-10006.
[http://dx.doi.org/10.1016/j.biomaterials.2013.09.040] [PMID: 24079891]

[23] Sekiya S, Shimizu T, Okano T. Vascularization in 3D tissue using cell sheet technology. Regen Med 2013; 8(3): 371-7.
[http://dx.doi.org/10.2217/rme.13.16] [PMID: 23627830]

[24] Sekiya S, Shimizu T, Yamato M, Okano T. Hormone supplying renal cell sheet in vivo produced by tissue engineering technology. Biores Open Access 2013; 2(1): 12-9.

[25] Kanai N, Yamato M, Okano T. Cell sheets engineering for esophageal regenerative medicine. Ann Transl Med 2014; 2(3): 28.
[PMID: 25333004]

[26] Sato M, Yamato M, Hamahashi K, Okano T, Mochida J. Articular cartilage regeneration using cell sheet technology. Anat Rec (Hoboken) 2014; 297(1): 36-43.
[http://dx.doi.org/10.1002/ar.22829] [PMID: 24293096]

[27] Wang J, Zhang R, Shen Y, *et al.* Recent advances in cell sheet technology for periodontal regeneration. Curr Stem Cell Res Ther 2014; 9(3): 162-73.
[http://dx.doi.org/10.2174/1574888X09666140213150218] [PMID: 24524797]

[28] Bardag-Gorce F, Oliva J, Wood A, *et al.* Carrier-free cultured autologous oral mucosa epithelial cell sheet (CAOMECS) for corneal epithelium reconstruction: a histological study. Ocul Surf 2015; 13(2): 150-63.
[http://dx.doi.org/10.1016/j.jtos.2014.12.003] [PMID: 25881998]

[29] Berner A, Henkel J, Woodruff MA, *et al.* Delayed minimally invasive injection of allogenic bone

marrow stromal cell sheets regenerates large bone defects in an ovine preclinical animal model. Stem Cells Transl Med 2015; 4(5): 503-12.
[http://dx.doi.org/10.5966/sctm.2014-0244] [PMID: 25834121]

[30] Hirabaru M, Kuroki T, Adachi T, *et al.* A Method for Performing Islet Transplantation Using Tissue-Engineered Sheets of Islets and Mesenchymal Stem Cells. Tissue Eng Part C Methods 2015; 21(12): 1205-15.
[http://dx.doi.org/10.1089/ten.tec.2015.0035] [PMID: 26066973]

[31] Imamura T, Ogawa T, Minagawa T, *et al.* Engineered bone marrow-derived cell sheets restore structure and function of radiation-injured rat urinary bladders. Tissue Eng Part A 2015; 21(9-10): 1600-10.
[http://dx.doi.org/10.1089/ten.tea.2014.0592] [PMID: 25669695]

[32] Yamaguchi Y, Kubo T, Tarutani M, *et al.* Epithelial-mesenchymal interactions in wounds: treatment of palmoplantar wounds by nonpalmoplantar pure epidermal sheet grafts. Arch Dermatol 2001; 137(5): 621-8.
[PMID: 11346340]

[33] Nishida K, Yamato M, Hayashida Y, *et al.* Corneal reconstruction with tissue-engineered cell sheets composed of autologous oral mucosal epithelium. N Engl J Med 2004; 351(12): 1187-96.
[http://dx.doi.org/10.1056/NEJMoa040455] [PMID: 15371576]

[34] Ohki T, Yamato M, Ota M, *et al.* Prevention of esophageal stricture after endoscopic submucosal dissection using tissue-engineered cell sheets. Gastroenterology 2012; 143(3): 582-8.e1, 2.
[http://dx.doi.org/10.1053/j.gastro.2012.04.050] [PMID: 22561054]

[35] Sawa Y, Miyagawa S, Sakaguchi T, *et al.* Tissue engineered myoblast sheets improved cardiac function sufficiently to discontinue LVAS in a patient with DCM: report of a case. Surg Today 2012; 42(2): 181-4.
[http://dx.doi.org/10.1007/s00595-011-0106-4] [PMID: 22200756]

[36] You HJ, Han SK, Lee JW, Chang H. Treatment of diabetic foot ulcers using cultured allogeneic keratinocytes--a pilot study. Wound Repair Regen 2012; 20(4): 491-9.
[PMID: 22672334]

[37] Okuda K, Kawase T, Nagata M, *et al.* Tissue-engineered cultured periosteum sheet application to treat infrabony defects: case series and 5-year results. Int J Periodontics Restorative Dent 2013; 33(3): 281-7.
[http://dx.doi.org/10.11607/prd.1545] [PMID: 23593621]

[38] Iwata T, Washio K, Yoshida T, *et al.* Cell sheet engineering and its application for periodontal regeneration. J Tissue Eng Regen Med 2015; 9(4): 343-56.
[http://dx.doi.org/10.1002/term.1785] [PMID: 23881816]

[39] Jiang Y, Jahagirdar BN, Reinhardt RL, *et al.* Pluripotency of mesenchymal stem cells derived from adult marrow. Nature 2002; 418(6893): 41-9.
[http://dx.doi.org/10.1038/nature00870] [PMID: 12077603]

[40] Paschos NK, Brown WE, Eswaramoorthy R, Hu JC, Athanasiou KA. Advances in tissue engineering through stem cell-based co-culture. J Tissue Eng Regen Med 2015; 9(5): 488-503.
[http://dx.doi.org/10.1002/term.1870] [PMID: 24493315]

[41] Osugi M, Katagiri W, Yoshimi R, Inukai T, Hibi H, Ueda M. Conditioned media from mesenchymal stem cells enhanced bone regeneration in rat calvarial bone defects. Tissue Eng Part A 2012; 18(13-14): 1479-89.
[http://dx.doi.org/10.1089/ten.tea.2011.0325] [PMID: 22443121]

[42] Piscioneri A, Morelli S, Mele M, *et al.* Neuroprotective effect of human mesenchymal stem cells in a compartmentalized neuronal membrane system. Acta Biomater 2015; 24: 297-308.
[http://dx.doi.org/10.1016/j.actbio.2015.06.013] [PMID: 26087109]

[43] Sasagawa T, Shimizu T, Yamato M, Okano T. Expression profiles of angiogenesis-related proteins in prevascular three-dimensional tissues using cell-sheet engineering. Biomaterials 2014; 35(1): 206-13.
[http://dx.doi.org/10.1016/j.biomaterials.2013.09.104] [PMID: 24119455]

[44] Wang K, Yu LY, Jiang LY, Wang HB, Wang CY, Luo Y. The paracrine effects of adipose-derived stem cells on neovascularization and biocompatibility of a macroencapsulation device. Acta Biomater 2015; 15: 65-76.
[http://dx.doi.org/10.1016/j.actbio.2014.12.025] [PMID: 25575852]

[45] Ishii M, Shibata R, Shimizu Y, *et al.* Multilayered adipose-derived regenerative cell sheets created by a novel magnetite tissue engineering method for myocardial infarction. Int J Cardiol 2014; 175(3): 545-53.
[http://dx.doi.org/10.1016/j.ijcard.2014.06.034] [PMID: 25023793]

[46] Ouyang HW, Cao T, Zou XH, *et al.* Mesenchymal stem cell sheets revitalize nonviable dense grafts: implications for repair of large-bone and tendon defects. Transplantation 2006; 82(2): 170-4.
[http://dx.doi.org/10.1097/01.tp.0000226232.79106.72] [PMID: 16858278]

[47] Imanishi Y, Miyagawa S, Maeda N, *et al.* Induced adipocyte cell-sheet ameliorates cardiac dysfunction in a mouse myocardial infarction model: a novel drug delivery system for heart failure. Circulation 2011; 124(11) (Suppl.): S10-7.
[http://dx.doi.org/10.1161/CIRCULATIONAHA.110.009993] [PMID: 21911798]

[48] Inukai T, Katagiri W, Yoshimi R, *et al.* Novel application of stem cell-derived factors for periodontal regeneration. Biochem Biophys Res Commun 2013; 430(2): 763-8.
[http://dx.doi.org/10.1016/j.bbrc.2012.11.074] [PMID: 23206704]

[49] Shudo Y, Miyagawa S, Ohkura H, *et al.* Addition of mesenchymal stem cells enhances the therapeutic effects of skeletal myoblast cell-sheet transplantation in a rat ischemic cardiomyopathy model. Tissue Eng Part A 2014; 20(3-4): 728-39.
[PMID: 24164292]

[50] Ishii M, Shibata R, Numaguchi Y, *et al.* Enhanced angiogenesis by transplantation of mesenchymal stem cell sheet created by a novel magnetic tissue engineering method. Arterioscler Thromb Vasc Biol 2011; 31(10): 2210-5.
[http://dx.doi.org/10.1161/ATVBAHA.111.231100] [PMID: 21757660]

[51] Zhao J, Liu L, Wei J, *et al.* A novel strategy to engineer small-diameter vascular grafts from marrow-derived mesenchymal stem cells. Artif Organs 2012; 36(1): 93-101.
[http://dx.doi.org/10.1111/j.1525-1594.2011.01231.x] [PMID: 21790675]

[52] Liu Y, Luo H, Wang X, *et al.* *In vitro* construction of scaffold-free bilayered tissue-engineered skin

containing capillary networks. Biomed Res Int 2013; 2013: 561410.

[53] Owaki T, Shimizu T, Yamato M, Okano T. Cell sheet engineering for regenerative medicine: current challenges and strategies. Biotechnol J 2014; 9(7): 904-14. [http://dx.doi.org/10.1002/biot.201300432] [PMID: 24964041]

[54] Xiao L, Nasu M. From regenerative dentistry to regenerative medicine: progress, challenges, and potential applications of oral stem cells. Stem Cells Cloning 2014; 7: 89-99. [http://dx.doi.org/10.2147/SCCAA.S51009] [PMID: 25506228]

[55] Langenbach F, Naujoks C, Smeets R, *et al.* Scaffold-free microtissues: differences from monolayer cultures and their potential in bone tissue engineering. Clin Oral Investig 2013; 17(1): 9-17. [http://dx.doi.org/10.1007/s00784-012-0763-8] [PMID: 22695872]

[56] Ma D, Ren L, Yao H, *et al.* Locally injection of cell sheet fragments enhances new bone formation in mandibular distraction osteogenesis: a rabbit model. J Orthop Res 2013; 31(7): 1082-8. [http://dx.doi.org/10.1002/jor.22336] [PMID: 23494761]

[57] Nakamura A, Akahane M, Shigematsu H, *et al.* Cell sheet transplantation of cultured mesenchymal stem cells enhances bone formation in a rat nonunion model. Bone 2010; 46(2): 418-24. [http://dx.doi.org/10.1016/j.bone.2009.08.048] [PMID: 19716454]

[58] Kawamura M, Miyagawa S, Fukushima S, *et al.* Xenotransplantation of bone marrow-derived human mesenchymal stem cell sheets attenuates left ventricular remodeling in a porcine ischemic cardiomyopathy model. Tissue Eng Part A 2015; 21(15-16): 2272-80. [http://dx.doi.org/10.1089/ten.tea.2014.0036] [PMID: 26046810]

[59] Ishida O, Hagino I, Nagaya N, *et al.* Adipose-derived stem cell sheet transplantation therapy in a porcine model of chronic heart failure. Transl Res 2015; 165(5): 631-9. [http://dx.doi.org/10.1016/j.trsl.2014.12.005] [PMID: 25613060]

[60] Kato Y, Iwata T, Morikawa S, Yamato M, Okano T, Uchigata Y. Allogeneic transplantation of an adipose-derived stem cell sheet combined with artificial skin accelerates wound healing in a rat wound model of type 2 diabetes and obesity. Diabetes 2015; 64(8): 2723-34. [http://dx.doi.org/10.2337/db14-1133] [PMID: 25795216]

[61] Liu J, Wang L, Liu W, Li Q, Jin Z, Jin Y. Dental follicle cells rescue the regenerative capacity of periodontal ligament stem cells in an inflammatory microenvironment. PLoS One 2014; 9(9): e108752. [http://dx.doi.org/10.1371/journal.pone.0108752] [PMID: 25275580]

[62] Burillon C, Huot L, Justin V, *et al.* Cultured autologous oral mucosal epithelial cell sheet (CAOMECS) transplantation for the treatment of corneal limbal epithelial stem cell deficiency. Invest Ophthalmol Vis Sci 2012; 53(3): 1325-31. [http://dx.doi.org/10.1167/iovs.11-7744] [PMID: 22064987]

[63] Sekine W, Haraguchi Y, Shimizu T, Yamato M, Umezawa A, Okano T. Chondrocyte differentiation of human endometrial gland-derived MSCs in layered cell sheets. Scientific World J 2013; 2013(11): 359109. [PMID: 24348153]

[64] Xie Q, Wang Z, Huang Y, *et al.* Characterization of human ethmoid sinus mucosa derived

mesenchymal stem cells (hESMSCs) and the application of hESMSCs cell sheets in bone regeneration. Biomaterials 2015; 66: 67-82.
[http://dx.doi.org/10.1016/j.biomaterials.2015.07.013] [PMID: 26196534]

[65] Diederichs S, Tuan RS. Functional comparison of human-induced pluripotent stem cell-derived mesenchymal cells and bone marrow-derived mesenchymal stromal cells from the same donor. Stem Cells Dev 2014; 23(14): 1594-610.
[http://dx.doi.org/10.1089/scd.2013.0477] [PMID: 24625206]

[66] Reinisch A, Etchart N, Thomas D, *et al.* Epigenetic and *in vivo* comparison of diverse MSC sources reveals an endochondral signature for human hematopoietic niche formation. Blood 2015; 125(2): 249-60.
[http://dx.doi.org/10.1182/blood-2014-04-572255] [PMID: 25406351]

[67] Liao HT, Chen CT. Osteogenic potential: Comparison between bone marrow and adipose-derived mesenchymal stem cells. World J Stem Cells 2014; 6(3): 288-95.
[http://dx.doi.org/10.4252/wjsc.v6.i3.288] [PMID: 25126378]

[68] Gao X, Usas A, Tang Y, *et al.* A comparison of bone regeneration with human mesenchymal stem cells and muscle-derived stem cells and the critical role of BMP. Biomaterials 2014; 35(25): 6859-70.
[http://dx.doi.org/10.1016/j.biomaterials.2014.04.113] [PMID: 24856105]

[69] Kim Y, Kim H, Cho H, Bae Y, Suh K, Jung J. Direct comparison of human mesenchymal stem cells derived from adipose tissues and bone marrow in mediating neovascularization in response to vascular ischemia. Cell Physiol Biochem 2007; 20(6): 867-76.
[http://dx.doi.org/10.1159/000110447] [PMID: 17982269]

[70] Monaco E, Bionaz M, Rodriguez-Zas S, Hurley WL, Wheeler MB. Transcriptomics comparison between porcine adipose and bone marrow mesenchymal stem cells during *in vitro* osteogenic and adipogenic differentiation. PLoS One 2012; 7(3): e32481.
[http://dx.doi.org/10.1371/journal.pone.0032481] [PMID: 22412878]

[71] Davies OG, Cooper PR, Shelton RM, Smith AJ, Scheven BA. A comparison of the *in vitro* mineralisation and dentinogenic potential of mesenchymal stem cells derived from adipose tissue, bone marrow and dental pulp. J Bone Miner Metab 2015; 33(4): 371-82.
[http://dx.doi.org/10.1007/s00774-014-0601-y] [PMID: 24997523]

[72] Tsumanuma Y, Iwata T, Washio K, *et al.* Comparison of different tissue-derived stem cell sheets for periodontal regeneration in a canine 1-wall defect model. Biomaterials 2011; 32(25): 5819-25.
[http://dx.doi.org/10.1016/j.biomaterials.2011.04.071] [PMID: 21605900]

[73] Shah FS, Li J, Dietrich M, *et al.* Comparison of stromal/stem cells isolated from human omental and subcutaneous adipose depots: differentiation and immunophenotypic characterization. Cells Tissues Organs 2015.
[http://dx.doi.org/10.1159/000430088] [PMID: 26089088]

[74] Yamahara K, Harada K, Ohshima M, *et al.* Comparison of angiogenic, cytoprotective, and immunosuppressive properties of human amnion- and chorion-derived mesenchymal stem cells. PLoS One 2014; 9(2): e88319.
[http://dx.doi.org/10.1371/journal.pone.0088319] [PMID: 24551087]

[75] Zhang ZY, Teoh SH, Chong MS, *et al.* Superior osteogenic capacity for bone tissue engineering of

fetal compared with perinatal and adult mesenchymal stem cells. Stem Cells 2009; 27(1): 126-37. [http://dx.doi.org/10.1634/stemcells.2008-0456] [PMID: 18832592]

[76] Marchesan JT, Scanlon CS, Soehren S, Matsuo M, Kapila YL. Implications of cultured periodontal ligament cells for the clinical and experimental setting: a review. Arch Oral Biol 2011; 56(10): 933-43. [http://dx.doi.org/10.1016/j.archoralbio.2011.03.003] [PMID: 21470594]

[77] Wu RX, Bi CS, Yu Y, Zhang LL, Chen FM. Age-related decline in the matrix contents and functional properties of human periodontal ligament stem cell sheets. Acta Biomater 2015; 22: 70-82. [http://dx.doi.org/10.1016/j.actbio.2015.04.024] [PMID: 25922305]

[78] Chuah YJ, Zhang Y, Wu Y, *et al.* Combinatorial effect of substratum properties on mesenchymal stem cell sheet engineering and subsequent multi-lineage differentiation. Acta Biomater 2015; 23: 52-62. [http://dx.doi.org/10.1016/j.actbio.2015.05.023] [PMID: 26026305]

[79] Zhong ZY, Jiang YD. Surface modification and characterization of indium-tin oxide for organic light-emitting devices. J Colloid Interface Sci 2006; 302(2): 613-9. [http://dx.doi.org/10.1016/j.jcis.2006.07.009] [PMID: 16890950]

[80] Palamà IE, D'Amone S, Coluccia AM, Biasiucci M, Gigli G. Cell self-patterning on uniform PDMS-surfaces with controlled mechanical cues. Integr Biol (Camb) 2012; 4(2): 228-36. [http://dx.doi.org/10.1039/C2IB00116K] [PMID: 22146870]

[81] Takahashi H, Nakayama M, Yamato M, Okano T. Controlled chain length and graft density of thermoresponsive polymer brushes for optimizing cell sheet harvest. Biomacromolecules 2010; 11(8): 1991-9.
[http://dx.doi.org/10.1021/bm100342e] [PMID: 20593758]

[82] Chuah YJ, Zhang Y, Wu Y, *et al.* Combinatorial effect of substratum properties on mesenchymal stem cell sheet engineering and subsequent multi-lineage differentiation. Acta Biomater 2015; 23: 52-62. [http://dx.doi.org/10.1016/j.actbio.2015.05.023] [PMID: 26026305]

[83] Akiyama Y, Kikuchi A, Yamato M, Okano T. Accelerated cell-sheet recovery from a surface successively grafted with polyacrylamide and poly(N-isopropylacrylamide). Acta Biomater 2014; 10(8): 3398-408.
[http://dx.doi.org/10.1016/j.actbio.2014.03.024] [PMID: 24681372]

[84] Becerra NY, López BL, Restrepo LM. Thermosensitive behavior in cell culture media and cytocompatibility of a novel copolymer: poly(N-isopropylacrylamide-co-butylacrylate). J Mater Sci Mater Med 2013; 24(4): 1043-52.
[http://dx.doi.org/10.1007/s10856-013-4861-1] [PMID: 23361967]

[85] Feng Y, Jin X, Dai G, Liu J, Chen J, Yang L. *In vitro* targeted magnetic delivery and tracking of superparamagnetic iron oxide particles labeled stem cells for articular cartilage defect repair. J Huazhong Univ Sci Technolog Med Sci 2011; 31(2): 204-9.
[http://dx.doi.org/10.1007/s11596-011-0253-2] [PMID: 21505986]

[86] Kito T, Shibata R, Ishii M, *et al.* iPS cell sheets created by a novel magnetite tissue engineering method for reparative angiogenesis. Sci Rep 2013; 3: 1418. [http://dx.doi.org/10.1038/srep01418] [PMID: 23475393]

[87] Idota N, Tsukahara T, Sato K, Okano T, Kitamori T. The use of electron beam lithographic graft-

polymerization on thermoresponsive polymers for regulating the directionality of cell attachment and detachment. Biomaterials 2009; 30(11): 2095-101.
[http://dx.doi.org/10.1016/j.biomaterials.2008.12.058] [PMID: 19157534]

[88] Tang Z, Kikuchi A, Akiyama Y, *et al.* Novel cell sheet carriers using polyion complex gel modified membranes for tissue engineering technology for cell sheet manipulation and transplantation. React Funct Polym 2007; 67(11): 1388-97.
[http://dx.doi.org/10.1016/j.reactfunctpolym.2007.07.058]

[89] Fukumori K, Akiyama Y, Yamato M, Kobayashi J, Sakai K, Okano T. Temperature-responsive glass coverslips with an ultrathin poly(N-isopropylacrylamide) layer. Acta Biomater 2009; 5(1): 470-6.
[http://dx.doi.org/10.1016/j.actbio.2008.06.018] [PMID: 18676189]

[90] Biazar E, Khorasani M, Daliri M. Cell sheet engineering: solvent effect on nanometric grafting of poly-N-isopropylacrylamide onto polystyrene substrate under ultraviolet radiation. Int J Nanomedicine 2011; 6: 295-302.
[http://dx.doi.org/10.2147/IJN.S15371] [PMID: 21383854]

[91] Hannachi IE, Yamato M, Okano T. Cell sheet technology and cell patterning for biofabrication. Biofabrication 2009; 1(2): 022002.
[http://dx.doi.org/10.1088/1758-5082/1/2/022002] [PMID: 20811100]

[92] Matsuura K, Utoh R, Nagase K, Okano T. Cell sheet approach for tissue engineering and regenerative medicine. J Control Release 2014; 190: 228-39.
[http://dx.doi.org/10.1016/j.jconrel.2014.05.024] [PMID: 24858800]

[93] Nagase K, Kimura A, Shimizu T, *et al.* Dynamically cell separating thermo-functional biointerfaces with densely packed polymer brushes. J Mater Chem 2012; 22(37): 19514.
[http://dx.doi.org/10.1039/c2jm31797d]

[94] Zhang H, Iwama M, Akaike T, *et al.* Human amniotic cell sheet harvest using a novel temperature-responsive culture surface coated with protein-based polymer. Tissue Eng 2006; 12(2): 391-401.
[http://dx.doi.org/10.1089/ten.2006.12.391] [PMID: 16548697]

[95] Rayatpisheh S, Li P, Chan-Park MB. Argon-plasma-induced ultrathin thermal grafting of thermoresponsive pNIPAm coating for contractile patterned human SMC sheet engineering. Macromol Biosci 2012; 12(7): 937-45.
[http://dx.doi.org/10.1002/mabi.201100477] [PMID: 22535772]

[96] Miyahara Y, Nagaya N, Kataoka M, *et al.* Monolayered mesenchymal stem cells repair scarred myocardium after myocardial infarction. Nat Med 2006; 12(4): 459-65.
[http://dx.doi.org/10.1038/nm1391] [PMID: 16582917]

[97] Gärtner A, Pereira T, Armada-da-Silva P, *et al.* Effects of umbilical cord tissue mesenchymal stem cells (UCX®) on rat sciatic nerve regeneration after neurotmesis injuries. J Stem Cells Regen Med 2014; 10(1): 14-26.
[PMID: 25075157]

[98] Yeh TS, Fang YH, Lu CH, *et al.* Baculovirus-transduced, VEGF-expressing adipose-derived stem cell sheet for the treatment of myocardium infarction. Biomaterials 2014; 35(1): 174-84.
[http://dx.doi.org/10.1016/j.biomaterials.2013.09.080] [PMID: 24120047]

[99] Wang Z, Weng Y, Lu S, *et al.* Osteoblastic mesenchymal stem cell sheet combined with Choukroun platelet-rich fibrin induces bone formation at an ectopic site. J Biomed Mater Res B Appl Biomater 2015; 103(6): 1204-16.
[http://dx.doi.org/10.1002/jbm.b.33288] [PMID: 25327691]

[100] Wei F, Qu C, Song T, *et al.* Vitamin C treatment promotes mesenchymal stem cell sheet formation and tissue regeneration by elevating telomerase activity. J Cell Physiol 2012; 227(9): 3216-24.
[http://dx.doi.org/10.1002/jcp.24012] [PMID: 22105792]

[101] Jin H, Zhang K, Qiao C, *et al.* Efficiently engineered cell sheet using a complex of polyethylenimine-alginate nanocomposites plus bone morphogenetic protein 2 gene to promote new bone formation. Int J Nanomedicine 2014; 9: 2179-90.
[PMID: 24855355]

[102] Guo P, Zeng JJ, Zhou N. A novel experimental study on the fabrication and biological characteristics of canine bone marrow mesenchymal stem cells sheet using vitamin C. Scanning 2015; 37(1): 42-8.
[http://dx.doi.org/10.1002/sca.21177] [PMID: 25588682]

[103] Gao LN, An Y, Lei M, *et al.* The effect of the coumarin-like derivative osthole on the osteogenic properties of human periodontal ligament and jaw bone marrow mesenchymal stem cell sheets. Biomaterials 2013; 34(38): 9937-51.
[http://dx.doi.org/10.1016/j.biomaterials.2013.09.017] [PMID: 24095254]

[104] Qi Y, Zhao T, Yan W, Xu K, Shi Z, Wang J. Mesenchymal stem cell sheet transplantation combined with locally released simvastatin enhances bone formation in a rat tibia osteotomy model. Cytotherapy 2013; 15(1): 44-56.
[http://dx.doi.org/10.1016/j.jcyt.2012.10.006] [PMID: 23260085]

[105] Kikuchi T, Shimizu T, Wada M, Yamato M, Okano T. Automatic fabrication of 3-dimensional tissues using cell sheet manipulator technique. Biomaterials 2014; 35(8): 2428-35.
[http://dx.doi.org/10.1016/j.biomaterials.2013.12.014] [PMID: 24370007]

[106] Jun I, Lee YB, Choi YS, Engler AJ, Park H, Shin H. Transfer stamping of human mesenchymal stem cell patches using thermally expandable hydrogels with tunable cell-adhesive properties. Biomaterials 2015; 54: 44-54.
[http://dx.doi.org/10.1016/j.biomaterials.2015.03.016] [PMID: 25907038]

[107] Kim SJ, Cho HR, Cho KW, *et al.* Multifunctional cell-culture platform for aligned cell sheet monitoring, transfer printing, and therapy. ACS Nano 2015; 9(3): 2677-88.
[http://dx.doi.org/10.1021/nn5064634] [PMID: 25687418]

[108] Shimizu T, Sekine H, Isoi Y, Yamato M, Kikuchi A, Okano T. Long-term survival and growth of pulsatile myocardial tissue grafts engineered by the layering of cardiomyocyte sheets. Tissue Eng 2006; 12(3): 499-507.
[http://dx.doi.org/10.1089/ten.2006.12.499] [PMID: 16579683]

[109] Tremblay PL, Hudon V, Berthod F, Germain L, Auger FA. Inosculation of tissue-engineered capillaries with the host's vasculature in a reconstructed skin transplanted on mice. Am J Transplant 2005; 5(5): 1002-10.
[http://dx.doi.org/10.1111/j.1600-6143.2005.00790.x] [PMID: 15816880]

[110] Sekiya S, Shimizu T, Yamato M, Kikuchi A, Okano T. Bioengineered cardiac cell sheet grafts have intrinsic angiogenic potential. Biochem Biophys Res Commun 2006; 341(2): 573-82. [http://dx.doi.org/10.1016/j.bbrc.2005.12.217] [PMID: 16434025]

[111] Saxena AK, Willital GH, Vacanti JP. Vascularized three-dimensional skeletal muscle tissue-engineering. Biomed Mater Eng 2001; 11(4): 275-81. [PMID: 11790859]

[112] Levenberg S, Rouwkema J, Macdonald M, *et al.* Engineering vascularized skeletal muscle tissue. Nat Biotechnol 2005; 23(7): 879-84. [http://dx.doi.org/10.1038/nbt1109] [PMID: 15965465]

[113] Lesman A, Koffler J, Atlas R, Blinder YJ, Kam Z, Levenberg S. Engineering vessel-like networks within multicellular fibrin-based constructs. Biomaterials 2011; 32(31): 7856-69. [http://dx.doi.org/10.1016/j.biomaterials.2011.07.003] [PMID: 21816465]

[114] L'Heureux N, Pâquet S, Labbé R, Germain L, Auger FA. A completely biological tissue-engineered human blood vessel. FASEB J 1998; 12(1): 47-56. [PMID: 9438410]

[115] Rayatpisheh S, Heath DE, Shakouri A, Rujitanaroj PO, Chew SY, Chan-Park MB. Combining cell sheet technology and electrospun scaffolding for engineered tubular, aligned, and contractile blood vessels. Biomaterials 2014; 35(9): 2713-9. [http://dx.doi.org/10.1016/j.biomaterials.2013.12.035] [PMID: 24411678]

[116] Tsuda Y, Shimizu T, Yamato M, *et al.* Cellular control of tissue architectures using a three-dimensional tissue fabrication technique. Biomaterials 2007; 28(33): 4939-46. [http://dx.doi.org/10.1016/j.biomaterials.2007.08.002] [PMID: 17709135]

[117] Isenberg BC, Tsuda Y, Williams C, *et al.* A thermoresponsive, microtextured substrate for cell sheet engineering with defined structural organization. Biomaterials 2008; 29(17): 2565-72. [http://dx.doi.org/10.1016/j.biomaterials.2008.02.023] [PMID: 18377979]

[118] Zhang R, Gao Z, Geng W, Yan X, Chen F, Liu Y. Engineering vascularized bone graft with osteogenic and angiogenic lineage differentiated bone marrow mesenchymal stem cells. Artif Organs 2012; 36(12): 1036-46. [http://dx.doi.org/10.1111/j.1525-1594.2012.01529.x] [PMID: 23020776]

[119] von Tell D, Armulik A, Betsholtz C. Pericytes and vascular stability. Exp Cell Res 2006; 312(5): 623-9. [http://dx.doi.org/10.1016/j.yexcr.2005.10.019] [PMID: 16303125]

[120] Mills SJ, Cowin AJ, Kaur P. Pericytes, mesenchymal stem cells and the wound healing process. Cells 2013; 2(3): 621-34. [http://dx.doi.org/10.3390/cells2030621] [PMID: 24709801]

[121] Mendes LF, Pirraco RP, Szymczyk W, *et al.* Perivascular-like cells contribute to the stability of the vascular network of osteogenic tissue formed from cell sheet-based constructs. PLoS One 2012; 7(7): e41051. [http://dx.doi.org/10.1371/journal.pone.0041051] [PMID: 22829909]

[122] Boumaza I, Srinivasan S, Witt WT, *et al.* Autologous bone marrow-derived rat mesenchymal stem

cells promote PDX-1 and insulin expression in the islets, alter T cell cytokine pattern and preserve regulatory T cells in the periphery and induce sustained normoglycemia. J Autoimmun 2009; 32(1): 33-42.
[http://dx.doi.org/10.1016/j.jaut.2008.10.004] [PMID: 19062254]

[123] Manabe T, Mizumoto K, Nagai E, *et al.* Cell-based protein delivery system for the inhibition of the growth of pancreatic cancer: NK4 gene-transduced oral mucosal epithelial cell sheet. Clin Cancer Res 2003; 9(8): 3158-66.
[PMID: 12912968]

[124] Kusumi T, Ishihara K, Mizumoto H, *et al.* Evaluation of a bioreactor with stacked sheet shaped organoids of primary hepatocytes. J Biosci Bioeng 2009; 107(5): 552-5.
[http://dx.doi.org/10.1016/j.jbiosc.2009.01.005] [PMID: 19393557]

[125] Matsuura K, Wada M, Konishi K, *et al.* Fabrication of mouse embryonic stem cell-derived layered cardiac cell sheets using a bioreactor culture system. PLoS One 2012; 7(12): e52176.
[http://dx.doi.org/10.1371/journal.pone.0052176] [PMID: 23284924]

[126] Takahashi Y, Tomotsune D, Takizawa S, *et al.* New model for cardiomyocyte sheet transplantation using a virus-cell fusion technique. World J Stem Cells 2015; 7(5): 883-93.
[http://dx.doi.org/10.4252/wjsc.v7.i5.883] [PMID: 26131319]

[127] Xu ZC, Zhang Q, Li H. Elastic large muscular vessel wall engineered with bone marrow□derived cells under pulsatile stimulation in a bioreactor. Mol Med Rep 2015; 12(4): 6005-12.
[PMID: 26238613]

[128] Kobayashi T, Kan K, Nishida K, Yamato M, Okano T. Corneal regeneration by transplantation of corneal epithelial cell sheets fabricated with automated cell culture system in rabbit model. Biomaterials 2013; 34(36): 9010-7.
[http://dx.doi.org/10.1016/j.biomaterials.2013.07.065] [PMID: 23978517]

[129] Oie Y, Nozaki T, Takayanagi H, *et al.* Development of a cell sheet transportation technique for regenerative medicine. Tissue Eng Part C Methods 2014; 20(5): 373-82.
[http://dx.doi.org/10.1089/ten.tec.2013.0266] [PMID: 24044382]

[130] Brede C, Friedrich M, Jordán-Garrote AL, *et al.* Mapping immune processes in intact tissues at cellular resolution. J Clin Invest 2012; 122(12): 4439-46.
[http://dx.doi.org/10.1172/JCI65100] [PMID: 23143304]

[131] Ishikane S, Hosoda H, Yamahara K, *et al.* Allogeneic transplantation of fetal membrane-derived mesenchymal stem cell sheets increases neovascularization and improves cardiac function after myocardial infarction in rats. Transplantation 2013; 96(8): 697-706.
[http://dx.doi.org/10.1097/TP.0b013e31829f753d] [PMID: 23912174]

[132] Zhao YH, Zhang M, Liu NX, *et al.* The combined use of cell sheet fragments of periodontal ligament stem cells and platelet-rich fibrin granules for avulsed tooth reimplantation. Biomaterials 2013; 34(22): 5506-20.
[http://dx.doi.org/10.1016/j.biomaterials.2013.03.079] [PMID: 23639531]

[133] Imanishi Y, Miyagawa S, Maeda N, *et al.* Induced adipocyte cell-sheet ameliorates cardiac dysfunction in a mouse myocardial infarction model: a novel drug delivery system for heart failure. Circulation 2011; 124(11) (Suppl.): S10-7.

[http://dx.doi.org/10.1161/CIRCULATIONAHA.110.009993] [PMID: 21911798]

[134] Amagai Y, Karasawa K, Kyungsook J, *et al.* Development of a novel carrier optimized for cell sheet transplantation. Biomatter 2015; 5(1): e1027846. [http://dx.doi.org/10.1080/21592535.2015.1027846] [PMID: 25869322]

[135] Chong MS, Chan J, Choolani M, Lee CN, Teoh SH. Development of cell-selective films for layered co-culturing of vascular progenitor cells. Biomaterials 2009; 30(12): 2241-51. [http://dx.doi.org/10.1016/j.biomaterials.2008.12.056] [PMID: 19200592]

[136] Ahn H, Ju YM, Takahashi H, *et al.* Engineered small diameter vascular grafts by combining cell sheet engineering and electrospinning technology. Acta Biomater 2015; 16: 14-22. [http://dx.doi.org/10.1016/j.actbio.2015.01.030] [PMID: 25641646]

[137] Mouthuy PA, El-Sherbini Y, Cui Z, Ye H. Layering PLGA-based electrospun membranes and cell sheets for engineering cartilage-bone transition. J Tissue Eng Regen Med 2016; 10(4): E263-74. [http://dx.doi.org/10.1002/term.1765] [PMID: 23754692]

[138] Gao Z, Chen F, Zhang J, *et al.* Vitalisation of tubular coral scaffolds with cell sheets for regeneration of long bones: a preliminary study in nude mice. Br J Oral Maxillofac Surg 2009; 47(2): 116-22. [http://dx.doi.org/10.1016/j.bjoms.2008.07.199] [PMID: 18992973]

[139] Zhou W, Han C, Song Y, *et al.* The performance of bone marrow mesenchymal stem cells for regeneration of long bones: a preliminary study in nude mice. Br J Oral Maxillofac Surg 2009.

[140] Chen YS, Tsou PC, Lo JM, Tsai HC, Wang YZ, Hsiue GH. Poly(N-isopropylacrylamide) hydrogels with interpenetrating multiwalled carbon nanotubes for cell sheet engineering. Biomaterials 2013; 34(30): 7328-34. [http://dx.doi.org/10.1016/j.biomaterials.2013.06.017] [PMID: 23827188]

[141] Geng W, Ma D, Yan X, *et al.* Engineering tubular bone using mesenchymal stem cell sheets and coral particles. Biochem Biophys Res Commun 2013; 433(4): 595-601. [http://dx.doi.org/10.1016/j.bbrc.2013.03.034] [PMID: 23523796]

[142] Chaudhuri B, Bhadra D, Moroni L, Pramanik K. Myoblast differentiation of human mesenchymal stem cells on graphene oxide and electrospun graphene oxide-polymer composite fibrous meshes: importance of graphene oxide conductivity and dielectric constant on their biocompatibility. Biofabrication 2015; 7(1): 015009. [http://dx.doi.org/10.1088/1758-5090/7/1/015009] [PMID: 25691492]

[143] Lee B, Jiao A, Yu S, You JB, Kim DH, Im SG. Initiated chemical vapor deposition of thermoresponsive poly(N-vinylcaprolactam) thin films for cell sheet engineering. Acta Biomater 2013; 9(8): 7691-8. [http://dx.doi.org/10.1016/j.actbio.2013.04.049] [PMID: 23665504]

[144] Akiyama Y, Kikuchi A, Yamato M, Okano T. Ultrathin poly(N-isopropylacrylamide) grafted layer on polystyrene surfaces for cell adhesion/detachment control. Langmuir 2004; 20(13): 5506-11. [http://dx.doi.org/10.1021/la036139f] [PMID: 15986693]

[145] Shimizu T, Yamato M, Kikuchi A, Okano T. Two-dimensional manipulation of cardiac myocyte sheets utilizing temperature-responsive culture dishes augments the pulsatile amplitude. Tissue Eng 2001; 7(2): 141-51.

[http://dx.doi.org/10.1089/107632701300062732] [PMID: 11304450]

[146] Idota N, Tsukahara T, Sato K, Okano T, Kitamori T. The use of electron beam lithographic graft-polymerization on thermoresponsive polymers for regulating the directionality of cell attachment and detachment. Biomaterials 2009; 30(11): 2095-101. [http://dx.doi.org/10.1016/j.biomaterials.2008.12.058] [PMID: 19157534]

[147] Nandkumar MA, Yamato M, Kushida A, *et al.* Two-dimensional cell sheet manipulation of heterotypically co-cultured lung cells utilizing temperature-responsive culture dishes results in long-term maintenance of differentiated epithelial cell functions. Biomaterials 2002; 23(4): 1121-30. [http://dx.doi.org/10.1016/S0142-9612(01)00225-3] [PMID: 11791915]

[148] Yamato M, Utsumi M, Kushida A, Konno C, Kikuchi A, Okano T. Thermo-responsive culture dishes allow the intact harvest of multilayered keratinocyte sheets without dispase by reducing temperature. Tissue Eng 2001; 7(4): 473-80. [http://dx.doi.org/10.1089/10763270152436517] [PMID: 11506735]

[149] Ohashi K, Yokoyama T, Yamato M, *et al.* Engineering functional two- and three-dimensional liver systems *in vivo* using hepatic tissue sheets. Nat Med 2007; 13(7): 880-5. [http://dx.doi.org/10.1038/nm1576] [PMID: 17572687]

[150] Canavan HE, Cheng X, Graham DJ, Ratner BD, Castner DG. Cell sheet detachment affects the extracellular matrix: a surface science study comparing thermal liftoff, enzymatic, and mechanical methods. J Biomed Mater Res A 2005; 75(1): 1-13. [http://dx.doi.org/10.1002/jbm.a.30297] [PMID: 16086418]

[151] Canavan HE, Cheng X, Graham DJ, Ratner BD, Castner DG. Surface characterization of the extracellular matrix remaining after cell detachment from a thermoresponsive polymer. Langmuir 2005; 21(5): 1949-55. [http://dx.doi.org/10.1021/la048546c] [PMID: 15723494]

[152] Cheng X, Canavan HE, Stein MJ, *et al.* Surface chemical and mechanical properties of plasma-polymerized N-isopropylacrylamide. Langmuir 2005; 21(17): 7833-41. [http://dx.doi.org/10.1021/la050417o] [PMID: 16089389]

[153] Pan YV, Wesley RA, Luginbuhl R, Denton DD, Ratner BD. Plasma polymerized N-isopropylacrylamide: synthesis and characterization of a smart thermally responsive coating. Biomacromolecules 2001; 2(1): 32-6. [http://dx.doi.org/10.1021/bm0000642] [PMID: 11749152]

[154] Rayatpisheh S, Li P, Chan-Park MB. Argon-plasma-induced ultrathin thermal grafting of thermoresponsive pNIPAm coating for contractile patterned human SMC sheet engineering. Macromol Biosci 2012; 12(7): 937-45. [http://dx.doi.org/10.1002/mabi.201100477] [PMID: 22535772]

[155] von Recum HA, Kim SW, Kikuchi A, Okuhara M, Sakurai Y, Okano T. Novel thermally reversible hydrogel as detachable cell culture substrate. J Biomed Mater Res 1998; 40(4): 631-9. [http://dx.doi.org/10.1002/(SICI)1097-4636(19980615)40:4<631::AID-JBM15>3.0.CO;2-I] [PMID: 9599040]

[156] Hatakeyama H, Kikuchi A, Yamato M, Okano T. Patterned biofunctional designs of thermoresponsive surfaces for spatiotemporally controlled cell adhesion, growth, and thermally induced detachment.

Biomaterials 2007; 28(25): 3632-43.
[http://dx.doi.org/10.1016/j.biomaterials.2007.04.019] [PMID: 17470377]

[157] Liu H, Ito Y. Gradient micropattern immobilization of a thermo-responsive polymer to investigate its effect on cell behavior. J Biomed Mater Res A 2003; 67(4): 1424-9.
[http://dx.doi.org/10.1002/jbm.a.20004] [PMID: 14624531]

[158] Moran MT, Carroll WM, Selezneva I, Gorelov A, Rochev Y. Cell growth and detachment from protein-coated PNIPAAm-based copolymers. J Biomed Mater Res A 2007; 81(4): 870-6.
[http://dx.doi.org/10.1002/jbm.a.31089] [PMID: 17236213]

[159] Moran MT, Carroll WM, Gorelov A, Rochev Y. Intact endothelial cell sheet harvesting from thermoresponsive surfaces coated with cell adhesion promoters. J R Soc Interface 2007; 4(17): 1151-7.
[http://dx.doi.org/10.1098/rsif.2007.1023] [PMID: 17504736]

[160] Reed JA, Lucero AE, Hu S, *et al.* A low-cost, rapid deposition method for “smart” films: applications in mammalian cell release. ACS Appl Mater Interfaces 2010; 2(4): 1048-51.
[http://dx.doi.org/10.1021/am900821t] [PMID: 20423125]

[161] Nash ME, Carroll WM, Nikoloskya N, *et al.* Straightforward, one-step fabrication of ultrathin thermoresponsive films from commercially available pNIPAm for cell culture and recovery. ACS Appl Mater Interfaces 2011; 3(6): 1980-90.
[http://dx.doi.org/10.1021/am200204j] [PMID: 21534571]

[162] Patel NG, Cavicchia JP, Zhang G, Zhang Newby BM. Rapid cell sheet detachment using spin-coated pNIPAAm films retained on surfaces by an aminopropyltriethoxysilane network. Acta Biomater 2012; 8(7): 2559-67.
[http://dx.doi.org/10.1016/j.actbio.2012.03.031] [PMID: 22475785]

[163] Yeo WS, Hodneland CD, Mrksich M. Electroactive monolayer substrates that selectively release adherent cells. ChemBioChem 2001; 2(7-8): 590-3.
[http://dx.doi.org/10.1002/1439-7633(20010803)2:7/8<590::AID-CBIC590>3.0.CO;2-D] [PMID: 11828494]

[164] Seto Y, Inaba R, Okuyama T, Sassa F, Suzuki H, Fukuda J. Engineering of capillary-like structures in tissue constructs by electrochemical detachment of cells. Biomaterials 2010; 31(8): 2209-15.
[http://dx.doi.org/10.1016/j.biomaterials.2009.11.104] [PMID: 20022631]

[165] Fukuda J, Kameoka Y, Suzuki H. Spatio-temporal detachment of single cells using microarrayed transparent electrodes. Biomaterials 2011; 32(28): 6663-9.
[http://dx.doi.org/10.1016/j.biomaterials.2011.05.068] [PMID: 21665269]

[166] Guillaume-Gentil O, Gabi M, Zenobi-Wong M, Vörös J. Electrochemically switchable platform for the micro-patterning and release of heterotypic cell sheets. Biomed Microdevices 2011; 13(1): 221-30.
[http://dx.doi.org/10.1007/s10544-010-9487-1] [PMID: 21057978]

[167] Hong Y, Yu M, Weng W, Cheng K, Wang H, Lin J. Light-induced cell detachment for cell sheet technology. Biomaterials 2013; 34(1): 11-8.
[http://dx.doi.org/10.1016/j.biomaterials.2012.09.043] [PMID: 23069710]

[168] Higuchi A, Hamamura A, Shindo Y, *et al.* Photon-modulated changes of cell attachments on poly(spiropyran-co-methyl methacrylate) membranes. Biomacromolecules 2004; 5(5): 1770-4.

[http://dx.doi.org/10.1021/bm049737x] [PMID: 15360286]

[169] Edahiro J, Sumaru K, Tada Y, *et al.* *In situ* control of cell adhesion using photoresponsive culture surface. Biomacromolecules 2005; 6(2): 970-4.
[http://dx.doi.org/10.1021/bm0493382] [PMID: 15762667]

[170] Auernheimer J, Dahmen C, Hersel U, Bausch A, Kessler H. Photoswitched cell adhesion on surfaces with RGD peptides. J Am Chem Soc 2005; 127(46): 16107-10.
[http://dx.doi.org/10.1021/ja053648q] [PMID: 16287297]

[171] Wirkner M, Alonso JM, Maus V, *et al.* Triggered cell release from materials using bioadhesive photocleavable linkers. Adv Mater 2011; 23(34): 3907-10.
[http://dx.doi.org/10.1002/adma.201100925] [PMID: 21618293]

[172] Kolesnikova TA, Kohler D, Skirtach AG, Möhwald H. Laser-induced cell detachment, patterning, and regrowth on gold nanoparticle functionalized surfaces. ACS Nano 2012; 6(11): 9585-95.
[http://dx.doi.org/10.1021/nn302891u] [PMID: 23066742]

[173] Byambaa B, Konno T, Ishihara K. Detachment of cells adhered on the photoreactive phospholipid polymer surface by photoirradiation and their functionality. Colloids Surf B Biointerfaces 2013; 103(1): 489-95.
[http://dx.doi.org/10.1016/j.colsurfb.2012.11.008] [PMID: 23261571]

[174] Guillaume-Gentil O, Semenov OV, Zisch AH, Zimmermann R, Vörös J, Ehrbar M. pH-controlled recovery of placenta-derived mesenchymal stem cell sheets. Biomaterials 2011; 32(19): 4376-84.
[http://dx.doi.org/10.1016/j.biomaterials.2011.02.058] [PMID: 21458856]

[175] Yeh HY, Lin JC. Surface characterization and *in vitro* platelet compatibility study of surface sulfonated chitosan membrane with amino group protection-deprotection strategy. J Biomater Sci Polym Ed 2008; 19(3): 291-310.
[http://dx.doi.org/10.1163/156856208783720985] [PMID: 18325232]

[176] Lin CW, Lin JC. Surface characterization and platelet compatibility evaluation of surface-sulfonated chitosan membrane. J Biomater Sci Polym Ed 2001; 12(5): 543-57.
[http://dx.doi.org/10.1163/156856201300194270] [PMID: 11469784]

[177] Chen YH, Chung YC, Wang IJ, Young TH. Control of cell attachment on pH-responsive chitosan surface by precise adjustment of medium pH. Biomaterials 2012; 33(5): 1336-42.
[http://dx.doi.org/10.1016/j.biomaterials.2011.10.048] [PMID: 22078808]

[178] Hasegawa M, Yamato M, Kikuchi A, Okano T, Ishikawa I. Human periodontal ligament cell sheets can regenerate periodontal ligament tissue in an athymic rat model. Tissue Eng 2005; 11(3-4): 469-78.
[http://dx.doi.org/10.1089/ten.2005.11.469] [PMID: 15869425]

[179] Nishida K, Yamato M, Hayashida Y, *et al.* Corneal reconstruction with tissue-engineered cell sheets composed of autologous oral mucosal epithelium. N Engl J Med 2004; 351(12): 1187-96.
[http://dx.doi.org/10.1056/NEJMoa040455] [PMID: 15371576]

[180] Haraguchi Y, Shimizu T, Yamato M, Okano T. Regenerative therapies using cell sheet-based tissue engineering for cardiac disease. Cardiology Research and Practice 2011; 2011: 8. Article ID 845170
[http://dx.doi.org/10.4061/2011/845170]

[181] Harimoto M, Yamato M, Hirose M, *et al.* Novel approach for achieving double-layered cell sheets co-

culture: overlaying endothelial cell sheets onto monolayer hepatocytes utilizing temperature-responsive culture dishes. J Biomed Mater Res 2002; 62(3): 464-70.
[http://dx.doi.org/10.1002/jbm.10228] [PMID: 12209933]

[182] Nishida K, Yamato M, Hayashida Y, *et al.* Functional bioengineered corneal epithelial sheet grafts from corneal stem cells expanded *ex vivo* on a temperature-responsive cell culture surface. Transplantation 2004; 77(3): 379-85.
[http://dx.doi.org/10.1097/01.TP.0000110320.45678.30] [PMID: 14966411]

[183] Hayashida Y, Nishida K, Yamato M, *et al.* Ocular surface reconstruction using autologous rabbit oral mucosal epithelial sheets fabricated *ex vivo* on a temperature-responsive culture surface. Invest Ophthalmol Vis Sci 2005; 46(5): 1632-9.
[http://dx.doi.org/10.1167/iovs.04-0813] [PMID: 15851562]

[184] Yaji N, Yamato M, Yang J, Okano T, Hori S. Transplantation of tissue-engineered retinal pigment epithelial cell sheets in a rabbit model. Biomaterials 2009; 30(5): 797-803.
[http://dx.doi.org/10.1016/j.biomaterials.2008.10.045] [PMID: 19036433]

[185] Yamamoto K, Hama T, Yamato M, *et al.* The effect of transplantation of nasal mucosal epithelial cell sheets after middle ear surgery in a rabbit model. Biomaterials 2015; 42: 87-93.
[http://dx.doi.org/10.1016/j.biomaterials.2014.11.037] [PMID: 25542796]

[186] Yaguchi Y, Murakami D, Yamato M, *et al.* Middle ear mucosal regeneration with three-dimensionally tissue-engineered autologous middle ear cell sheets in rabbit model. J Tissue Eng Regen Med 2016; 10(3): E188-94.
[PMID: 23894137]

[187] Liu Y, Zheng Y, Ding G, *et al.* Periodontal ligament stem cell-mediated treatment for periodontitis in miniature swine. Stem Cells 2008; 26(4): 1065-73.
[http://dx.doi.org/10.1634/stemcells.2007-0734]

[188] Hasegawa M, Yamato M, Kikuchi A, Okano T, Ishikawa I. Human periodontal ligament cell sheets can regenerate periodontal ligament tissue in an athymic rat model. Tissue Eng 2005; 11(3-4): 469-78.
[http://dx.doi.org/10.1089/ten.2005.11.469]

[189] Akizuki T, Oda S, Komaki M, *et al.* Application of periodontal ligament cell sheet for periodontal regeneration: a pilot study in beagle dogs.. J Periodont Res 2005; 40(3): 245-51.

[190] Bai Y, Bai Y, Matsuzaka K, *et al.* Cementum- and periodontal ligament-like tissue formation by dental follicle cell sheets co-cultured with Hertwig's epithelial root sheath cells. Bone 2011; 48(6): 1417-26.
[http://dx.doi.org/10.1016/j.bone.2011.02.016] [PMID: 21376148]

[191] Tsumanuma Y, Iwata T, Washio K, *et al.* Comparison of different tissue-derived stem cell sheets for periodontal regeneration in a canine 1-wall defect model. Biomaterials 2011; 32(25): 5819-25.
[http://dx.doi.org/10.1016/j.biomaterials.2011.04.071] [PMID: 21605900]

[192] Zhou Y, Li Y, Mao L, Peng H. Periodontal healing by periodontal ligament cell sheets in a teeth replantation model. Arch Oral Biol 2012; 57(2): 169-76.
[http://dx.doi.org/10.1016/j.archoralbio.2011.08.008] [PMID: 21907971]

[193] Iwata T, Yamato M, Tsuchioka H, *et al.* Periodontal regeneration with multi-layered periodontal ligament-derived cell sheets in a canine model. Biomaterials 2009; 30(14): 2716-23.

[http://dx.doi.org/10.1016/j.biomaterials.2009.01.032] [PMID: 19201461]

[194] Yang B, Chen G, Li J, *et al.* Tooth root regeneration using dental follicle cell sheets in combination with a dentin matrix - based scaffold. Biomaterials 2012; 33(8): 2449-61. [http://dx.doi.org/10.1016/j.biomaterials.2011.11.074] [PMID: 22192537]

[195] Ma D, Ren L, Chen F, *et al.* Reconstruction of rabbit critical-size calvarial defects using autologous bone marrow stromal cell sheets. Ann Plast Surg 2010; 65(2): 259-65. [http://dx.doi.org/10.1097/SAP.0b013e3181c9c3f5] [PMID: 20606584]

[196] Li G, Wang X, Cao J, *et al.* Coculture of peripheral blood CD34+ cell and mesenchymal stem cell sheets increase the formation of bone in calvarial critical-size defects in rabbits. Br J Oral Maxillofac Surg 2014; 52(2): 134-9. [http://dx.doi.org/10.1016/j.bjoms.2013.10.004] [PMID: 24210781]

[197] Yao C, Bu L, Wang K, *et al.* A study of repairing mandibular defect using tissue engineering bone with bone marrow stem cells cell sheets in dog. Hua xi kouqiangyixuezazhi= Huaxikouqiangyixuezazhi= West China journal of stomatology 2012; 30(3): 229-33, 242.

[198] Oyama T, Tomori A, Hotta K, *et al.* Endoscopic submucosal dissection of early esophageal cancer. ClinGastroenterolHepatol 2005; 3(7) (Suppl. 1): S67n do. [http://dx.doi.org/10.1016/S1542-3565(05)00291-0]

[199] Soetikno R, Kaltenbach T, Yeh R, Gotoda T. Endoscopic mucosal resection for early cancers of the upper gastrointestinal tract. J Clin Oncol 2005; 23(20): 4490-8. [http://dx.doi.org/10.1200/JCO.2005.19.935] [PMID: 16002839]

[200] Ohki T, Yamato M, Murakami D, *et al.* Treatment of oesophageal ulcerations using endoscopic transplantation of tissue-engineered autologous oral mucosal epithelial cell sheets in a canine model. Gut 2006; 55(12): 1704-10. [http://dx.doi.org/10.1136/gut.2005.088518] [PMID: 16709659]

[201] Kanai N, Yamato M, Ohki T, Yamamoto M, Okano T. Fabricated autologous epidermal cell sheets for the prevention of esophageal stricture after circumferential ESD in a porcine model. Gastrointest Endosc 2012; 76(4): 873-81. [http://dx.doi.org/10.1016/j.gie.2012.06.017] [PMID: 22867446]

[202] Kanzaki M, Yamato M, Hatakeyama H, *et al.* Tissue engineered epithelial cell sheets for the creation of a bioartificial trachea. Tissue Eng 2006; 12(5): 1275tion. [http://dx.doi.org/10.1089/ten.2006.12.ft-70]

[203] Arauchi A, Shimizu T, Yamato M, Obara T, Okano T. Tissue-engineered thyroid cell sheet rescued hypothyroidism in rat models after receiving total thyroidectomy comparing with nontransplantation models. Tissue Engineering Part A 2009; 15(12): 39433mi. [http://dx.doi.org/10.1089/ten.tea.2009.0119]

[204] Kaneshiro N, Sato M, Ishihara M, Mitani G, Sakai H, Mochida J. Bioengineered chondrocyte sheets may be potentially useful for the treatment of partial thickness defects of articular cartilage. Biochem Biophys Res Commun 2006; 349(2): 723-31. [http://dx.doi.org/10.1016/j.bbrc.2006.08.096] [PMID: 16949051]

[205] Sato M, Ishihara M, Furukawa K, *et al.* Recent technological advancements related to articular

cartilage regeneration. Med Biol Eng Comput 2008; 46(8): 735-43. [M].
[http://dx.doi.org/10.1007/s11517-008-0360-9] [PMID: 18546028]

[206] Ebihara G, Sato M, Yamato M, *et al.* Cartilage repair in transplanted scaffold-free chondrocyte sheets using a minipig model. Biomaterials 2012; 33(15): 3846-51.
[http://dx.doi.org/10.1016/j.biomaterials.2012.01.056] [PMID: 22369960]

[207] Kokubo M, Sato M, Yamato M, *et al.* Characterization of chondrocyte sheets prepared using a co-culture method with temperature-responsive culture inserts. J Tissue Eng Regen Med 2013.
[http://dx.doi.org/10.1002/term.1764] [PMID: 23868865]

[208] Kanzaki M, Yamato M, Yang J, *et al.* Dynamic sealing of lung air leaks by the transplantation of tissue engineered cell sheets. Biomaterials 2007; 28(29): 4294-302.
[http://dx.doi.org/10.1016/j.biomaterials.2007.06.009] [PMID: 17602737]

[209] Kanzaki M, Yamato M, Yang J, *et al.* Functional closure of visceral pleural defects by autologous tissue engineered cell sheets. Eur J Cardiothorac Surg 2008; 34(4): 864-9.
[http://dx.doi.org/10.1016/j.ejcts.2008.05.048] [PMID: 18586511]

[210] Porrello ER, Mahmoud AI, Simpson E, *et al.* Transient regenerative potential of the neonatal mouse heart. Science 2011; 331(6020): 1078-80.
[http://dx.doi.org/10.1126/science.1200708] [PMID: 21350179]

[211] Sano M, Minamino T, Toko H, *et al.* p53-induced inhibition of Hif-1 causes cardiac dysfunction during pressure overload. Nature 2007; 446(7134): 444-8.
[http://dx.doi.org/10.1038/nature05602] [PMID: 17334357]

[212] Sekine H, Shimizu T, Dobashi I, *et al.* Cardiac cell sheet transplantation improves damaged heart function *via* superior cell survival in comparison with dissociated cell injection. Tissue Eng Part A 2011; 17(23-24): 2973-80.
[http://dx.doi.org/10.1089/ten.tea.2010.0659] [PMID: 21875331]

[213] Matsuura K, Honda A, Nagai T, *et al.* Transplantation of cardiac progenitor cells ameliorates cardiac dysfunction after myocardial infarction in mice. J Clin Invest 2009; 119(8): 2204-17.
[PMID: 19620770]

[214] Hamdi H, Planat-Benard V, Bel A, *et al.* Epicardial adipose stem cell sheets results in greater post-infarction survival than intramyocardial injections. Cardiovasc Res 2011; 91(3): 483-91.
[http://dx.doi.org/10.1093/cvr/cvr099] [PMID: 21565832]

[215] Memon IA, Sawa Y, Fukushima N, *et al.* Repair of impaired myocardium by means of implantation of engineered autologous myoblast sheets. J Thorac Cardiovasc Surg 2005; 130(5): 1333-41.
[http://dx.doi.org/10.1016/j.jtcvs.2005.07.023] [PMID: 16256786]

[216] Sekiya N, Matsumiya G, Miyagawa S, *et al.* Layered implantation of myoblast sheets attenuates adverse cardiac remodeling of the infarcted heart. J Thorac Cardiovasc Surg 2009; 138(4): 985-93.
[http://dx.doi.org/10.1016/j.jtcvs.2009.02.004] [PMID: 19769884]

[217] Kondoh H, Sawa Y, Miyagawa S, *et al.* Longer preservation of cardiac performance by sheet-shaped myoblast implantation in dilated cardiomyopathic hamsters. Cardiovasc Res 2006; 69(2): 466-75.
[http://dx.doi.org/10.1016/j.cardiores.2005.11.005] [PMID: 16423569]

[218] Hoashi T, Matsumiya G, Miyagawa S, *et al.* Skeletal myoblast sheet transplantation improves the

diastolic function of a pressure-overloaded right heart. J Thorac Cardiovasc Surg 2009; 138(2): 460-7. [http://dx.doi.org/10.1016/j.jtcvs.2009.02.018] [PMID: 19619796]

[219] Hata H, Matsumiya G, Miyagawa S, *et al.* Grafted skeletal myoblast sheets attenuate myocardial remodeling in pacing-induced canine heart failure model. J Thorac Cardiovasc Surg 2006; 132(4): 918-24.
[http://dx.doi.org/10.1016/j.jtcvs.2006.01.024] [PMID: 17000305]

[220] Miyagawa S, Saito A, Sakaguchi T, *et al.* Impaired myocardium regeneration with skeletal cell sheets--a preclinical trial for tissue-engineered regeneration therapy. Transplantation 2010; 90(4): 364-72. [http://dx.doi.org/10.1097/TP.0b013e3181e6f201] [PMID: 20555308]

[221] Imanishi Y, Miyagawa S, Maeda N, *et al.* Induced adipocyte cell-sheet ameliorates cardiac dysfunction in a mouse myocardial infarction model: a novel drug delivery system for heart failure. Circulation 2011; 124(11) (Suppl.): S10-7.
[http://dx.doi.org/10.1161/CIRCULATIONAHA.110.009993] [PMID: 21911798]

[222] Miyagawa S, Sawa Y, Sakakida S, *et al.* Tissue cardiomyoplasty using bioengineered contractile cardiomyocyte sheets to repair damaged myocardium: their integration with recipient myocardium. Transplantation 2005; 80(11): 1586-95.
[http://dx.doi.org/10.1097/01.tp.0000181163.69108.dd] [PMID: 16371930]

[223] Miyahara Y, Nagaya N, Kataoka M, *et al.* Monolayered mesenchymal stem cells repair scarred myocardium after myocardial infarction. Nat Med 2006; 12(4): 459-65.
[http://dx.doi.org/10.1038/nm1391] [PMID: 16582917]

[224] Sekine H, Shimizu T, Hobo K, *et al.* Endothelial cell coculture within tissue-engineered cardiomyocyte sheets enhances neovascularization and improves cardiac function of ischemic hearts. Circulation 2008; 118(14) (Suppl.): S145-52.
[http://dx.doi.org/10.1161/CIRCULATIONAHA.107.757286] [PMID: 18824746]

[225] Kobayashi H, Shimizu T, Yamato M, *et al.* Fibroblast sheets co-cultured with endothelial progenitor cells improve cardiac function of infarcted hearts. J Artif Organs 2008; 11(3): 141-7.
[http://dx.doi.org/10.1007/s10047-008-0421-8] [PMID: 18836875]

[226] Bel A, Planat-Bernard V, Saito A, *et al.* Composite cell sheets: a further step toward safe and effective myocardial regeneration by cardiac progenitors derived from embryonic stem cells. Circulation 2010; 122(11) (Suppl.): S118-23.
[http://dx.doi.org/10.1161/CIRCULATIONAHA.109.927293] [PMID: 20837902]

[227] Masumoto H, Matsuo T, Yamamizu K, *et al.* Pluripotent stem cell-engineered cell sheets reassembled with defined cardiovascular populations ameliorate reduction in infarct heart function through cardiomyocyte-mediated neovascularization. Stem Cells 2012; 30(6): 1196-205.
[http://dx.doi.org/10.1002/stem.1089] [PMID: 22438013]

[228] Hida N1. Nishiyama N, Miyoshi S, Kira S, Segawa K, Uyama T, Mori T, Miyado K, Ikegami Y, Cui C, Kiyono T, Kyo S, Shimizu T, Okano T, Sakamoto M, Ogawa S, Umezawa A. Novel cardiac precursor-like cells fromhumanmenstrual blood-derived mesenchymal cells. Stem Cells 2008; 26(7): 1695-704.
[PMID: 18420831]

[229] Kawamura M, Miyagawa S, Miki K, *et al.* Feasibility, safety, and therapeutic efficacy of human

induced pluripotent stem cell-derived cardiomyocyte sheets in a porcine ischemic cardiomyopathy model. Circulation 2012; 126(11) (Suppl. 1): S29-37.
[http://dx.doi.org/10.1161/CIRCULATIONAHA.111.084343] [PMID: 22965990]

[230] Perez-Ilzarbe M, Agbulut O, Pelacho B, *et al.* Feasibility, safety, and therapeutic efficacy of human induced pluripotent stem cell-derived cardiomyocyte sheets in a port Fail. 2008; 10(11): 1065-72.

[231] Kerbel RS. Tumor angiogenesis. N Engl J Med 2008; 358: 2039Ener.

[232] Nakagami H, Morishita R, Yamamoto K, *et al.* Mitogenic and antiapoptotic actions of hepatocyte growth factor through ERK, STAT3, and AKT in endothelial cells. Hypertension 2001; 37(2 Pt 2): 581-6.
[http://dx.doi.org/10.1161/01.HYP.37.2.581] [PMID: 11230338]

[233] Shudo Y, Miyagawa S, Ohkura H, *et al.* Addition of mesenchymal stem cells enhances the therapeutic effects of skeletal myoblast cell-sheet transplantation in a rat ischemic cardiomyopathy model. Tissue Eng Part A 2014; 20(3-4): 728-39.
[PMID: 24164292]

[234] Shudo Y, Miyagawa S, Fukushima S, *et al.* Novel regenerative therapy using cell-sheet covered with omentum flap delivers a huge number of cells in a porcine myocardial infarction model. J Thorac Cardiovasc Surg 2011; 142(5): 1188-96.
[http://dx.doi.org/10.1016/j.jtcvs.2011.07.002] [PMID: 21924436]

[235] Burchfield JS, Xie M, Hill JA. Pathological ventricular remodeling: mechanisms: part 1 of 2. Circulation 2013; 128(4): 388-400.
[http://dx.doi.org/10.1161/CIRCULATIONAHA.113.001878] [PMID: 23877061]

[236] Frangogiannis NG. Regulation of the incal ventricular remodeling: mechanisms: part 1 of 2. Circulation 2

[237] Chung ES, Packer M, Lo KH, Fasanmade AA, Willerson JT. Anti-TNF Therapy Against Congestive Heart Failure Investigators. Randomized, double-blind, placebo-controlled, pilot trial of infliximab, a chimeric monoclonal antibody to tumor necrosis factor-alpha, in patients with moderate-to-severe heart failure: results of the anti-TNF Therapy Against Congestive Heart Failure (ATTACH) trial. Circulation 2003; 107(25): 3133-40.
[http://dx.doi.org/10.1161/01.CIR.0000077913.60364.D2] [PMID: 12796126]

[238] Mann DL, McMurray JJ, Packer M, *et al.* Targeted anti cytokine therapy in patients with chronic heart failure: results of the Randomized Etanercept Worldwide Evaluation (RENEWAL). Circulation 2004; 109(13): 1594-602.
[http://dx.doi.org/10.1161/01.CIR.0000124490.27666.B2] [PMID: 15023878]

[239] Ray WA, Stein CM, Hall K, Daugherty JR, Griffin MR. Non-steroidal anti-inflammatory drugs and risk of serious coronary heart disease: an observational cohort study. Lancet 2002; 359(9301): 118-23.
[http://dx.doi.org/10.1016/S0140-6736(02)07370-1] [PMID: 11809254]

[240] Le Blanc K, Frassoni F, Ball L, *et al.* Developmental Committee of the European Group for Blood and Marrow Transplantation. Mesenchymal stem cells for treatment of steroid-resistant, severe, acute graft-*versus*-host disease: a phase II study. Lancet 2008; 371(9624): 1579-86.
[http://dx.doi.org/10.1016/S0140-6736(08)60690-X] [PMID: 18468541]

[241] Egawa H, Tanabe K, Fukushima N, Date H, Sugitani A, Haga H. Current status of organ transplantation in Japan. Am J Transplant 2012; 12(3): 523-30.
[http://dx.doi.org/10.1111/j.1600-6143.2011.03822.x] [PMID: 22054061]

[242] Joseph B, Malhi H, Bhargava KK, Palestro CJ, McCuskey RS, Gupta S. Kupffer cells participate in early clearance of syngeneic hepatocytes transplanted in the rat liver. Gastroenterology 2002; 123(5): 1677-85.
[http://dx.doi.org/10.1053/gast.2002.36592] [PMID: 12404242]

[243] Gupta S, Rajvanshi P, Sokhi R, *et al.* Entry and integration of transplanted hepatocytes in rat liver plates occur by disruption of hepatic sinusoidal endothelium. Hepatology 1999; 29(2): 509-19.
[http://dx.doi.org/10.1002/hep.510290213] [PMID: 9918929]

[244] Ohashi K, Yokoyama T, Yamato M, *et al.* Engineering functional two-and three-dimensional liver systems in vivo using hepatic tissue sheets. NatMed 2007; 13(7): 880e g.

[245] Ohashi K, Tatsumi K, Tateno C, *et al.* Liver tissue engineering utilizing hepatocytes propagated in mouse livers in vivo. Cell Transplant 2012; 21(2-3): 429 hep.

[246] Tatsumi K, Sugimoto M, Lillicrap D, *et al.* A novel cell-sheet technology that achieves durable factor VIII delivery in a mouse model of hemophilia A. PLoS One 2013; 8(12): e83280.
[http://dx.doi.org/10.1371/journal.pone.0083280] [PMID: 24358271]

[247] Matsui H, Shibata M, Brown B, *et al.* *Ex vivo* gene therapy for hemophilia A that enhances safe delivery and sustained *in vivo* factor VIII expression from lentivirally engineered endothelial progenitors. Stem Cells 2007; 25(10): 2660-9.
[http://dx.doi.org/10.1634/stemcells.2006-0699] [PMID: 17615271]

[248] Shapiro AM. Islet transplantation in type 1 diabetes: ongoing challenges, refined procedures, and long-term outcome. Rev Diabet Stud 2012; 9(4): 385-406.
[http://dx.doi.org/10.1900/RDS.2012.9.385] [PMID: 23804275]

[249] Shapiro AM, Lakey JR, Ryan EA, *et al.* Islet transplantation in seven patients with type 1 diabetes mellitus using a glucocorticoid-free immunosuppressive regimen. N Engl J Med 2000; 343(4): 230-8.
[http://dx.doi.org/10.1056/NEJM200007273430401] [PMID: 10911004]

[250] Shimizu H, Ohashi K, Utoh R, *et al.* Bioengineering of a functional sheet of islet cells for the treatment of diabetes mellitus. Biomaterials 2009; 30(30): 5943-9.
[http://dx.doi.org/10.1016/j.biomaterials.2009.07.042] [PMID: 19674781]

[251] Sekiya S, Shimizu T, Yamato M, Okano T. Hormone supplying renal cell sheet in vivo produced by tissue engineering technology. Bioresearch Open Access 2013; 2(1): 12als.

[252] Shiroyanagi Y, Yamato M, Yamazaki Y, Toma H, Okano T. Transplantable urothelial cell sheets harvested noninvasively from temperature-responsive culture surfaces by reducing temperature. Tissue Eng 2003; 9(5): 1005-12.
[http://dx.doi.org/10.1089/107632703322495646] [PMID: 14633384]

[253] Shiroyanagi Y, Yamato M, Yamazaki Y, Toma H, Okano T. Urothelium regeneration using viable cultured urothelial cell sheets grafted on demucosalized gastric flaps. BJU Int 2004; 93(7): 1069sive.

[254] Nishida K, Yamato M, Hayashida Y, *et al.* Corneal reconstruction with tissue-engineered cell sheets composed of autologous oral mucosal epithelium. New England J Med 2004; 351(12): 1187of M.

[http://dx.doi.org/10.1056/NEJMoa040455]

[255] Li H, Zhang D, Wen Y, *et al.* Growth of compound layer tissue engineered oral mucosa and its clinical application in hetero-transplantation. Chinese Journal of Reparative and Reconstructive Surgery 2006; 20(2): 177-80.

[256] Ohki T, Yamato M, Ota M, *et al.* Prevention of esophageal stricture after endoscopic submucosal dissection using tissue-engineered cell sheets. Gastroenterology 2012; 143(3): 582-8.e1, 2. [http://dx.doi.org/10.1053/j.gastro.2012.04.050] [PMID: 22561054]

[257] Egami M, Haraguchi Y, Shimizu T, Yamato M, Okano T. Latest status of the clinical and industrial applications of cell sheet engineering and regenerative medicine. Arch Pharm Res 2014; 37(1): 96-106.
[http://dx.doi.org/10.1007/s12272-013-0299-8] [PMID: 24293063]

[258] Sawa Y, Miyagawa S, Sakaguchi T, *et al.* Tissue engineered myoblast sheets improved cardiac function sufficiently to discontinue LVAS in a patient with DCM: report of a case. Surg Today 2012; 42(2): 181-4.
[http://dx.doi.org/10.1007/s00595-011-0106-4] [PMID: 22200756]

CHAPTER 6

New Methods of Scaffold Fabrication: Cell-instructive Scaffolds, Structural Scaffolds, Scaffolds Fabricated from Smart Materials Able to Respond Sensitively to Environmental Cues

Jing Wang[1,*], Yuehua Liu[2], Jing Chen[3], Yun Shen[1], Yiming Zhang[1], Jinglin Zhou[4], Shu Zhang[4], Chenyuan Xu[1] and **Raorao Wang[1]**

[1] *Department of Stomatology, Shanghai Tenth People's Hospital, Tongji University School of Medicine, Shanghai, China*

[2] *Department of Orthodontics, Shanghai Stomatological Hospital, Shanghai, China*

[3] *Department of Orthodontics, School & Hospital of Stomatology, Tongji University, Shanghai Engineering Research Center of Tooth Restoration and Regeneration, Shanghai, China*

[4] *State Key Laboratory of Oral Diseases, West China School of Stomatology, Sichuan University, Chengdu, China*

Abstract: In the field of regenerative medicine and tissue engineering, biomaterials are critical for the encouragement and maintenance of biological activities. Scaffolds can be cell-instructive, structural and fabricated from smart materials. Design considerations include the creation of well-defined materials, cell-material interactions, and controlled-release bioactive agents.

Keywords: Cell-instructive, Controlled release, Extracellular matrix, Scaffold.

1. INTRODUCTION

Biomaterials are critical components that encourage and sustain biological activities in regenerative medicine and tissue engineering to provide an appro-

* **Corresponding author Jing Wang:** Department of Stomatology, Shanghai Tenth People's Hospital, Tongji University School of Medicine, No. 301, Middle Yanchang Road, Shanghai 200072, P.R. China; Tel: 86-21-66301726; Fax: 86-21-66301725; E-mail: dentist_tracy@hotmail.com

Jing Wang & YunFeng Lin (Eds.)

priate environment for cell behavior. Mimicking the extracellular environment should be considered in cellular scaffold design to render them capable of facilitating cell behavior, such as proliferation, adhesion and differentiation. Scaffolds can be cell-instructive, structural and fabricated from smart materials which could respond sensitively to environmental cues. The fabrication of materials and interactions between cells and materials should be taken into consideration when the materials are designed. Obviously, the interaction of materials with cells or local tissue has important effects in the regeneration process. Knowledge of materials characteristics and cell-material interactions is necessary for guiding cell phenotypes and tissue formation.

2. CELL-INSTRUCTIVE SCAFFOLDS

Scaffolds for tissue engineering should match some physical and biochemical characteristics. Cell-instructive scaffolds that may mimic the extracellular environment offer a favorable environment for the natural behavior of cells. In this review, we discuss fabrication strategies such as spinning and self-assembly techniques to provide well-designed scaffolds, related factors of cell-instructive scaffold for cell behavior, and methods for loading of bioactive molecules.

2.1. Synthesis and Fabrication of Cell-instructive Scaffolds

Both natural and synthetic materials are available for the construction of cell-instructive scaffolds [1]. Natural materials frequently employed include polysaccharides (*e.g.*, alginate, chitosan) and proteins (*e.g.*, collagen, gelatin) with biological activity and biocompatibility [2]. However, natural materials have the disadvantages of rapid degradation rates and high variability, as well as obstacles in sterilization and purification [3, 4]. In contrast, synthetic materials provide control for their physico-chemical properties and can be modified [5]. Moreover, composite materials overcome the limitations of single-component materials that lack the desired mechanical properties or cell–matrix interactions. Kuo *et al.* [6] promoted cell adhesion and viability on poly(lactide-co-glycolide) (PLGA) /chitosan scaffolds with type I collagen. Polyesters such as Poly(ε-caprolactone)(PCL),poly(glycolic acid) (PGA), poly(lacticacid) (PLA)and their copolymers are usually applied in synthetic polymers [7]. However, their acidic degradation products may result in a strong inflammatory response [8], rendering them unfavorable for cell adherence to the surfaces of polyesters due to their hydrophobicity [9].

Various approaches, such as spinning and self-assembly, have been used in the creation of polymeric fibers for the synthesis and fabrication of scaffolds with a three-dimensional network [10 - 12]. Other techniques have also been introduced for the fabrication of nanofibrous or microfibrous assemblies, such as thermally-induced phase separation, printing, and UV light and photomask application.

2.1.1. Spinning

Nanofiber or microfiber assembly scaffolds have been made by the physical bonding of fibers through spinning, such as wet-, electro- and melt-spinning. As a phase inversion technique, wet-spinning produces microscale polymeric fibers by an immersion precipitation process [13]. Electro-spinning that is able to mimic the physical functions of the extracellular matrix (ECM) has been used to produce polymeric nanofibrous assemblies for tissue regeneration (Fig. **1**). This technique formed fibers with a diameter from nanometers to micrometers by a polymeric solution through a capillary tube to a highly electric field [14,15]. For instance, periodontal ligament cells (PDLSCs) seeded on an electro-spun gelatin membrane showed good attachment and proliferation after 7 days [16]. Recently, live cell electro-spinning was used for the direct preparation of scaffolds with living cells. Jayasinghe and his colleagues [17] developed multiple core-vessel-like structures by the cell electro-spinning technique. This structure was comprised of various cell types around the core. Contrary to the electro-spinning method, melt-spinning is free of solvents and thus avoids the volatility and toxicity issues caused by solvents. Melt-electro-spinning writing (MEW) generates accurate fiber deposition and scaffolds with specific designs and shapes by layer-by-layer fabrication. Scaffolds designed by MEW have been demonstrated to support cell attachment, proliferation and infiltration [18].

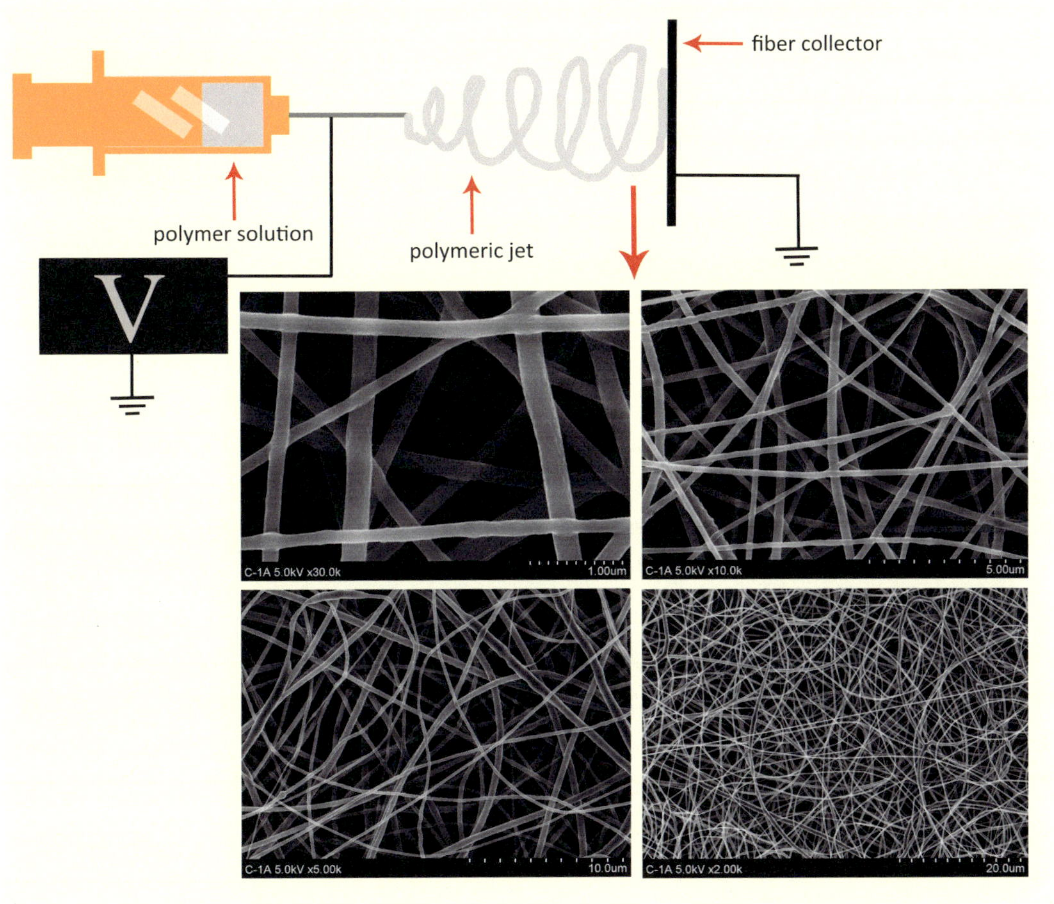

Fig. (1). Scanning electron microscopy (SEM) image of scaffold prepared by the electro-spinning technique.

2.1.2. Self-assembly

Bioactive molecules, especially proteins, tend to self-assemble and form supramolecular structures by various stimuli [19]. The major driving forces for self-assembly are hydrophobic and other non-covalent bonding interactions, such as Van der Waals forces, hydrophobic interactions, hydrogen bonds, and electrostatic interactions [20 - 22].

These interactions, although individually weak, have strong cooperative action that dominates the scaffold structure [23]. Compared with electrospun fibers, molecular self-assembly can obtain hydrogel formation with nanofibers of a much smaller diameter [24]. One study encapsulated basic fibroblast growth factors (bFGFs) into peptide amphiphile(PA) nanofibrous hydrogels by the self-assembly method and found that hydrogel with bFGF showed significant angiogenesis, in contrast that achieved with bFGF or PA injection alone in the backs of mice [25]. Injection application seems a viable alternative for use in self-assembly. However, there are some unstable factors that may affect the degradation rate of scaffolds, such as uncontrolled internal pore shape, poor mechanical strength and susceptibility to enzymatic degradation.

2.1.3. Thermally Induced Phase Separation (TIPS)

The phase separation method involves the application of a single-phase homogenous polymer solution, rapidly cooled below the solvent's freezing point (Tk), after which the solution is separated into two phases [26]. This method obtains both polymer-rich and solvent-rich phases, valuable for internal pore structure preparation and complex scaffold geometries [27, 28] (Fig. **2**). The quenching temperature and rate affect the scaffold structure and the process of solvent crystallization [29]. Porogens, such as sugar and salt, are applied to form pores of material with leaching out following phase separation. Obviously, structures with internal pores are beneficial for cell migration, nutrient distribution, and nutrient/waste exchange [30].

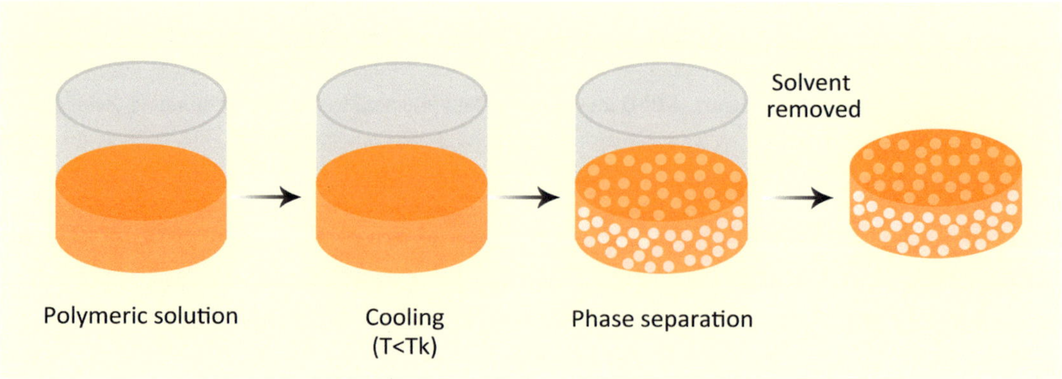

Fig. (2). Thermallyinduced phase separation.

Niu prepared composite scaffolds with chitosan microspheres (CMs) through thermally-induced phase separation and found that they stimulated the alkaline phosphatase (ALP) activity of mesenchymal stem cells(MSCs) [31]. Another study reported that MSCs seeded on a PLGA/nano-hydroxyapatite (n-HA) scaffold showed significantly higher alkaline phosphatase activity compared with that on PLGA scaffolds without NHA [32]. Compared with the PLGA-MS--treated group, cartilage with glycosaminoglycan and collagen type II deposition was found in osteochondral defects of rats in the PLGA/NHA-MSC-treated group after 12 weeks [33].

2.1.4. Printing

Printer technology has been applied to achieve high spatial control over surface properties of biomaterials to guide cell behavior. Inkjet printing is the most common application for this field. As a non-contact technique, the inkjet releases ink droplets triggered by thermal, piezoelectrics or magnetics at a controlled position [34]. Moreover, a more complex signaling environment can be created by multiple printed layers, such as growth factor printing. Researchers have already patterned fibroblast growth factor-2 (FGF-2) onto fibrin hydrogels by inkjet printing, depending on the affinity between the fibrin and FGF-2 [35]. Another new method of bioactive factor fabrication is a lithographic technique to create micro- and nano-patterned biomaterials. Islands of fibronectin were printed onto non-adhesive hard materials by microcontact printing. Results showed that the sizes of the adhesive islands could limit cell spreading [36]. Gao developed a polymerized poly(ethylene glycol) dimethacrylate-gelatin methacrylate (PEG-GelMA) scaffold by the inkjet bioprinting method, in which hMSCs were distributed layer-by-layer. With this approach, 80% of the hMSCs survived during the printing process [37].

2.1.5. Ultraviolet (UV) Photolithography

UV light and photomasks have been used to immobilize bioactive molecules in controlled regions of the bio-scaffold surface. Song applied this technique to prepare various poly(vinyl alcohol) (PVA) micropatterns and discovered that hMSCs formed single-cell arrays with different geometric micropatterns. Small hMSCs seeded with different shapes had similar adipogenic differentiation potential [38]. One study used a thin layer of PEG-diacrylate (PEGDA) solution containing BMP-2 on a base of cross-linked PCL/gelatin nanofibers. For PEGDA cross-linked through a photomask, the uncross-linked monomer was removed, where a solution containing FGF-2 was adsorbed. Thus, both the FGF-2 and BMP-2 were released in different regions of the nanofibers over a period of several days. hMSCs seeded onto the scaffold suggested a better osteogenic differentiation [39]. Another application of UV light through photomasks was the creation of controlled regions of covalently coupled adhesion ligands with alginate solutions of acrylate GGGGRGDSP peptide and calcium ions cross-linked. These materials could direct MC3T3 cell adhesion and proliferation [40].

2.2. Related Factors of Cell-instructive Scaffolds for Cell Behavior

Interactions between cells and materials are of increasing importance in tissue engineering, due to the strong requirement for phenotype control or differentiation for stem cells. The key related factor is a native extracellular matrix (ECM) mimic. Therefore, biomaterials would offer structural support and intrinsic properties to the cells, such as substrate mechanics, surface topography, stiffness and viscoelasticity [41]. These factors present micro- and nano-scopic organization and can regulate cell adhesion, proliferation, differentiation and apoptosis.

2.2.1. Scaffold Mechanics

The physical properties of scaffolds significantly affect MSC differentiation. Scaffold mechanics, such as mechanical strength and intrafiber cross-link density, may have an important effect on the interactions and gene expression of hMSCs. In their research, Lee's group found that the expression of osteogenesis markers of MSCs cultured on protein-coated hydrogels depended on stiffness, with a maximum at ~30 kPa [42]. However, the problem of how to balance the mechanical strength and porous nature of a material still exists. Composite materials have been proposed as a solution to this problem. The mechanical properties of materials can be modulated by the use of composite materials combined with nanoscale fibers, tubes or particulates.

2.2.2. Scaffold Composition

Some research has demonstrated scaffold composition as an important factor for stem cell differentiation [43, 44]. Naturally derived polymers such as collagen, fibrinogen and chitosan, *etc.*, can help cells to attach to the surfaces of materials and control cell function compared with that achievable in synthetic materials [4]. For example, human MSCs on cross linked collagen scaffolds had greater alkaline phosphatase activity than those on unmodified scaffolds [45]. Chemical composition of the scaffold can also be considered as a determining factor for stem cell differentiation. Researchers found that n-HA) deposited on the surfaces of electro-spinning nanofibers *via* a calcium phosphate dipping method promoted higher osteogenic differentiation of ADSCs [46].

2.2.3. Scaffold Topography

Surface topography with micro- and nano-scale structures is another key feature of materials influencing cell proliferation, alignment, migration and gene expression, *etc.* [47]. The study by Christopherson *et al.* suggested that more rat neural stem/progenitor cells (rNSCs) spread and proliferate when the fiber diameter decreased [48]. Moreover, the order of nano-scale surface roughness directs cell function. Surface topographical features contain grooves, pillars, lattices and other shapes that can be generated by different micro-fabrication and nanofabrication methods, such as micro-contact printing, electron beam lithography, photolithography, soft lithography, electro-spinning, and polymer phase separation [49]. Compared with tissue culture polystyrene (TCPS), plain PCL nanofibers have been demonstrated to up-regulate the expressions of neural markers of MSCs at both the mRNA and protein levels [50].

2.3. Methods for the Loading of Bioactive Molecules

Controlled delivery of specific protein or growth factors to the proper site may support the host or transplanted cells and thus guide tissue regeneration [51]. In a recent study, non-covalent and covalent growth factor immobilization represents distinct strategies for such direct delivery [52]. Both physical encapsulation and adsorption on the surfaces of materials are examples of non-covalent binding with growth factors. The types of mobilization for growth factors determine the release mechanism. Generally, diffusion-controlled release and swelling-controlled release were exhibited by non-covalent immobilization, which exhibits an initial burst release phase followed by a slower drug release. However, covalently immobilized proteins usually display the chemical/enzymatic reaction-controlled release mechanism [52, 53]. Although the delivery of a single growth factor has been studied, the methods involved in the release of two or more growth factors have not been well-examined. The bioactive factors delivered include growth factors, genetic material and drugs or small molecules; of the genetic material, nucleic acids may change gene expression at the transcriptional or post-transcriptional level. Candidate genes play an important part in the processes of bone development and repair, such as bone regeneration [54]. Antibiotics are commonly used to control infections at a bone regeneration site for anti-inflammatory effect [55, 56].

Biomimetic and nanotechnology-based scaffolds can be exploited to deliver bioactive molecules at specific concentrations, temporally and spatially. For the controlled-release system to deliver biomolecules at appropriate concentrations both temporally and spatially, the release kinetics of the bioactive factors should be first considered. Different controlled system strategies may lead to different presentations of bioactive factors. For instance, the factor *via* covalent attachment to the material will release when the bond is broken or the material degrades. This section will review some methods for the loading of bioactive agents in cell-instructive scaffolds.

2.3.1. Adsorption

For bioactive polymeric materials to be obtained, a straightforward approach involves bioactive molecules immobilized on the surfaces of scaffolds by physical or chemical methods. As mentioned previously, nano/microfibrous scaffolds with large surface areas adsorb bioactive molecules through non-covalent bonding interactions. One study reported that 3D ceramic porous scaffolds could be fabricated with BMP-2 loaded by direct absorption, although it is difficult for the adsorbed agents to be released for a long period of time [57]. Cai explored a multilayer with morphogenetic protein 2 (BMP-2) in the films using poly(allylamine hydrochloride) (PAH) and poly(sodium 4-styrenesulfonate) (PSS) by a layer-by-layer technique [58]. The adsorbed BMP-2 in the films enhanced early osteogenic differentiation of MSCs (Fig. **3**).

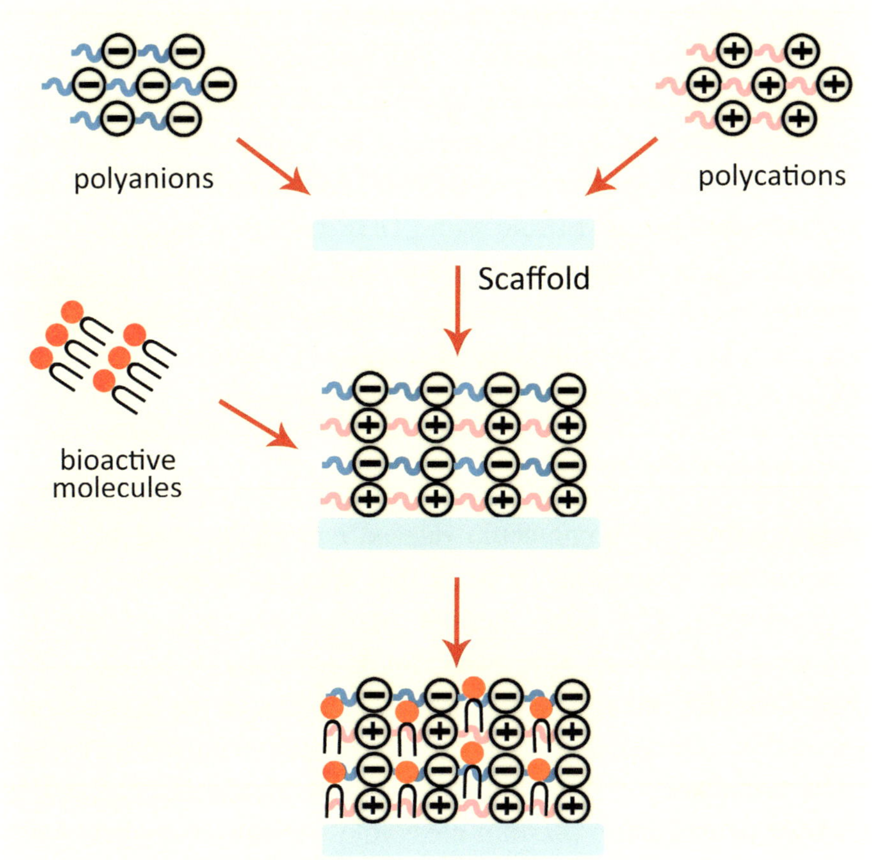

Fig. (3). The layer-by-layer (LbL) technique.

In addition to physical adsorption, chemical immobilization is another approach for biomolecule immobilization, which can promote coating and long-time release of bioactive agents [12]. The mechanism for covalent attachment may be through targeting the thiol, amino or carboxylic groups of peptides, protein fragments or full-length proteins [59]. Initially, integrin-mediated cell attachment was applied to promote cell adhesion. Arginine-glycine-aspartate (RGD) sequences were attached to the synthetic materials' surfaces to improve cell seeding.

2.3.2. Controlled-release Particles

Particles containing cytokine encapsulation have a 'burst' release, while microparticulates and nanoparticulates show sustained or delayed release [60, 61].

For easy incorporation with drugs or molecules, hydrogels are often used as carriers [62]. However, compared with hydrogels, PLGA nanospheres incorporated by a double-emulsion approach showed a significantly longer controlled release. Wei *et al.* found initial bone formation by scaffolds with a PLGA nanosphere delivery system of BMP-7 [63]. Another study showed that a phase-separated nanofibrous scaffold with PDGF-releasing nanospheres promoted angiogenesis [64]. It is suggested that controlled delivery of PLGA nanospheres could be applied in craniofacial tissue regeneration. PLGA microspheres delivered into porous scaffolds showed prolonged release of vascular endothelial growth factor (VEGF) and significantly enhanced angiogenesis [65].

2.3.3. Stimulus-based Delivery

The temporal control of biomolecule release can be realized by biomaterials systems responding to external stimuli that can be controlled in space. As described previously, UV light through photomasks makes light-dependent reactions to triggered release of growth factors or other molecules [66]. In this delivery, the chemical linkages with bioactive factors or the biomaterial loaded with encapsulations are degraded to release the bioactive molecules by leveraging cell secretion of enzymes which break down the ECM. Zisch, *et al.* [67] reported coupling VEGF containing a plasmin-cleavable site into hydrogels with MMP-degradable peptides, in which VEGF covalently immobilized in the materials, could be released triggered by cell-secreted plasmin and MMPs. Compared with soluble VEGF or VEGF-bound hydrogels, these hydrogel delivery system improved angiogenesis in rats.

3. STRUCTURAL SCAFFOLDS IN TISSUE ENGINEERING

Structural scaffolds have been widely used in tissue engineering to maintain the constructs of implants and sustain the cells before a mature tissue can be regenerated and exert its function. To achieve superior performance, four properties are of primary concern during the development of structural scaffolds, including: mechanical behavior, bio-inductive properties, biodegradation rate, and micro-/nano-structure and porosity. As a replacement for missing tissue, scaffolds should possess enough strength to resist the tension from surrounding tissue, maintaining suitable room for regeneration. Bio-inductive properties enable the scaffolds to encourage the attachment, proliferation and differentiation of the stem/progenitor cells, which are usually achieved by carrying growth factors or relative genes to the scaffolds. Ideal scaffolds should be degradable, leaving no foreign substance in the body when a biological regeneration is achieved. Last but not least, the micro-/nano-structure of scaffolds regulates the interaction between biomaterials and cells and directly interferes with the cytoskeleton spreading through focal adhesion, which play important roles in the control of cell behavior. However, an optimized material that exhibits unchallenged superiority in all aspects remains to be developed. For instance, a titanium implant possesses the best mechanical properties, but it is osteo-conductive instead of osteo-inductive. Titanium implants facilitate cell attachment and ECM deposition on the surface, but cannot encourage cell differentiation and proliferation [68]. Moreover, the manufacture of porous 3D titanium frameworks remains challenging. Comparatively, natural ECM molecules such as collagen provide attractive bio-inductive ability by promoting cell adhesion, proliferation, migration and differentiation, where as they have insufficient structural stability and can barely maintain the shape of the implant. Recently, with rapid progress in hybrid materials and nanotechnology, increasing numbers of novel biomaterials have been developed. These methods promise to lead to breakthroughs in bio-scaffold construction.

The term 'biomimetic scaffold' refers to the bioengineered scaffold that mimics the 3D structure and chemical properties of natural ECM. Given the differences between ECMs in bone, cartilage, cutaneous and other tissues, various scaffolds have been selected. These materials share several common features but should be optimized specifically to fit their host.

3.1. Features of Different Tissue Matrices

3.1.1. Bone Matrix

Bone matrix is composed of organic collagen components, predominantly type I collagen and inorganic bone mineral salt, mainly hydroxyapatite. In normal bone tissue, a basic element of bone matrix, lamellar bone, is formed by parallel type I collagen fibers and hydroxyapatite crystals aligned intensively along the collagen fibrils. These fibers are attached to each other by glycosaminoglycans (GAGs) and proteins. In alternating layers of lamellar bone, the fibers are aligned perpendicularly, enhancing mechanical strength and resistance to torsion forces. It has been implied by Bettinger, *et al.* that the highly ordered nano topography contributes to cell matrix signaling [69].

The research team led by Robert E. Guldberg has reported that the orientation of nanofibers in scaffolds dramatically influences the behavior of hMSCs [70]. Enhanced cell migration was revealed in those scaffolds with aligned fiber mesh, compared with those with random meshes. It can be ascribed to the mechanism of 'contact guidance', which implies that the cells could align by themselves along the fibers after spreading on the nanofibers. Similar phenomena were also reported by Basher *et al.* [71] and Yang *et al.* [72]. As a result, optimized biomimetic scaffolds for bone tissue engineering are supposed to be assembled by highly ordered aligned fibers, just like the fibers in the lamellar bone.

Another essential feature of bone ECM is its mechanical properties. Bone is a hard tissue with high mechanical strength and moderate Young's modulus. As a result, bone scaffolds should provide similar properties. As an example, SiO_2-based scaffolds are not suitable because of their high Young's modulus. It has been demonstrated that the cells in SiO_2-based scaffolds fail to receive physiologic mechanical signals, which decreases cell bioactivity [73]. In contrast, soft polymer substrates with the same 3D structure fulfilled the function and enhanced the adhesion, proliferation and differentiation of the cells. It is suggested that the properties of mechanical strength and elasticity should be finely adjusted on the scaffolds.

3.1.2. Cartilage Matrix

Regeneration of cartilage is even more challenging than that of bone. Unlike bone tissue, cartilage has poor ability for spontaneous repair. The matrix of cartilage tissue has its unique features: cells encapsulated in self-produced extracellular matrix. The matrix is helpful in maintaining the spherical shapes of the chondrocytes, which is extremely important in chondrogenic differentiation. It has been reported that primary chondrocytes will undergo dedifferentiation during *in vitro* culture [74]. However, during *in vitro* chondrogenesis, MSCs will eventually differentiate into hypertrophic chondrocytes, which are unstable and are most likely to be replaced by blood vessels and fibrocartilage after transplantation. It was hypothesized that the cartilage-like ECM could maximally support the chondrocytes and contributed to the maintenance of a stable hyaline phenotype. The ECM was in abundance with cartilage-related matrix molecules, such as collagen type II, aggrecan or decorin. These molecules have also been proved to contribute in supporting the production and maintenance of the hyaline cartilage matrix [75]. Therefore, the scaffold should be optimized to mimic the geometric shape and chemical composition of the cartilage ECM.

Shear stresses loaded on articular cartilage are also helpful to the maintenance of the hyaline phenotype. It has been reported that fluid flow of 1dyne/cm^2 promoted the production of collagen type II and improved the tensile strength of the cartilage constructs [76, 77]. This research highlights the importance of an elastic, tough and tensile scaffold for cartilage engineering.

3.1.3. Skin Matrix

Natural skin is composed of two layers: the epidermis, derived from the ectoderm, and dermis, derived from the mesoderm. Severe trauma usually results in the defection of both layers, which calls for full-thickness regeneration. In skin engineering, the epidermis is usually regenerated by keratinocytes or epidermal stem cells encapsulated in lipid-rich matrix, and the dermis is usually regenerated by synthetic polymers or natural matrix proteins. Therefore, a typical skin scaffold requires a bilayer structure. As an example, Integra(Integra Life Sciences Corp), a dual-layer wound matrix which has been approved by the FDA for clinical application, has proved its efficiency in burn wounds, diabetic foot ulcers and other types of skin wounds [78 - 81].

Another critical factor of skin engineering scaffolds is the water uptake ratio (WUR),which represents the ability of the scaffolds to exchange the nutrients and exudate from the adjacent tissue, which is extremely important to maintain transplant survival and prevent swelling. Scaffolds with high WUR can be applied in full-thickness wounds with excessive exudates and inflammation, whereas a lower WUR is optimized for partial-thickness wounds [82].

3.1.4. Scaffolds in Tooth Tissue Engineering

Tooth tissue engineering is even more challenging than that of other tissues involving the interaction of multiple cells. Fully functional tooth engineering requires complete regeneration of enamel, dentin and periodontal tissue. Enamel and dentin regeneration is still in its initial stage, due to the inaccessibility of odontogenic stem cells. Only collagen gels were used in recent bioengineered tooth technology [83, 84].

Compared with enamel and dentin, periodontal regeneration seems more promising. Scaffolds for periodontal regeneration should be osteo-inductive and osteo-conductive, which can facilitate the regeneration and integration of alveolar bone. Moreover, the scaffolds should degrade at an appropriate pace to match the regeneration of neo-bone instead of the mucosa, which functions as a barrier to prevent epithelial growth.

3.2. Materials and Features of Biomimicry Scaffolds

3.2.1. Essential Features of the Scaffolds

As summarized previously, these scaffold materials have both advantages and disadvantages. It is impossible to build an ideal scaffold with only one material. At the same time, regeneration of different tissues requires distinct scaffolds. Scientists are investigating the combination of several biomaterials, attempting to use the advantages of each while avoiding accompanying complications. Additionally, by adjustments to the amount of each substance and the methods of combination, the features of the hybrid materials can be modified to meet the needs of different tissues. The essential properties requiring special concern include: biocompatibility, porous properties, mechanical properties and degradation properties.

3.2.1.1. Biocompatibility

There are four outcomes of implantation. If the material is toxic, it will elicit local inflammation and necrosis of the tissue, leading to rejection of the implant. If the material is degradable, it will be absorbed and replaced by neotissues. If the material is bio-active, integration between the implants and adjacent tissues will be obtained. If the material is biologically inactive, proliferation of the fibroblasts will be induced and finally encapsulate the implant in the fibrous capsule. Biocompatibility refers to the property of the material to support the attachment, proliferation and differentiation of surrounding tissues without any toxicity or extra activation of immune-responses. It is a basic requirement of all implantations.

3.2.1.2. Porous Properties

The scaffolds should also be highly porous to facilitate cell attachment and the exchange of nutrients and metabolism products. Pore size is one of the most important properties of the biomaterials that may accelerate tissue reconstruction [85]. It has been reported that different pore sizes are needed in optimized scaffolds [86]. Microscopic pores benefit cell proliferation and attachment, and in contrast, larger pores play roles in accelerating the ingrowth of cells and blood capillaries [87].

Various synthetic techniques result in distinct porous properties in the scaffolds. Parameters such as temperature, saturation and particle size can be finely modulated to create different porous scaffolds.

3.2.1.3. Structural Strength

Structural strength is essential for load-bearing structures in the body, specifically, bone and cartilage. To efficiently support the attached cells and maintain a designed shape, scaffolds must resist relevant stresses without obvious deformation. Conversely, moderate strain should be allowed, which can apply beneficial fluid shear stress to the cells and initiate a series of reactions. Generally, several mechanical parameters, such as tensile strength, stiffness, Young's modulus, *etc.*, should be taken into account. Besides the inherent characteristics of the materials, chemical bonding among the molecules dramatically influences the mechanical strength of the scaffolds. As a result, hybrids and cross-linking techniques containing massive chemical bonding have been common in this field for decades.

3.2.1.4. Biodegradation Rate

Ideal scaffolds should be biodegradable at an appropriate rate, gradually replaced by regenerated tissue and leaving only natural organics. The degradation rate should match the regeneration rate of neotissues. If absorption occurs before neotissue regeneration, the scaffold is likely to collapse, while if the scaffold degrades too slowly, it will occupy the space and interfere with the growth of neotissues.

The degradation rate depends mainly on the chemical ingredients of the scaffolds. The biodegradation rate is also associated with the microstructure of the scaffold and the surface coating technique, as well as the environment of surrounding tissues. The ingredient is an essential factor that determine how the polymers are depolymerized. As an example, ECM-derived polymers are easily decomposed by enzymes, which make them degrade at a high rate. Similarly, polyanhydrides also degrade at a remarkable pace, because their chemical activity facilitates the process by surface erosion and does not require the involvement of macrophages and enzymes.

Building interactions and cross-linking also contributes to the modulation of the degradation rate. Massive chemical bonding increases the stability of the macromolecules and protects the inorganic particles (usually hydrophilic) from hydrolysis, result in resistance to biodegradation.

Furthermore, the amounts and sizes of inorganic particles, and the pore properties, are also involved in the control of degradation rates. These factors should be taken into account during the construction of a synthetic scaffold.

However, a challenge is that the porous, strength, and biodegradation properties contradict each other. A high porosity rate usually implies low mechanical strength. Meanwhile, biodegradation processing may increasingly impair the mechanical strength of scaffolds. Conversely, scaffolds with high mechanical strength, including metal alloys and ceramics, are non-degradable or minimally degradable. These characteristics should be designed and modulated according to the features of regenerated tissue.

3.2.2. Biomaterials used in Scaffold Manufacture

Structural scaffolds can be produced from different materials, the most common of which can be divided into four categories, namely, metal alloys, bioceramics, natural polymers and synthetic polymers. Their properties will be briefly reviewed in the following section.

1. Titanium is the most widely used metal material in regenerative medicine. Its excellent bio-compatibility as well as its mechanical properties have been recognized for decades. However, Ti-based scaffolds are osteo-conductive instead of osteo-inductive, suggesting that the surfaces of implants may support osteoblast and osteoprogenitor cell attachment and subsequent bone matrix deposition and formation, but can barely encourage osteogenic differentiation and cell proliferation [88]. Additionally, Ti scaffolds are not good carriers for stem cells, growth factors and genes. Finally, craniofacial bone defects require individual model building through computer-aided design and manufacture, but insufficient machinability of Ti materials restrains their application. To overcome the shortage of Ti and incorporate osteo-inductive properties into Ti scaffolds, a combination of Ti and other biodegradable materials was proposed. Datta, *et al.* optimized Ti mesh by culturing MSCs on a Ti scaffold, to generate bone-like ECM on the surface of Ti. It was suggested that the Ti scaffold has significant osteo-inductivity because of the presence of the ECM coating [89, 90].
2. Bioceramics used in tissue engineering are different from traditional ceramic materials, which are composed mainly of silicon compound. Bioceramics, including hydroxyapatite (HA), tricalcium phosphate (TCP) and calcium sulfate, are biocompatible materials composed of calcium salts. A favorable characteristic of these bioceramics is the calcium ion, which might facilitate the mineralization of bone matrix. Hydroxyapatite is the predominant ingredient of human hard tissues. It has been reported that HA scaffolds are osteo-conductive and osteo-inductive, which implies that osteogenesis is encouraged on the surfaces of the scaffolds [88, 91]. However, unlike their natural isomers, synthetic HA scaffolds have limited mechanical strength. Combination

approaches are proposed to enhance the properties of HA scaffolds. Hybrids composed of HA particles and PCL have been proved to achieve higher tensile strength than non-modified HA [92, 93]. HA particles are also combined with hydrogels such as gelatine, chitosan or collagen to increase the stiffness of the hydrogels and sometimes modify HA crystallization [94 - 96].

Two distinct crystal types of TCP material, α-TCP and β-TCP, can be used in tissue engineering scaffolds. α-TCP scaffolds are supposed to have higher tensile strength than β-TCP because their crystals are more thermally stable. In contrast, β-TCP material is degraded at a lower rate than α-TCP, which matches the regeneration of neobone, whereas α-TCP is not suitable for bone engineering unless the degradation rate is modulated.

TCP scaffold also takes advantage of their remarkable osteo-inductive capacity. It has been revealed by several teams that TCP scaffolds are helpful in accelerating bone formation and vascularity without additional growth factors [97, 98]. However, this characteristic is species-specific. For instance, highly purified β-TCP scaffolds with 75% porosity are usually considered as osteo-inductive in dogs [99, 100], but failed to initiate new bone formation in FVB mice [101].In contrast, β-TCP scaffolds with 60% porosity show remarkable osteo-inductive potential in FVB mice. These distinct results could be attributed to the different pore size distribution.

3. 'Natural polymers' refers to the subset of multiple naturally derived materials, including collagen, fibrin, hyaluronic acid, silk fibroin, gelatin, chitosan and alginate. One of the similarities of these materials is that most of them can form hydrogel through molecular cross-linking. These hydrogels are excellent cell encapsulation materials, being soft, porous, biocompatible, biodegradable and able to deliver drugs, proteins or genes. The three-dimensional hydrogel scaffolds are cartilage-biomimetic materials, which makes them favorable to chondrocytes in the maintenance of their round shape and hyaline phenotype. For instance, hybrid scaffolds of alginate cross-linked with Ca^{2+} were found to be an appropriate environment for hyaline formation [102, 103]. In particular, the cross-linked alginate hydrogels supported the chondrocytes and maintained spherical cell morphology by encapsulating the cells in the hydrogel form and

porous construct [104, 105]. Hence, chondrogenesis was promoted and the cell phenotype was maintained, since the micro-environment simulated the embryonic development of cartilage. Alginate is not easily degraded by mammalian metabolism; however, its poor mechanical properties and high degradability have limited its application. The hydrogel form also facilitates the injection of cell-encapsulated materials and the application of bioprinting.

Another advantage of natural polymers is their excellent biocompatibility. Most of these polymers are derived from natural ECM, facilitating a maximal mimic of ECM chemical properties. Further, as a component of normal tissue, natural polymers can easily be degraded by enzymolysis. As a result, natural polymers are usually combined with other materials to enhance the biocompatibility and modulate the biodegradation of the entire scaffold.

4. Synthetic polymers are comprised of a wide range of macromolecules composed of repeated small molecules. These materials will be detailed below.

3.3. Strategy for Fabricating and Assembling Scaffolds which Mimic Natural ECM

3.3.1. Materials Available for Fabrication

Different categories of materials can be utilized as structured scaffolds for tissue engineering, but synthetic biodegradable polymers may be the most machinable material. The mechanical and biological properties of scaffolds derived from polymers can be finely modulated through adjustment of the amounts of monomer ingredients, the minimal polymer units and the additional molecules or steps during the process. The latest research is focused on designing and fabricating polymer-based scaffolds with the aid of computers, in attempts to develop optimized scaffolds for tissue engineering [106, 107].

Varieties of polymers are used as scaffolds in tissue engineering. However, most of them can be divided into 5 categories, namely, poly(α-hydroxy acids), polylactone, poly(anhydrides), polyhydroxyalkanoates and poly(propylene fumarates).

1. Poly(α-hydroxy acids) may be the most classic polymers in regenerative medicine. As a member of the family of poly(α-hydroxy acids), PLA (Fig. **4**) and PGA (Fig. **5**) and their copolymers, PLGA (Fig. **6**) has been intensively investigated, and some of the products have been approved for clinical utilization by the FDA. A common property of this kind of polymer is that they can biodegrade through hydrolysis of the ester bonds and release acid during the process. As a result, one of the disadvantages of these polymers is inflammation in the implantation region, induced by low pH.
2. PCL is a frequently used polymer and shares a molecular structure similar to that of the poly(α-hydroxy acids), but it degrades at a lower rate than PLA, PGA and PLGA. Typical degradation time of a PCL scaffold is more than 24 months, which makes PCL an attractive polymer for bone regeneration, which calls for long-term stability and mechanical strength. The degradation rate can also be increased by copolymerization with dilactide [108].

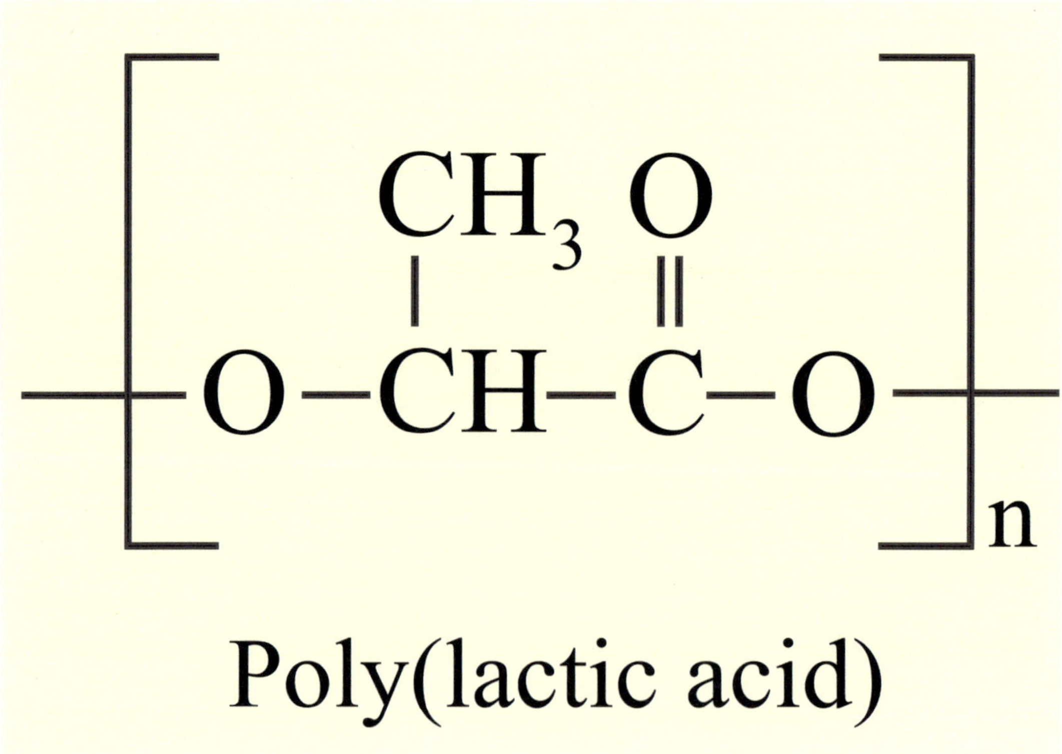

Fig. (4). Structure of poly(lactic acid).

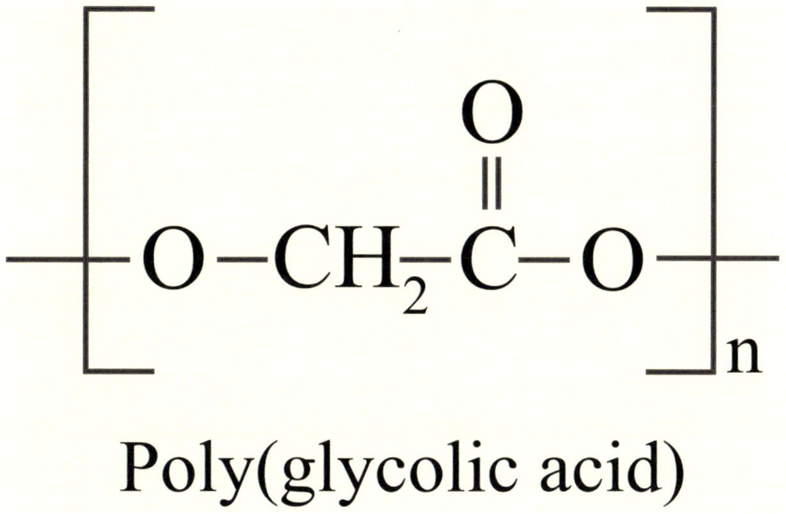

Fig. (5). Structure of poly(glycolic acid).

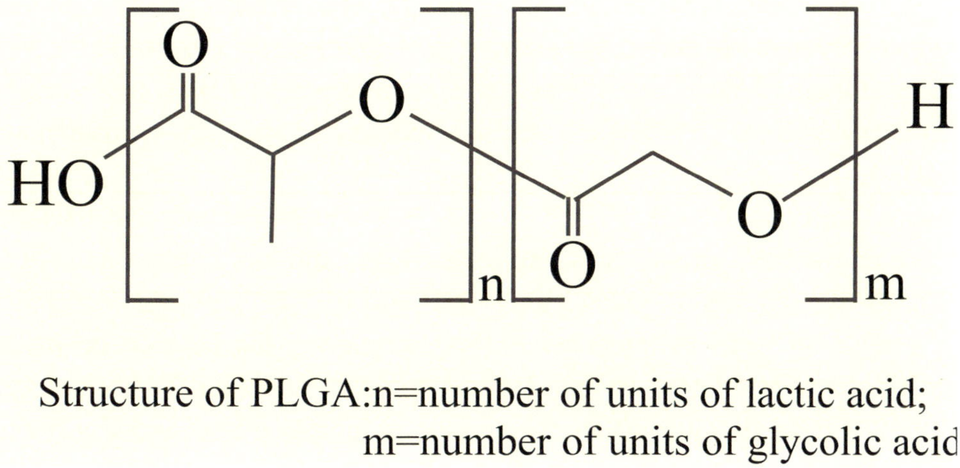

Fig. (6). Structure of PLGA (n=number of units of lactic acid; m=number of units of glycolic acid).

Another modulation of PCL can be achieved by incorporation with tricalcium phosphate (TCP), which improves the biocompatibility and also regulates the degradation kinetics of the scaffold. Additionally, TCP could contribute in buffering the low pH value resulting from degradation products [109, 110].

Poly(p-dioxanone) (PDS) is another of the polylactones. Sutures made of PDS have been approved for clinical application and have shown positive effects. The

PDS suture is superior to PLA or PGA, not only due to the non-toxic degradation products, but also because of its flexible property. PDS also holds the potential to act in the repair of bone fractures, as an alternative to metal materials. However, PDS scaffolds should be incorporated with other osteo-inductive materials, to modulate and enhance the bioactivity of the scaffolds.

3. The most significant characteristic of the polyanhydrides (Fig. **7**) is their high chemical activity, which enables them to be easily hydrolyzed without additional catalysts. Because of their high degradation rate, polyanhydrides have low mechanical properties, and their Young's modulus is near 1.3 MPa. As a result, they are not suitable for bone scaffolds. Copolymerizing polyanhydride and other monomers such as sebacic acid (SA) or carboxy-phenoxypropane (CPP) can change the hydrophobicity of the matrix and modulate the degradation rate [111].
4. Polyhydroxyalkanoates, or PHAs (Fig. **8**), are linear aliphatic polyesters produced by bacteria during unbalanced growth (excessive carbon) as a form of energy storage. Several copolymers are also members of PHAs, including poly-β-hydroxybutyrate (PHB), poly(3-hydroxybutyrate-co-hydroxyvalerate) (PHBV), and poly(3-hydroxybutyrate-co-3-hydroxyhexanoate) (PHBHHx). Among these copolymers, PHBHHx may be the most attractive. PHBHHx can be fabricated through salt leaching technology, and has better flexibility and lower brittleness and toughness than PHB and PHVB. Through adjustment of the amount of hydroxyhexanoate (HHx) monomer, the balance between the flexibility and toughness of the scaffold could be achieved. Additionally, the amount of HHx also influences the surface properties and therefore affects the adhesion and proliferation of attached cells. It has been implied that each cell line favors PHBHHx with different amounts of HHx: 20% HHx is suitable for fibroblasts, while 12% HHx is suitable for osteoblasts [112]. It has also been reported that the fibroblasts and osteoblasts seeded on a PHBHHx scaffold reproduced much more rapidly than those on PLA and PHB scaffolds [113]. Similar results were observed with chondrocytes and MSCs [112, 114, 115]. Although the investigations of PHBHHx for scaffolds are only in the initial stage, existing evidence has shown promising prospects for this polymer.
5. Fumarate-based polymer is also known as polypropylene fumarate (PPF)

(Fig. **9**). The most attractive feature of PPF is its ability to cross-link through the reactive double bond. The cross-linking reaction dramatically increases its mechanical strength and makes it a candidate for bone scaffolds. PPF materials can be applied by injection to cure bone defects *in vivo*. This property facilitates the shaping of the scaffold and enables minimally invasive surgery to be performed. Another advantage of a PPF scaffold is that its compressive strength perfectly matches that of human trabecular bone, ranging from 20-130 MPa, and can be modulated by adjustments in the cross-linking level and the molecular weight [116].

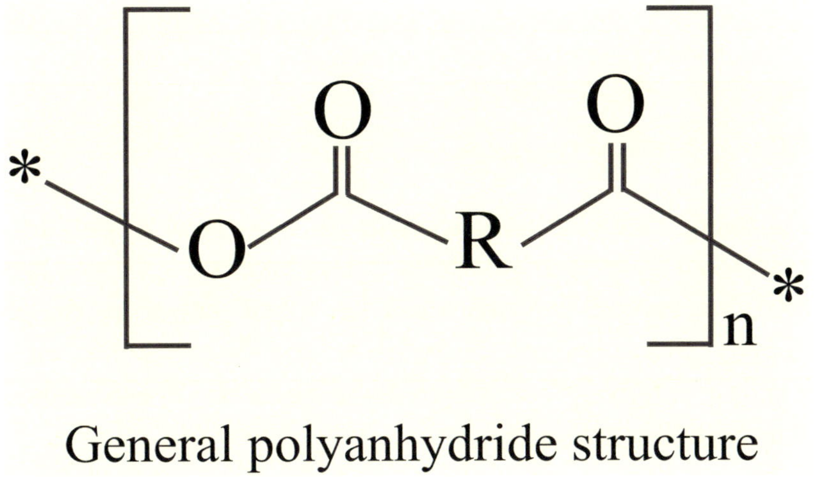

Fig. (7). Structure of a general polyanhydride.

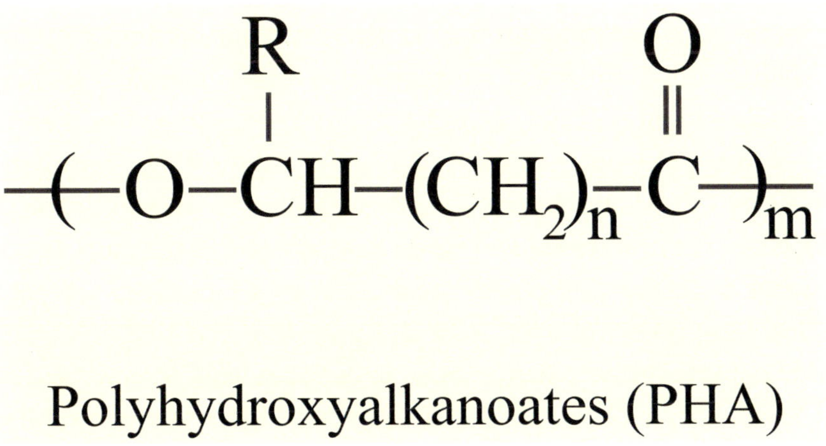

Fig. (8). Structure of polyhydroxyalkanoates (PHA).

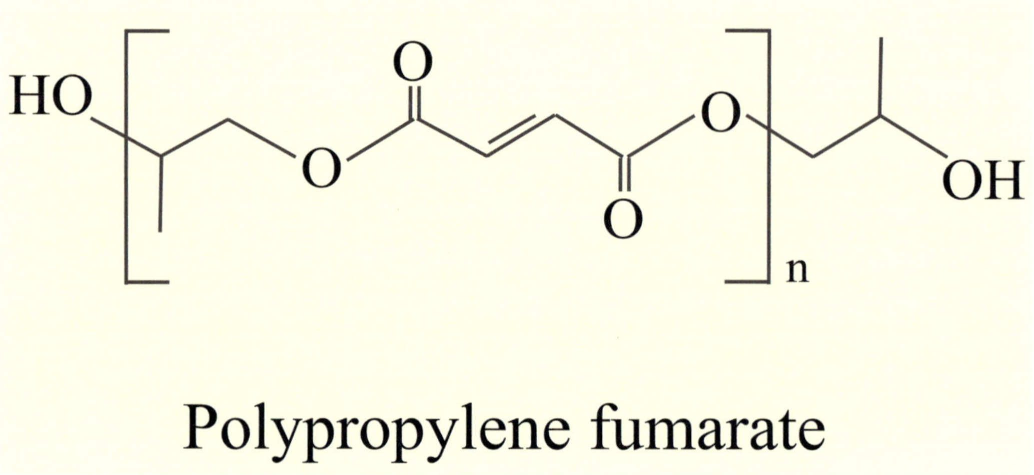

Fig. (9). Structure of polypropylene fumarate.

3.3.2. Synthesis of Polymer Scaffolds

To mimic the highly ordered microstructure of living tissues, polymer particles should be assembled in the same pattern. In this process, appropriate pore sizes can be formed, mechanical properties can be 'tuned' and a hybrid of multiple particles and cells can be obtained. There are several strategies that can be utilized to fabricate porous scaffolds from micro-/nano-particles. These methods will be summarized in the following section.

3.3.2.1. Particulate Leaching

Particulate leaching is a convenient way to product porous polymer scaffolds with particular shapes. Two different strategies are used, namely, melt molding-particulate leaching and solvent casting-particulate leaching. For melt molding-particulate leaching, the polymers are cast with water-soluble salt particles and shaped into a mold. Heat and pressure are then applied to the mixture in the mold. After the reaction, the composite is washed to dissolve the salt particles, leaving a porous polymer with the designated shape. As an alternative, solvent casting-particulate leaching does not need heat and pressure, and polymer particles are dissolved in a particularly volatile organic solvent and then mixed with salt particles. These salt particles are also washed out after the evaporation of the organic solvent.

3.3.2.2. Phase Separation

The phase separation technique is based on the thermally induced phase separation (TIPS) technique. Two strategies are proposed to induce phase separation in the polymer solution, by adding an immiscible solvent or by cooling the solution and inducing the crystallization of the solvent. Thereby, a polymer-rich phase and a polymer-lean phase are separated, leaving a porous scaffold. The characteristic of phase separation is its high porosity, up to 97% of the product. The types and concentrations of the polymers, the solvent component and the quenching temperature can be balanced to modulate pore size and morphology.

3.3.2.3. Gas Foaming

The gas foaming technique produces porous scaffolds by releasing the gas porogen generated by the polymerization reactions or by changes in heat or pressure. It takes advantages of the absence of the organic solvents that may be harmful to the cells. However, this approach requires special equipment and is limited to certain polymers.

3.3.2.4. Electro-spinning

Electro-spinning is a unique technique for the production of nanofibers with a highly ordered, aligned orientation. During fabrication, polymer particles are dissolved in the solution and supplied to a spinneret. The electric voltage is then applied to create electrostatic repulsion in the solution, which pushes a thin filament from the spinneret. These filaments are collected by a plate with an opposite charge and form a porous scaffold. The nanostructure of the scaffold can be modulated by varying the voltage, the ingredient of the solution and the distance from the spinneret to the collector. Electro-spinning technology is easy accessible and facilitates control of the scaffold architecture, which makes it a popular method in the laboratory.

3.3.2.5. Solid Free-form Fabrication

Solid free-form fabrication represents one of the most attractive techniques in materials science. It is also called rapid prototyping, based on a layer-by-layer three-dimensional manufacture. Several strategies, including fused deposition modeling, stereolithography, selective laser sintering and digital light processing, are included in this group. These techniques are always combined with CAD/CAM and enable the precisely controlled manufacture of the scaffolds. It has been reported that a cell-encapsulating scaffold can be printed with a custom-built device (Fig. **10**) [106]. Furthermore, scientists are developing scaffold-free systems which can build up vascular tissue with smooth-muscle cells, fibroblasts and agarosetemplate [107]. With the promising progress in this field, bioprinting of large-scale and fully functional tissues can likely be attained in the future.

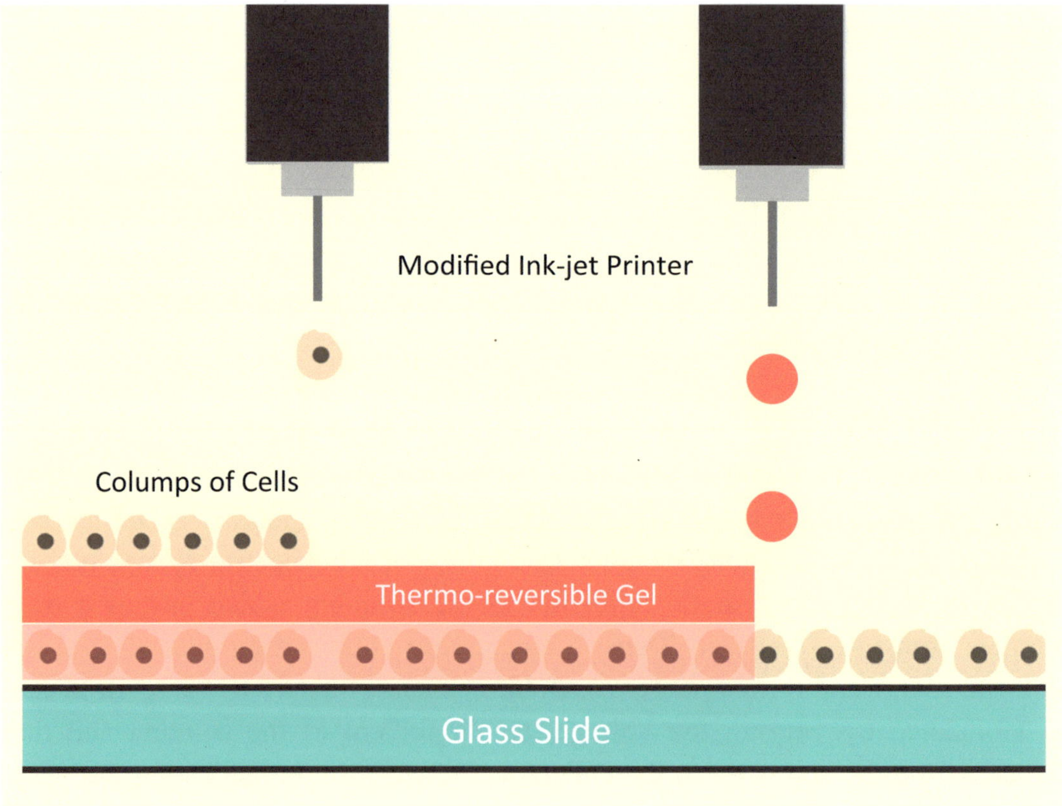

Fig. (10). Cell 3D printing.

3.3.2.6. Nanoparticle Self-assembly

In the self-assembly process, nanoparticles can organize spontaneously and form specific architecture without extra intervention. The implementation of self-assembly is initiated by interactions between the monomers, either covalent or non-covalent. The process of assembly is somehow similar to the formation of protein tertiary structures. Predominant self-assembling particles are derivations of peptides, including α-helical-forming peptides, β-sheet-and β-hairpin-forming peptides and peptide-amphiphiles [117].

The self-assembly scaffold takes advantage of skipping the process of synthesis, and encapsulated cells can be injected directly into the wounds. Further investigations are planned to focus on the acceleration of the gelation reaction and the enhancement of the mechanical properties.

CONFLICT OF INTEREST

The authors declare that they have no financial or other relationships with other people or organizations that might lead to a conflict of interest.

ACKNOWLEDGEMENTS

We express our deepest gratitude to colleagues in the Department of Stomatology, Shanghai Tenth People's Hospital, whose kind support has made this work possible. This work is financially supported by the National Natural Science Foundation of China (81500808,81201410,81470768) and the Natural Science Foundation of Shanghai (15ZR1432700).

REFERENCES

[1] Hubbell JA. Biomaterials in tissue engineering. Biotechnology (N Y) 1995; 13(6): 565-76. [PMID: 9634795]

[2] Puppi D, Chiellini F, Piras AM, *et al.* Polymeric materials for bone and cartilage repair. Prog Polym Sci 2010; 35(4): 403-40.

[3] Saha K, Pollock JF, Schaffer DV, Healy KE. Designing synthetic materials to control stem cell phenotype. Curr Opin Chem Biol 2007; 11(4): 381-7. [PMID: 17669680]

[4] Dawson E, Mapili G, Erickson K, Taqvi S, Roy K. Biomaterials for stem cell differentiation. Adv Drug Deliv Rev 2008; 60(2): 215-28. [PMID: 17997187]

[5] Zhu J. Bioactive modification of poly(ethylene glycol) hydrogels for tissue engineering. Biomaterials 2010; 31(17): 4639-56. [PMID: 20303169]

[6] Kuo YC, Yeh CF. Effect of surface-modified collagen on the adhesion, biocompatibility and

differentiation of bone marrow stromal cells in poly(lactide-co-glycolide)/chitosan scaffolds. Colloids Surf B Biointerfaces 2011; 82(2): 624-31.
[PMID: 21074381]

[7] Kohn J, Langer R. Bioresorbable and bioerodible materials. An Introduction. New York 1996; pp. 64-72.

[8] Martin C, Winet H, Bao JY. Acidity near eroding polylactide-polyglycolide *in vitro* and *in vivo* in rabbit tibial bone chambers. Biomaterials 1996; 17(24): 2373-80.
[PMID: 8982478]

[9] Place ES, George JH, Williams CK, Stevens MM. Synthetic polymer scaffolds for tissue engineering. Chem Soc Rev 2009; 38(4): 1139-51.
[PMID: 19421585]

[10] Mota C, Puppi D, Chiellini F, Chiellini E. Additive manufacturing techniques for the production of tissue engineering constructs. J Tissue Eng Regen Med 2015; 9(3): 174-90.
[PMID: 23172792]

[11] Beachley V, Wen X. Polymer nanofibrous structures: Fabrication, biofunctionalization, and cell interactions. Prog Polym Sci 2010; 35(7): 868-92.
[PMID: 20582161]

[12] Puppi D, Dinucci D, Bartoli C, *et al.* Development of 3D wet-spun polymericscaffolds loaded with antimicrobial agents for bone engineering. J Bioact Compat Polym 2011; 26(5): 478-92.

[13] Puppi D, Piras AM, Chiellini F, *et al.* Optimized electro- and wet-spinning techniques for the production of polymeric fibrous scaffolds loaded with bisphosphonate and hydroxyapatite. J Tissue Eng Regen Med 2011; 5(4): 253-63.
[PMID: 20661867]

[14] Puppi D, Detta N, Piras AM, *et al.* Development of electrospun three-arm star poly(ε-caprolactone) meshes for tissue engineering applications. Macromol Biosci 2010; 10(8): 887-97.
[PMID: 20376838]

[15] Puppi D, Piras AM, Detta N, *et al.* Poly(vinyl alcohol)-based electrospun meshes as potential candidate scaffolds in regenerative medicine. J Bioact Compat Polym 2011; 26(1): 20-34.

[16] Zhang S, Huang Y, Yang X, *et al.* Gelatin nanofibrous membrane fabricated by electrospinning of aqueous gelatin solution for guided tissue regeneration. J Biomed Mater Res A 2009; 90(3): 671-9.
[PMID: 18563816]

[17] Jayasinghe SN. Cell electrospinning: a novel tool for functionalising fibres, scaffolds and membranes with living cells and other advanced materials for regenerative biology and medicine. Analyst (Lond) 2013; 138(8): 2215-23.
[PMID: 23457706]

[18] Muerza-Cascante ML, Haylock D, Hutmacher DW, Dalton PD. Melt electrospinning and its technologization in tissue engineering. Tissue Eng Part B Rev 2015; 21(2): 187-202.
[PMID: 25341031]

[19] Rajagopal K, Schneider JP. Self-assembling peptides and proteins for nanotechnological applications. Curr Opin Struct Biol 2004; 14(4): 480-6.

[PMID: 15313243]

[20] Zhang S. Fabrication of novel biomaterials through molecular self-assembly. Nat Biotechnol 2003; 21(10): 1171-8.
[PMID: 14520402]

[21] Zhang S. Building from bottom-up. Mater Today 2003; 6: 20-7.

[22] Mao HQ, Lim SH, Zhang S, *et al.* The Nanofiber Matrix as an Artificial Stem Cell Niche. StudyMechanobiol Tissue EngBiomater 2010; 2: 89-118.

[23] Zhang S. Emerging biological materials through molecular self-assembly. Biotechnol Adv 2002; 20(5-6): 321-39.
[PMID: 14550019]

[24] Hartgerink JD, Beniash E, Stupp SI. Self-assembly and mineralization of peptide-amphiphile nanofibers. Science 2001; 294(5547): 1684-8.
[PMID: 11721046]

[25] Hosseinkhani H, Hosseinkhani M, Khademhosseini A, Kobayashi H, Tabata Y. Enhanced angiogenesis through controlled release of basic fibroblast growth factor from peptide amphiphile for tissue regeneration. Biomaterials 2006; 27(34): 5836-44.
[PMID: 16930687]

[26] Maquet V, Boccaccini AR, Pravata L, Notingher I, Jérôme R. Preparation, characterization, and *in vitro* degradation of bioresorbable and bioactive composites based on Bioglass-filled polylactide foams. J Biomed Mater Res A 2003; 66(2): 335-46.
[PMID: 12889004]

[27] Chen VJ, Smith LA, Ma PX. Bone regeneration on computer-designed nano-fibrous scaffolds. Biomaterials 2006; 27(21): 3973-9.
[PMID: 16564086]

[28] Heijkants RG, Van Tienen TG, De Groot JH, *et al.* Preparation of a polyurethane scaffold for tissue engineering made by a combination of salt leaching and freeze-drying of dioxane. J Mater Sci 2006; 41(8): 2423-8.

[29] Rowlands AS, Lim SA, Martin D, Cooper-White JJ. Polyurethane/poly(lactic-co-glycolic) acid composite scaffolds fabricated by thermally induced phase separation. Biomaterials 2007; 28(12): 2109-21.
[PMID: 17258315]

[30] Ma PX. Biomimetic materials for tissue engineering. Adv Drug Deliv Rev 2008; 60(2): 184-98.
[PMID: 18045729]

[31] Niu X, Feng Q, Wang M, Guo X, Zheng Q. Porous nano-HA/collagen/PLLA scaffold containing chitosan microspheres for controlled delivery of synthetic peptide derived from BMP-2. J Control Release 2009; 134(2): 111-7.
[PMID: 19100794]

[32] Huang YX, Ren J, Chen C, *et al.* Preparation and Properties of Poly(lactide-co-glycolide) (PLGA)/Nano-Hydroxyapatite (NHA)Scaffolds by Thermally Induced Phase Separation and Rabbit MSCs Culture on Scaffolds. J biomarerials app 2008; 22(2): 409-32.

[33] Xue D, Zheng Q, Zong C, *et al.* Osteochondral repair using porous poly(lactide-co-glycolide)/nao-hydroxyapatite hybrid scaffolds with undifferentiated mesenchymal stem cells in a rat model. J Biomed Mater Res A 2010; 94(1): 259-70.
[PMID: 20166224]

[34] Cui X, Boland T, D'Lima DD, Lotz MK. Thermal inkjet printing in tissue engineering and regenerative medicine. Recent Pat Drug Deliv Formul 2012; 6(2): 149-55.
[PMID: 22436025]

[35] Campbell PG, Miller ED, Fisher GW, Walker LM, Weiss LE. Engineered spatial patterns of FGF-2 immobilized on fibrin direct cell organization. Biomaterials 2005; 26(33): 6762-70.
[PMID: 15941581]

[36] Chen CS, Mrksich M, Huang S, Whitesides GM, Ingber DE. Geometric control of cell life and death. Science 1997; 276(5317): 1425-8.
[PMID: 9162012]

[37] Gao GF, Schilling A, Hubbell K, *et al.* Improved properties of bone and cartilage tissue from 3Dinkjet-bioprinted human mesenchymal stem cells by simultaneous deposition and photocrosslinking in PEG-GelMA. Biotechnol Lett 2015.

[38] Song W, Lu H, Kawazoe N, Chen G. Adipogenic differentiation of individual mesenchymal stem cell on different geometric micropatterns. Langmuir 2011; 27(10): 6155-62.
[PMID: 21486006]

[39] Lee HJ, Koh WG. Hydrogel micropattern-incorporated fibrous scaffolds capable of sequential growth factor delivery for enhanced osteogenesis of hMSCs. ACS Appl Mater Interfaces 2014; 6(12): 9338-48.
[PMID: 24915062]

[40] Samorezov JE, Morlock CM, Alsberg E. Dual ionic and photo crosslinked alginate hydrogels for micropatterned spatial control of material properties and cell behavior. Bioconjug Chem 2015; 26(7): 1339-47.
[PMID: 25799217]

[41] Hollister SJ. Scaffold design and manufacturing: from concept to clinic. Adv Mater 2009; 21(32-33): 3330-42.
[PMID: 20882500]

[42] Lee J, Abdeen AA, Huang TH, Kilian KA. Controlling cell geometry on substrates of variable stiffness can tune the degree of osteogenesis in human mesenchymal stem cells. J Mech Behav Biomed Mater 2014; 38: 209-18.
[PMID: 24556045]

[43] Elisseeff J, Ferran A, Hwang S, Varghese S, Zhang Z. The role of biomaterials in stem cell differentiation: applications in the musculoskeletal system. Stem Cells Dev 2006; 15(3): 295-303.
[PMID: 16846368]

[44] Ayala R, Zhang C, Yang D, *et al.* Engineering the cell-material interface for controlling stem cell adhesion, migration, and differentiation. Biomaterials 2011; 32(15): 3700-11.
[PMID: 21396708]

[45] Meinel L, Karageorgiou V, Fajardo R, *et al.* Bone tissue engineering using human mesenchymal stem cells: effects of scaffold material and medium flow. Ann Biomed Eng 2004; 32(1): 112-22.
[PMID: 14964727]

[46] Ravichandran R, Venugopal JR, Sundarrajan S, Mukherjee S, Ramakrishna S. Precipitation of nanohydroxyapatite on PLLA/PBLG/Collagen nanofibrous structures for the differentiation of adipose derived stem cells to osteogenic lineage. Biomaterials 2012; 33(3): 846-55.
[PMID: 22048006]

[47] Yim EK, Pang SW, Leong KW. Synthetic nanostructures inducing differentiation of human mesenchymal stem cells into neuronal lineage. Exp Cell Res 2007; 313(9): 1820-9.
[PMID: 17428465]

[48] Christopherson GT, Song H, Mao HQ. The influence of fiber diameter of electrospun substrates on neural stem cell differentiation and proliferation. Biomaterials 2009; 30(4): 556-64.
[PMID: 18977025]

[49] Prabhakaran MP, Venugopal J. Stem cells and nanostructures for advanced tissue regeneration. Biomedical applications of polymeric nanofibers 2012; 246: 21-62.

[50] Jiang X, Cao HQ, Shi LY, Ng SY, Stanton LW, Chew SY. Nanofiber topography and sustained biochemical signaling enhance human mesenchymal stem cell neural commitment. Acta Biomater 2012; 8(3): 1290-302.
[PMID: 22154861]

[51] Kaigler D, Cirelli JA, Giannobile WV. Growth factor delivery for oral and periodontal tissue engineering. Expert Opin Drug Deliv 2006; 3(5): 647-62.
[PMID: 16948560]

[52] Chen FM, Shelton RM, Jin Y, Chapple IL. Localized delivery of growth factors for periodontal tissue regeneration: role, strategies, and perspectives. Med Res Rev 2009; 29(3): 472-513.
[PMID: 19260070]

[53] Vasita R, Katti DS. Growth factor-delivery systems for tissue engineering: a materials perspective. Expert Rev Med Devices 2006; 3(1): 29-47.
[PMID: 16359251]

[54] Evans CH, Ghivizzani SC, Robbins PD. Orthopedic gene therapy--lost in translation? J Cell Physiol 2012; 227(2): 416-20.
[PMID: 21948071]

[55] Hetrick EM, Schoenfisch MH. Reducing implant-related infections: active release strategies. Chem Soc Rev 2006; 35(9): 780-9.
[PMID: 16936926]

[56] Mouriño V, Boccaccini AR. Bone tissue engineering therapeutics: controlled drug delivery in three-dimensional scaffolds. J R Soc Interface 2010; 7(43): 209-27.
[PMID: 19864265]

[57] Abarrategi A, Moreno-Vicente C, Martínez-Vázquez FJ, *et al.* Biological properties of solid free form designed ceramic scaffolds with BMP-2: *in vitro* and *in vivo* evaluation. PLoS One 2012; 7(3): e34117.

[PMID: 22470527]

[58] Cai P, Xue ZY, Qi W, *et al.* Adsorbed BMP-2 in polyelectrolyte multilayer films for enhanced early osteogenic differentiation of mesenchymal stem cells. Colloids and Surfaces A: Physicochem Eng 2013; 434(19): 110-7.

[59] Jabbari E. Bioconjugation of hydrogels for tissue engineering. Curr Opin Biotechnol 2011; 22(5): 655-60.
[PMID: 21306888]

[60] Zhang S, Uludağ H. Nanoparticulate systems for growth factor delivery. Pharm Res 2009; 26(7): 1561-80.
[PMID: 19415467]

[61] Wang X, Wenk E, Zhang X, Meinel L, Vunjak-Novakovic G, Kaplan DL. Growth factor gradients *via* microsphere delivery in biopolymer scaffolds for osteochondral tissue engineering. J Control Release 2009; 134(2): 81-90.
[PMID: 19071168]

[62] Elisseeff J, McIntosh W, Fu K, Blunk BT, Langer R. Controlled-release of IGF-I and TGF-β1 in a photopolymerizing hydrogel for cartilage tissue engineering. J Orthop Res 2001; 19(6): 1098-104.
[PMID: 11781011]

[63] Wei G, Jin Q, Giannobile WV, Ma PX. The enhancement of osteogenesis by nano-fibrous scaffolds incorporating rhBMP-7 nanospheres. Biomaterials 2007; 28(12): 2087-96.
[PMID: 17239946]

[64] Jin Q, Wei G, Lin Z, *et al.* Nanofibrous scaffolds incorporating PDGF-BB microspheres induce chemokine expression and tissue neogenesis *in vivo*. PloS One 2008; 3(3): e1729.

[65] Ennett AB, Kaigler D, Mooney DJ. Temporally regulated delivery of VEGF *in vitro* and *in vivo*. J Biomed Mater Res A 2006; 79(1): 176-84.
[PMID: 16788907]

[66] Brieke C, Rohrbach F, Gottschalk A, Mayer G, Heckel A. Light-controlled tools. Angew Chem Int Ed Engl 2012; 51(34): 8446-76.
[PMID: 22829531]

[67] Zisch AH, Lutolf MP, Ehrbar M, *et al.* Cell-demanded release of VEGF from synthetic, biointeractive cell ingrowth matrices for vascularized tissue growth. FASEB J 2003; 17(15): 2260-2.
[PMID: 14563693]

[68] Gala-Garcia A, Teixeira KIR, Wykrota FHL, Sinisterra RD, Cortés ME. Bioceramic/poly (glycolic)-poly (lactic acid) composite induces mineralized barrier after direct capping of rat tooth pulp tissue. Brazilian oral research 2010; 24(1): 08-14.

[69] Bettinger CJ, Langer R, Borenstein JT. Engineering substrate topography at the micro- and nanoscale to control cell function. Angew Chem Int Ed Engl 2009; 48(30): 5406-15.
[PMID: 19492373]

[70] Kolambkar YM, Bajin M, Wojtowicz A, Hutmacher DW, García AJ, Guldberg RE. Nanofiber orientation and surface functionalization modulate human mesenchymal stem cell behavior *in vitro*. Tissue Eng Part A 2014; 20(1-2): 398-409.

[PMID: 24020454]

[71] Bashur CA, Dahlgren LA, Goldstein AS. Effect of fiber diameter and orientation on fibroblast morphology and proliferation on electrospun poly(D,L-lactic-co-glycolic acid) meshes. Biomaterials 2006; 27(33): 5681-8.
[PMID: 16914196]

[72] Yang F, Murugan R, Wang S, Ramakrishna S. Electrospinning of nano/micro scale poly(L-lactic acid) aligned fibers and their potential in neural tissue engineering. Biomaterials 2005; 26(15): 2603-10.
[PMID: 15585263]

[73] Dickinson LE, Rand DR, Tsao J, Eberle W, Gerecht S. Endothelial cell responses to micropillar substrates of varying dimensions and stiffness. J Biomed Mater Res A 2012; 100(6): 1457-66.
[PMID: 22389314]

[74] Benya PD, Shaffer JD. Dedifferentiated chondrocytes reexpress the differentiated collagen phenotype when cultured in agarose gels. Cell 1982; 30(1): 215-24.
[PMID: 7127471]

[75] Dharmavaram RM, Liu G, Tuan RS, Stokes DG, Jiménez SA. Stable transfection of human fetal chondrocytes with a type II procollagen minigene: expression of the mutant protein and alterations in the structure of the extracellular matrix *in vitro*. Arthritis Rheum 1999; 42(7): 1433-42.
[PMID: 10403271]

[76] Gemmiti CV, Guldberg RE. Fluid flow increases type II collagen deposition and tensile mechanical properties in bioreactor-grown tissue-engineered cartilage. Tissue Eng 2006; 12(3): 469-79.
[PMID: 16579680]

[77] Guilak F, Fermor B, Keefe FJ, *et al.* The role of biomechanics and inflammation in cartilage injury and repair. Clin Orthop Relat Res 2004; 423(423): 17-26.
[PMID: 15232421]

[78] Heimbach DM, Warden GD, Luterman A, *et al.* Multicenter postapproval clinical trial of Integra dermal regeneration template for burn treatment. J Burn Care Rehabil 2003; 24(1): 42-8.
[PMID: 12543990]

[79] Jeschke MG, Rose C, Angele P, Füchtmeier B, Nerlich MN, Bolder U. Development of new reconstructive techniques: use of Integra in combination with fibrin glue and negative-pressure therapy for reconstruction of acute and chronic wounds. Plast Reconstr Surg 2004; 113(2): 525-30.
[PMID: 14758212]

[80] Yao M, Attalla K, Ren Y, French MA, Driver VR. Ease of use, safety, and efficacy of integra bilayer wound matrix in the treatment of diabetic foot ulcers in an outpatient clinical setting: a prospective pilot study. J Am Podiatr Med Assoc 2013; 103(4): 274-80.
[PMID: 23878379]

[81] Lagus H, Sarlomo-Rikala M, Böhling T, Vuola J. Prospective study on burns treated with Integra®, a cellulose sponge and split thickness skin graft: comparative clinical and histological study--randomized controlled trial. Burns 2013; 39(8): 1577-87.
[PMID: 23880091]

[82] Guptaa B, Agarwal R, Alan M. Textile-based smart wound dressings. Indian J Fibre. 2010; pp. 174-87.

[83] Ikeda E, Morita R, Nakao K, *et al.* Fully functional bioengineered tooth replacement as an organ replacement therapy. Proc Natl Acad Sci USA 2009; 106(32): 13475-80.
[PMID: 19666587]

[84] Nakao K, Morita R, Saji Y, *et al.* The development of a bioengineered organ germ method. Nat Methods 2007; 4(3): 227-30.
[PMID: 17322892]

[85] Yoon DM, Fisher JP. Chondrocyte signaling and artificial matrices for articular cartilage engineering. Tissue engineering. Springer 2007; pp. 67-86.

[86] Yoon DM, Fisher JP. Effects of exogenous IGF-1 delivery on the early expression of IGF-1 signaling molecules by alginate embedded chondrocytes. Tissue Eng Part A 2008; 14(7): 1263-73.
[PMID: 18491951]

[87] Ma PX. Biomimetic materials for tissue engineering. Adv Drug Deliv Rev 2008; 60(2): 184-98.
[PMID: 18045729]

[88] Burg KJ, Porter S, Kellam JF. Biomaterial developments for bone tissue engineering. Biomaterials 2000; 21(23): 2347-59.
[PMID: 11055282]

[89] Lin Z, Solomon KL, Zhang X, *et al. In vitro* evaluation of natural marine sponge collagen as a scaffold for bone tissue engineering. Int J Biol Sci 2011; 7(7): 968-77.
[PMID: 21850206]

[90] Datta N, Holtorf HL, Sikavitsas VI, Jansen JA, Mikos AG. Effect of bone extracellular matrix synthesized *in vitro* on the osteoblastic differentiation of marrow stromal cells. Biomaterials 2005; 26(9): 971-7.
[PMID: 15369685]

[91] Zhou H, Lee J. Nanoscale hydroxyapatite particles for bone tissue engineering. Acta Biomater 2011; 7(7): 2769-81.
[PMID: 21440094]

[92] Wang T, Chow LC, Frukhtbeyn SA, *et al.* Improve the Strength of PLA/HA Composite Through the Use of Surface Initiated Polymerization and Phosphonic Acid Coupling Agent. J Res Natl Inst Stand Technol 2011; 116(5): 785-96.
[PMID: 22399838]

[93] Gonçalves EM, Oliveira FJ, Silva RF, *et al.* Three-dimensional printed PCL-hydroxyapatite scaffolds filled with CNTs for bone cell growth stimulation. J Biomed Mater Res B Appl Biomater 2015.
[PMID: 26089195]

[94] Kumar R, Prakash KH, Cheang P, Gower L, Khor KA. Chitosan-mediated crystallization and assembly of hydroxyapatite nanoparticles into hybrid nanostructured films. J R Soc Interface 2008; 5(21): 427-39.
[PMID: 17698476]

[95] Huang JW, Lin SS, Chen LH, *et al.* The use of fluorescence-labeled mesenchymal stem cells in poly(lactide-co-glycolide)/hydroxyapatite/collagen hybrid graft as a bone substitute for posterolateral spinal fusion. J Trauma 2011; 70(6): 1495-502.

[PMID: 21817989]

[96] Zuo Y, Liu X, Wei D, *et al.* Photo-cross-linkable methacrylated gelatin and hydroxyapatite hybrid hydrogel for modularly engineering biomimetic osteon. ACS Appl Mater Interfaces 2015; 7(19): 10386-94.
[PMID: 25928732]

[97] Yang W, Both SK, van Osch GJ, Wang Y, Jansen JA, Yang F. Performance of different three-dimensional scaffolds for in vivo endochondral bone generation. Eur Cell Mater 2014; 27: 350-64.
[PMID: 24913441]

[98] Seol YJ, Park JY, Jung JW, *et al.* Improvement of bone regeneration capability of ceramic scaffolds by accelerated release of their calcium ions. Tissue Eng Part A 2014; 20(21-22): 2840-9.
[PMID: 24784792]

[99] Nasu T, Takemoto M, Akiyama N, Fujibayashi S, Neo M, Nakamura T. EP4 agonist accelerates osteoinduction and degradation of β-tricalcium phosphate by stimulating osteoclastogenesis. J Biomed Mater Res A 2009; 89(3): 601-8.
[PMID: 18437696]

[100] Fukuda C, Akiyama N, Takemoto M, *et al.* Local application of alendronate on β-tricalcium phosphate accelerated induction of osteogenesis with formation of giant osteoclast-like cell. J Biomater Nanobiotechnol 2012; 3(2): 169-77.

[101] Tsukanaka M, Fujibayashi S, Otsuki B, Takemoto M, Matsuda S. Osteoinductive potential of highly purified porous β-TCP in mice. J Mater Sci Mater Med 2015; 26(3): 132.
[PMID: 25698341]

[102] Steinert A, Weber M, Dimmler A, *et al.* Chondrogenic differentiation of mesenchymal progenitor cells encapsulated in ultrahigh-viscosity alginate. J Orthop Res 2003; 21(6): 1090-7.
[PMID: 14554223]

[103] Weber M, Steinert A, Jork A, *et al.* Formation of cartilage matrix proteins by BMP-transfected murine mesenchymal stem cells encapsulated in a novel class of alginates. Biomaterials 2002; 23(9): 2003-13.
[PMID: 11996042]

[104] Oh SH, Park IK, Kim JM, Lee JH. *In vitro* and *in vivo* characteristics of PCL scaffolds with pore size gradient fabricated by a centrifugation method. Biomaterials 2007; 28(9): 1664-71.
[PMID: 17196648]

[105] Ma PX, Choi JW. Biodegradable polymer scaffolds with well-defined interconnected spherical pore network. Tissue Eng 2001; 7(1): 23-33.
[PMID: 11224921]

[106] Kolesky DB, Truby RL, Gladman AS, Busbee TA, Homan KA, Lewis JA. 3D bioprinting of vascularized, heterogeneous cell-laden tissue constructs. Adv Mater 2014; 26(19): 3124-30.
[PMID: 24550124]

[107] Norotte C, Marga FS, Niklason LE, Forgacs G. Scaffold-free vascular tissue engineering using bioprinting. Biomaterials 2009; 30(30): 5910-7.
[PMID: 19664819]

[108] Kim JI, Kim Y, Kwon DY, *et al.* An injectable biodegradable temperature-responsive gel with an

adjustable persistence window. Biomaterials 2012; 33(10): 2823-34.
[PMID: 22261098]

[109] Shim JH, Yoon MC, Jeong CM, *et al.* Efficacy of rhBMP-2 loaded PCL/PLGA/β-TCP guided bone regeneration membrane fabricated by 3D printing technology for reconstruction of calvaria defects in rabbit. Biomed Mater 2014; 9(6): 065006.
[PMID: 25384105]

[110] Jensen J, Kraft DC, Lysdahl H, *et al.* Functionalization of polycaprolactone scaffolds with hyaluronic acid and β-TCP facilitates migration and osteogenic differentiation of human dental pulp stem cells in vitro. Tissue Eng Part A 2015; 21(3-4): 729-39.
[PMID: 25252795]

[111] Delgado-Rivera R, Rosario-Meléndez R, Yu W, Uhrich KE. Biodegradable salicylate-based poly(anhydride-ester) microspheres for controlled insulin delivery. J Biomed Mater Res A 2014; 102(8): 2736-42.
[PMID: 24027012]

[112] Sun M, Zhou P, Pan LF, Liu S, Yang HX. Enhanced cell affinity of the silk fibroin- modified PHBHHx material. J Mater Sci Mater Med 2009; 20(8): 1743-51.
[PMID: 19333570]

[113] Wu S, Liu YL, Cui B, Qu XH, Chen GQ. Study on decellularized porcine aortic valve/poly (3-hydroxybutyrate-co-3-hydroxyhexanoate) hybrid heart valve in sheep model. Artif Organs 2007; 31(9): 689-97.
[PMID: 17725696]

[114] Ye C, Hu P, Ma MX, Xiang Y, Liu RG, Shang XW. PHB/PHBHHx scaffolds and human adipose-derived stem cells for cartilage tissue engineering. Biomaterials 2009; 30(26): 4401-6.
[PMID: 19481254]

[115] Deng Y, Lin XS, Zheng Z, *et al.* Poly(hydroxybutyrate-co-hydroxyhexanoate) promoted production of extracellular matrix of articular cartilage chondrocytes *in vitro.* Biomaterials 2003; 24(23): 4273-81.
[PMID: 12853259]

[116] Nair LS, Laurencin CT. Polymers as biomaterials for tissue engineering and controlled drug delivery. Adv Biochem Eng Biotechnol 2006; 102: 47-90.
[PMID: 17089786]

[117] Wang J, Zheng J, Zheng Q, *et al.* FGL-functionalized self-assembling nanofiber hydrogel as a scaffold for spinal cord-derived neural stem cells. Mater Sci Eng C Mater Biol Appl 2015; 46(46): 140-7.
[PMID: 25491970]

CHAPTER 7

Tooth Regeneration: Dentin Regeneration, Periodontal Ligament Regeneration, Dental Pulp Regeneration

Guo Li, Na Fu, Xueqin Wei, Tao Zhang, Xiaoru Shao, Qian Huang, Ke Sun, Shiyu Lin, Sirong Shi and **Yunfeng Lin**[*]

State Key Laboratory of Oral Diseases, West China Hospital of Stomatology, Sichuan University, Chengdu, China

Abstract: A tooth is a compound organ, which is composed of calcified tissues of enamel, dentin and cementum, and a soft connective tissue of dental pulp in which blood vessels and nerves are protected. Periodontal ligaments anchor teeth into the alveolar bone in the jaw to ensure the proper function of teeth. In humans, tooth loss can not only lead to physical and mental suffering, but also affect the aesthetics, which compromise an individual's quality of life and self-esteem. With the development of tissue engineering, regenerating a whole tooth for clinical tooth replacement is now considered to be an acceptable scientific objective. This subject has a number of challenges for investigators over complicated disciplines including biology, dental medicine and biomaterial science. This chapter will summarize the current knowledge related to tooth regeneration, especially focus on dentin regeneration, periodontal ligament regeneration, and dental pulp regeneration.

Keywords: Growth factors, Regulation mechanism, Scaffolds, Stem cells, Tooth regeneration.

1. INTRODUCTION

"A tooth is a compound organ, which is composed of calcified tissues of enamel, dentin and cementum, and a soft connective tissue of dental pulp in which blood

[*] **Corresponding author Yunfeng Lin:** State Key Laboratory of Oral Diseases, West China Hospital of Stomatology, Sichuan University, Chengdu, P. R. China; Tel/Fax: 86-28-85503487; E-mail: yunfenglin@scu.edu.cn

Jing Wang & YunFeng Lin (Eds.)

vessels and nerves are protected". Periodontal ligaments anchor teeth into the alveolar bone in the jaw to ensure the proper function of teeth [1, 2]. Teeth are formed by epithelial–mesenchymal interactions [3]. The mesenchyme originates from neural crest cells and the epithelium originates from pharynx or stomodeum. During early development, the ectomesenchyme of the first brachial arch and the frontonasal process is formed by the lateral and anterior migration of cranial neural crest cells [4]. "Because the interactions between cells and tissues constitute a central mechanism regulating the development of all multicellular organisms, the development of a tooth is a very similar process to that for other skin derivatives such as feathers, scales and hairs" [2]. A thickening of the oral epithelium is the first morphological sign of tooth development. And it yields the vestibular lamina on the vestibular side and the dental lamina on the lingual side. The dental lamina induces the teeth and the vestibular lamina forms a groove between the cheek and the teeth. The dental lamina proliferates greatly in a specific location corresponding to the future teeth. And then oval or round structures known as placodes invade the mesenchyme, and will form the tooth buds. Mesenchyme will start to condense around the buds after the penetration of the buds into the under lying mesenchyme. The mechanism of cell growth and morphogenesis in the following cap stage and bell stage is very self-controlled and complicated, and gives rise to the establishment of the form of the tooth crown in the exact location (Fig. **1**). The dental papilla will be formed by the mesenchymal cells, and eventually will differentiate and form the pulp tissue and the odontoblasts. The enamel-forming ameloblasts and the dentine-forming odontoblasts differentiate terminally during the bell stage of tooth development and both of them are unique to teeth. "This takes place at the interface of the epithelium and mesenchyme and is regulated by interactions between the two tissues" [5 - 7]. Dentin, a bone-like hard tissue, is mineralized by collagenous extracellular matrix which is secreted by the odontoblasts. The mineralization of the enamel, which is the hardest tissue in the body, is directed by the enamel matrix deposited by the ameloblasts. The periodontal ligament including cementoblasts will be generated by the dental follicle that is formed by enamel organ, the outer cells of the condensed mesenchyme around the epithelium. The roots of the teeth develop after crown morphogenesis, and subsequently the teeth erupt into the oral cavity [8].

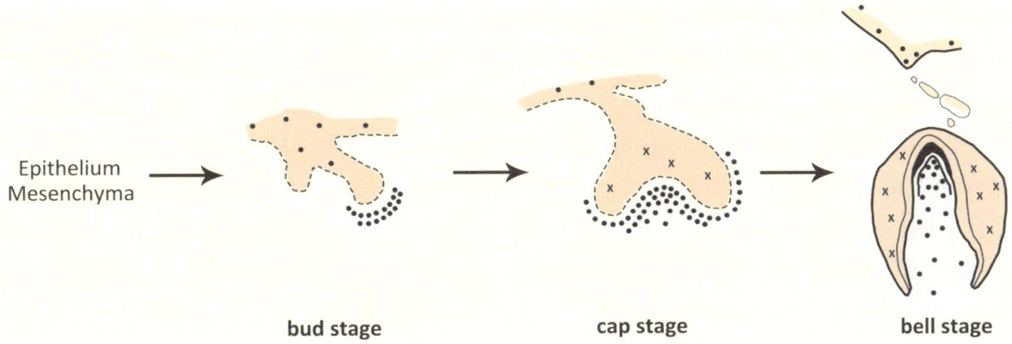

Fig. (1). The stage of tooth.

In humans, tooth loss cannot only lead to physical and mental suffering, but also affect the aesthetics, which compromise an individual's quality of life and self-esteem [4].There are several dental diseases that can eventually give rise to tooth loss. "Dental caries, known as tooth decay or cavity, is one of the most common disorders in humans, second only to common cold. It is an infectious disease primarily caused by bacterial colonies, which breakdown hard tissues of the tooth such as enamel and dentin, as well as soft tissue of the tooth known as dental pulp". Periodontal disease is another major cause for tooth loss. Bacterial plaque damages the periodontal tissues around the teeth and makes it unstable. Resection of oro-maxillo-facial tumors may include the extraction of teeth. In myriads craniofacial anomalies such as cleft palate, teeth can be congenitally missing [5 - 8]. Besides diseases, facial trauma may also lead to tooth loss in both children and adults. In clinical work, it is a great challenge for contemporary dentistry or stomatology to restore dentition defect or deficiency. Nowadays, dentures or dental implants are the most usual ways to restore missing teeth in contemporary dentistry or stomatology. Comparing with dentures, dental implants do not need to cause any damages of the consecutive tooth, and become favorite choices in many countries. However, it is too expensive for a large segment of the population to afford dental implants. And it can fail and will not adapt with surrounding bone that necessarily remodels throughout life, despite being the currently preferred treatment modality [9].

With the development of tissue engineering, regenerating a whole tooth for clinical tooth replacement is now considered to be an acceptable scientific objective. The constant knowledge of tooth developmental biology, progress in the understanding of the tooth biomineralization principles, developments in stem cells research, and practice in tissue engineering approaches and biomimetic technology have provided a solid foundation for such efforts. This subject has a number of challenges for investigators over complicated disciplines including biology, dental medicine and biomaterial science. Among them, there are three challenging areas of research, according to tissue engineering principles, attracting more and more attention: (1) how to identify, isolate and manipulate the suitable cells or stem cells to be used to regenerate the tissues or the whole organ; (2) how to identify, characterize and apply the suitable signaling molecules to be used for cell proliferation, differentiation and organogenesis and (3)how to design, develop and characterize the suitable carriers or scaffolds to be used to load the cells and/or signaling molecules, advance cell growth and facilitate tissue formation and organogenesis [10 - 12]. This chapter will summarize the current knowledge related to tooth regeneration, especially focus on dentin regeneration, periodontal ligament regeneration, and dental pulp regeneration.

2. DENTIN REGENERATION

2.1. The Stem Cells Used in Dentin Regeneration

Stem cell is a kind of multiple potential cell with self-renewing ability. Under certain conditions, it can differentiate into cells with different and more restricted functions. According to the developmental stages of stem cell, it can be divided into two basic categories: embryonic stem cells (ESC) and somatic stem cells, also called mesenchymal stem cells (MSC). For the limitation of ethical issues by using ESC, MSC becomes a more popular cellular source to be used in tissue engineering. So far, MSC have been isolated from several tissues, including brain, skin, hair follicles, skeletal muscle, bone marrow, adipose tissue and dental tissue.

2.1.1. Dental Pulp Stem Cells

Dental pulp stem cells (DPSCs), which are isolated from the dental pulp, can regenerate new stem cells or differentiate into functioning cells on the basis of specific signals from their environment [121]. Several different progenitor cell subpopulations exist in the dental pulp. And the proliferation rate, self-renewal ability, and differentiation potential are different among them [13]. Dental pulp can be gained from third molars or anterior molars extracted for the need of orthodontic, and pulpectomized teeth left *in situ*. Even after temporary storage in liquid nitrogen, DPSCs still keep their multipotent differentiation ability [14]. *In vitro*, DPSCs can differentiate to several cells, such as odontoblasts, osteoblasts, endothelocytes, smooth muscle cells, adipocytes, chondrocytes, and neurons. Because DPSCs have the differentiation ability to odontoblasts, they can be used directly for dental therapy [15]. "The regeneration ability of pulp-dentin complex is achieved by formation of tertiary dentin, reactionary dentin and reparative dentin. There are two different methods applied in regeneration of dentin by the use of tissue engineering techniques. The first method comprises a device that can be adopted as a filling material into a deep cavity of tooth with local layer of dentin on top of the pulp. In this procedure, some growth factors or molecules are used to form reparative dentin". The second method includes a scaffold which has odontoblast-like cells or stem cells such as DPSCs on it, putting on open pulp. These cells will form reparative dentin [15]. The differentiation ability of DPSCs is limited *in vitro*, while they can regenerate more complex tissues *in vivo*. *In vitro*, DPSCs differentiate into osteoblasts producing living autologous fibrous bone tissue [16]. However, *in vivo*, DPSCs can form bone tissue with Haversian canals and osteocytes [16 - 18], and dentin-pulp-like tissue complexes [19, 20]. Besides, many other differentiation directions were investigated in animal experiments.

2.1.2. Stem Cells from Human Exfoliated Deciduous Teeth (SHEDs)

It is relatively easy to get stem cells from human exfoliated deciduous teeth. SHEDs can be isolated from the coronal pulp of exfoliated deciduous teeth. It is found that SHEDs not only play an important role in the eruption of permanent teeth, but also have some effects on the osteogenesis associated [21]. *In vitro* and *in vivo*, different induce factors can make them differentiate into odontoblasts, osteoblasts, adipocyte, chondrocytes, or neurons.

2.1.3. Stem Cells from the Dental Apical Papilla (SCAPs)

"SCAPs are stem cells isolated from the apical part of dental apical papilla, which is the precursor tissue of the dental pulp." SCAPs can be obtained from impacted third molars. *In vitro*, SCAPs can differentiate odontogenically, osteogenically, and adipogenically. *In vivo*, it has been found to differentiate into odontonblasts and osteoblasts.

2.1.4. Non-dental Stem Cells

In order to make odontogenesis possible in a minimally invasive manner, researchers have been moving to find an extraoral, easily accessible source of stem cells. Ohazama *et al.* found that embryonic oral epithelium can induce odontogenic differentiation in mesenchyme which does not have a dental origin [22]. Bone marrow derived mesenchymal stem cells (BMSCs) have the self-replication ability and can be differentiated into several kinds of functioning cells. Co-culture Embryonic oral epithelium and BMSCs could induce the cells to express odontogenic genes. And tooth-like tissues with bone were observed *in vivo* [22]. Ye L *et al.* evaluated the odontogenic differentiation potential of BMSCs and adipose-derived stem cells (ADSCs), two non-dental-derived stem cells, both *in vitro* and *in vivo*. The results showed that both ADSCs and BMSCs had odontogenic differentiation potency. While, comparing with ADSCs, the odontogenic potential of BMSCs was greater, proving that BMSCs are a more suitable cell source for dentin regeneration [23].

Chen Y *et al.* have found that human umbilical cord mesenchymal stem cells (hUCMSCs) have an odontogenic differentiation potency to differentiate to odontoblast-like cells with characteristic deposition of dentin-like matrix *in vivo* [24].

Follicle dermal papilla mesenchymal cells (FDPMCs) are stem cells from hair follicles. For easily accessible isolation, FDPMCs attract researchers' attention. It is reported that FDPMCs from mouse tactile hairs had been isolated and differentiated to odontoblast-like cells *in vitro* [25].

2.2. The Scaffolds Used in Dentin Regeneration

The extracellular matrix (ECM) of dentin provides a scaffold for adhesion, proliferation and differentiation of odontoblasts. Scaffold should simulate dentin

ECM and permit effective transportation of oxygen, nutrients and metabolic waste. The scaffold should have favorable biodegradation, and eventually be taken the place by regenerated tissue matrix while retaining the morphological feature of the final tissue architecture and organization. The scaffold also should have biocompatibility, non-toxicity and optimal physical characteristics and mechanical properties. It can carry growth/differentiation factors and molecules to support morphogenetic activities.

In dentin regeneration, the scaffolds suited for dentin ogenesis should benefit to form reparative and regenerative dentin. The materials used to fabricate scaffolds are usually natural polymers and synthetic polymers. Both of them have their own advantages. The natural polymers, such as collagen and fibronectin, process good bioactivity and cytocompatibility, and the synthetic polymers have better control of the physic-chemical properties such as porosity, microstructure, degradation rate and mechanical properties [26]. The calcium hydroxide pulpcapping can form collagenous matrix of osteodentin, which stimulates the differentiation of odontoblasts [27]. The implantation of dentin matrix in the dental pulp can form tubular matrix lined with odontoblasts [28 - 30]. The scaffold of rhBMP2 with 4Mguanidine choloride extracted inactivated de-mineralized dentin matrix induced tubular dentin formation in the cavity on the exposed pulp. But the scaffold of a tendon collagen did not stimulate tubular formation [31, 32]. Thus, the insoluble dentin matrix seems to play a key role for pulp cells to attach and differentiate into odontoblasts in dentin regeneration. Fibronectin regulates the interaction of these substrates and pulp cells, and rearrange the cytoskeleton in odontoblast polarization [33]. Collagen fibers in the intercellular space of odontoblasts may assist to conduct the migration, adhesion of preodontoblasts and arrangement and repair of dentin [34]. Some synthetic scaffolds can be applied to pulp regeneration [35 - 37]. Calcium alginate hydrogel has an environment resembling ECM, and the hydration properties are good for wound healing. It can be as a carrier of growth factors, such as TGFb1, to promote cell activities and improve pulp regeneration [38].Concerning about the effect of 3D environment on cell behavior, the design of scaffolds should be improved to guide the dentin regeneration [39].

2.3. The Cytokines/growth Factors Used in Dentin Regeneration

The Transforming growth factor-β (TGF-β) family comprises a group of diverse growth factors including bone morphogenetic proteins (BMPs), TGF-β, Anti-Mullerian hormone (AMH), growth and differentiation factors (GDFs), active in and nodal [40]. "TGF-β is composed of ~390 amino acids, which are released mainly from platelets, macrophages, and bone. This inactive polypeptide undergoes proteolytic cleavage to create the active C-terminal 112-amino acid form. The active form of TGF-β dimerizes to form 25 kD homodimers" [41]. The three isoforms present in mammals, which are TGF-β1, TGF-β2, TGF-β3, are detected in human dentin [42]. Inactive TGF-β exists as a large latent complex [43]. "After proteolytic cleavage, the active TGFβ binds to the Type II receptor (TGFβRII) and recruits Type I receptor (TGFβRI) to dimerize. TGFβRI, in turn, phosphorylates the intracellular proteins SMAD (homologs of drosophila proteins including Caenorhabditis elegans protein, SMA, and mothers against decapentaplegic, MAD), in particular, SMAD2 and SMAD3" [44, 45]." The activated SMAD complex translocases to the nucleus and activates downstream TGF-βgene transcription. The effect of TGF-β is highly variable and dependent on the type of cells and tissues. TGF-β1 regulates a wide range of cellular activities, such as cell migration, cell proliferation, cell differentiation, and extracellular matrix synthesis" [46 - 50]." TGF-β1 has been shown to increase cell proliferation and production of the extracellular matrix in dental pulp tissue culture, and promotes odontoblastic differentiation of dental pulp cells" [51]." The effect of TGF-β1 can be synergistically upregulated by fibroblast growth fatocr-2 (FGF2), as evidenced by the increased ALP activity, the formation of mineralized nodule, and the expression of DSP and dentin matrix protein-1" [51]. The dentinogenic ability of dental pulp cells in the mechanically exposed dental pulp of dog teeth is shown to be induced by exogenous TGF-β1 [52]. TGF-β is chemotactic on dental pulp cells *in vitro*. [53, 120].

BMP family members are a subgroup of the TGF-β superfamily, and play an important role in the whole progress of tooth formation and development. Among them, BMP2, BMP4 and BMP6 were found in primary culture of human dental pulp cells [54]. In the dental pulp tissue of rat incisor, there are seven more BMP family members have been identified, including BMP7, BMP8, growth/

differentiation factor1 (Gdf1), Gdf5, Gdf6, Gdf7, Gdf11 and glial cell line-derived neurotrophic factor (GDNF) [55, 56]. The epithelial–mesenchymal interactions are crucial for tooth morphogenesis. It is reported that Bmp2, Bmp4, Bmp6, Bmp7 and Gdf11 were expressed during odontoblast differentiation [56]. BMP2 can direct pulp bovine and human progenitor/stem cell differentiation into odontoblasts in monolayer cultures and in three-dimensional pellet cultures [57]. Human recombinant BMP7 stimulated ECM secretion by odontoblasts in cultured tooth slices [58]. Recombinant human GDF11 transferred into mouse dental papilla mesenchymal cells also stimulates odontoblast differentiation and reparative dentin formation [59]. Recombinant human BMP2, BMP4 and BMP7 have been reported to induce reparative/regenerative dentin formation [60, 120].

Platelet-Derived Growth Factor (PDGF) is an alkaline protein stored in a platelet alpha granule. It can stimulate cell proliferation and promote angiogenesis [61, 62]. There are 4 isoform homodimers of PDGF, including AA, BB, CC, and DD, in addition to a heterodimer, PDGF-AB. "PDGF dimers bind to two cell surface receptors known as PDGFRα and PDGFRβ. The receptors form dimers before binding to different isoforms of PDGF. PDGF-AA, -BB, and -CC bind to PDGFR α/α, whereas PDGF-AB, -BB, -CC, and -DD bind to PDGFR α/β.PDGF-BB and DD bind to PDGFR β/β" [63]. Therefore, the biological effect of PDGF depends on the expression level of PDGFR dimer on target cells. However, "PDGFs appear to have little effects on the formation of the dentin-like nodule in dental pulp cells from rat lower incisors although PDGF-AB and -BB isoforms stimulate the expression of dentin sialoprotein (DSP)" [64]. PDGFs stimulate cell proliferation and dentin matrix protein synthesis but appear to inhibit alkaline phosphatase (ALP) activity in dental pulp cells inculture [65, 66]." DSP expression is inhibited by PDGF-AA, but enhanced by PDGF-AB and PDGF-BB although the mineralized tissue formation is inhibited, suggesting diverging effects of PDGFs on odontoblastic differentiation depending on dimeric form" [64]. *In vivo*, PDGF can stimulate renewal formation of dental-pulp like tissues in endodontically treated root canals of native human teeth [67, 120].

Fibroblast Growth Factor (FGF) plays an important role in cell migration, proliferation and differentiation during embryonic development and wound healing. Currently 22 members have been found in humans, of which FGF2 is significant in regeneration of the pulp-dentin complex. "Four FGF receptors, FGFR1 through FGFR4 are expressed in humans. Signal transduction is mediated by interaction between FGFs with the ability to bind to haparan sulfate and heparan sulfate proteoglycans on cell surface. FGF2 is a basic FGF, while FGF1 is acidic. FGF2 regulates tooth morphogenesis by controlling cell proliferation and differentiation" [68]. FGF2 is a potent angiogenic factor stimulating new blood vessel formation in the dental pulp along with PDGF and VEGF. FGF2 induces the migration of dental pulp cells. Using a transwell migration assay, significantly more dental pulp cells are recruited by bFGF (FGF2) into a 3D collagen gel than controls without cytokines and BMP7. FGF2 also promotes the proliferation of dental pulp cells without differentiation, whereas FGF2 combined with TGF-β1 stimulates dental pulp cells to differentiate into odotoblast-like cells, and synergistically upregulates the effect of TGF-β1 on odontoblast differentiation [69]. "The FGF2 on exposed dental pulp in rat molars induces vascular invasion and cell proliferation early in wound healing. Also, FGF2 stimulates reparative dentin formation or dentin-particles in the exposed pulp" [70 - 72, 120].

Insulin-like Growth Factors (IGFs), as multifunctional cell proliferation regulatory factors, have an important role in the differentiation, proliferation, and growth of the cells. IGFs, comprising IGF-1 and IGF-2, contribute to odontogenesis and dental tissue repair by cell proliferation and differentiation. "There are two known IGF receptors, of which IGF-1R has tyrosine kinase activity phosphorylating the insulin receptor substrates and activating MAP kinase and the phosphatidylinositol 3-kinase (PI3K) cascades. However, IGF-2R has no intrinsic kinase activity. IGF-1R binds to both isoforms of IGFs, but IGF-2R only binds to IGF-2.Of the two isoforms, IGF-1, also known as somatomedin C, has potency in proliferation and differentiation of dental pulp cells. IGF-1 induces proliferation and differentiation of dog dental pulp cells into odontoblast-like cells in serum-free culture conditions" [73]. "IGF-1 with PDGF-BB has a synergetic action on the proliferation of dental pulp cells *in vitro*.IGF-1 and IGF-1Rhave a higher level of expression in dental pulp tissue from teeth with complete root development than teeth with incomplete root formation, suggesting that IGF-1 stimulates mineralization and cell differentiation" [74 - 76, 120].

Nerve Growth Factor (NGF), also known as neutrophins, is a neurotrophic factor that plays an important role in the regulation of neuronal development, differentiation, growth, and regeneration. NGFs bind to two receptors, a p75 low-affinity neutrophin receptor (p75LANR) and a high-affinity tyrosine kinase receptor (trk) [77]. NGFs are related to the survival and differentiation of neuronal and non-neuronal cells through high-affinity trkA, but NGFs regulates apoptosis through p75 LANR [78]. The expression of NGF and p75 LANR increases in dental pulp cells at the damage site [79]. NGFs play a role in regulating tooth morphogenesis and tooth innervation in rat tooth development [80]. NGFs induce the differentiation of odontoblast-lineage cells with subsequent biomineralization *in vitro*, suggesting that NGFs act as a stimulant for mineralization [81, 120].

2.4. The Regulation Mechanism of Dentin Regeneration

The BMP signaling networks are complicated during morphogenesis. The regulation of BMP signaling pathway can be divided into 3 levels: intracellular domains, membrane site and extracellular sites [82]. The bioavailability of the morphogensis regulated by many factors, such as BMP antagonists, follistatin, noggin, DAN, chordin, Gremlin and Cerberus. Two transmembrane receptors, type I and type II with serine–threonine kinase activity are expressed in dental pulp [83]. BMP signals transduction from the membrane to the nucleus is *via* a few Smad proteins, receptor activated Smads (R-Smads), common mediator Smads (co-Smads) and inhibitory Smads (I-Smads). The number of Smad-interacting proteins are increasing by which the outcome of the signal is determined [84]. The BMP signals are repeatedly used in the communication between cells and tissues during tooth development [85].

LIM homeobox 8 (Lhx8) regulates dental mesenchyme differentiation and function *via* Wnt and TGF-β pathways. "Lhx8 expression was restricted to dental mesenchyme from E11.5 to a peak at E14.5, and absent in dental epithelium. By reconstituting dental epithelium and mesenchyme in an E16.5 tooth organ, Lhx8 knockdown accelerated dental mesenchyme differentiation; conversely, Lhx8 overexpression attenuated dentin formation. Lhx8 overexpressed adult human dental pulp stem/progenitor cells in β-tricalcium phosphate cubes attenuated mineralized matrix production *in vivo*. Gene profiling revealed that postnatal dental pulp stem/progenitor cells upon Lhx8 overexpression modified matrix related gene expression including dentin sialophosphoprotein (DSPP), Cola1 and osteocalcin. Lhx8 transcriptionally activated Wnt and TGF-β pathways and its attenuation unregulated multiple dentinogenesis genes. Together, Lhx8 regulates dentin development and regeneration by fine-turning Wnt and TGF-β signaling" [86].

Calcium ion is a major component of many pulp capping materials used to stimulate the regeneration of tertiary dentin. It can increase the expression of the DSPP, odontoblastic differentiation marker gene, and stimulate odontoblastic differentiation and mineralization of DPSCs, but inhibit ALP activity. Endogenous BMP2 expression increased by calcium ion activates the Smad1/5/8 pathway and stimulates the Smad1-Runx2 and Runx2-DSPP interaction in DPSCs. Dorsomorphin inhibits Smad1/5/8 pathway that leads to partially block of Runx2 activity; however, Noggin inhibits the BMP2 receptor that leads to nearly fully suppression of Runx2 activity. The results above demonstrate that calcium ion stimulates cell differentiation mainly *via* BMP2-mediated Smad-dependent and Smad-independent pathways. With calcium ion stimulation, the phosphorylation level of Erk1/2 was significantly increased. And when Erk1/2 is blocked, the Runx2 activity will be inhibited. So researchers presume that calcium ion triggers the Erk1/2 pathway, which subsequently adjusts Runx2 activity. Besides, block of Erk1/2 pathway differentially decreased the phosphorylation levels of Smad1/5/8 and Smad2/3. In summarize, calcium ion activates the BMP2-mediated Smad1/5/8 and Erk1/2 pathways in DPSCs and that Smad1/5/8 and Erk1/2 signaling converge at Runx2 to control the odontoblastic differentiation of DPSCs [87].

3. PERIODONTAL LIGAMENT REGENERATION

3.1. The Stem Cells Used in Periodontal Ligament Regeneration

3.1.1. Periodontal Ligament Stem Cells (PDLSCs)

The periodontal ligament is many fibers arranged in bundles, of which one end embedded in root cementum and the other embedded in alveolar bone to suspend the tooth in its alveolus. There are several elements in the periodontal ligament containing nerves, blood vessels, lymph, epithelial cells and stem cells. PDLSCs have the potency to regenerate periodontal tissues such as ligament and cementum. Under specified conditions, PDLSCs can differentiate into osteoblasts, cementoblasts, and adipocytes *in vitro*. In animal studies, PDLSCs was transplanted into mice and form analogical tissues like cementum, bone, cartilage, and PDL. PDLSC combined with SCAPs from impacted third molars were seeded on a hydroxyapatite-tricalcium scaffold and transplanted into the alveolar bone of juvenile miniature pigs. And the results found that a root and a periodontal complex were formed which could sustain a ceramic crown and perform the function of a natural tooth [88]. Besides, PDLSCs were also used to treat periodontal disease [89].

3.1.2. Dental Follicle Stem Cells (DFSCs)

The dental follicle is the structure surrounding the developing tooth, which plays an important role in the origin of cementum, periodontal ligament, and alveolar bone. DFSCs can be obtained from the follicles of impacted third molars [90]. DFSCs have characteristics of cementoblasts and osteoblasts when cultured *in vitro*. *In vivo*, tissue resembling dental cementum and differentiation into PDL progenitor cells have been observed.

Besides dental stem cells, periodontal tissue can be regenerated using ADSCs *in vivo* after transplantation into rat [91].

3.2. The Scaffolds Used in Periodontal Ligament Regeneration

With the development of hydrogels, a number of materials have been synthesized to fabricate membrane for GBR and GTR. There are several prerequisites for adequate biomaterials used as membrane, including biocompatibility, non-immunogenicity and non-toxicity. Biodegradability of materials is also important, which can refrain from second surgery to remove the membrane after healing. Besides, other elements such as cell occlusivity, space maintaining and tissue integration should not be ignored. The membranes used for GTR in the beginning, such as expanded polytetrafluoroethylene (ePTFE) and dense-polytetra-fluoroethylene, cannot be absorbable. For the second generation of membranes has good biodegradability, and there is minimal membrane exposure and no need to remove membrane, which make them a favorite choice in GTR procedure nowadays [92]. However, comparing with non-resorbable membrane, the biodegradable membrane may lead to a larger bone area.

PLA is a kind of biodegradable material for GTR membrane. In a 3-year controlled clinical experiment, curative effects of GTR were studied using a synthetic degradable PLA membrane in intrabony defects, and the control group were treated with open flap debridement (OFD). The indexes including probing pocket depth (PPD), gingival recession (GR), clinical attachment level (CAL), and linear alveolar bone level (ABL) were recorded at different time points. The results suggested that the curative effects between two groups have no significant difference [93]. A 10-year randomized-controlled trial follow-up demonstrated that biodegradable PLA membrane (Atrisorb1) with autogenous bone grafting led to poor regeneration of PDL [94]. The compositions of Atrisorb, a commercialized PLA biomaterial (DL-lactide polymer, Atrix Laboratories Inc.) are 37% of a liquid PLA resolved in 63% N-methyl-2-pyrrolidone (NMP). The results of PPD, GR, CAL demonstrated the potency of periodontal regeneration in both animal and human class II furcation defects [95, 96]. However, there is a lack of long-term study and histologic observations of tissue regeneration to evaluate Atrisorb membrane. In order to ameliorate the properties, the membranes of PLA combined with other polymers a refabricated, and get promising outcomes. A biodegradable GTR membrane, constituted by a biodegradable PLA/poly (glycolide-co-lactide) copolymer (PLGA) membrane with polyglycolic acid (PGA) mesh, shows ability of retaining space, good biocompatibility, suitable permeability of nutrients and non-cytotoxicity in animal study [97].

PLGA is composed of two monomers, PLA and hydroxy acetic acid at random polymerization. It is another biodegradable functional polymer, which has biocompatibility, non-toxicity and good film forming properties. PLGA has been certificated by FDA of United States, and is officially included into the United States Pharmacopoeia. The potency of periodontal regeneration of cloned cementoblasts (OCCMs), periodontal ligament fibroblasts (SVPDLs), and dental follicle (SV-F) was evaluated by a 3D PLGA scaffold. The results showed that porous PLGA scaffold could stimulate cementoblasts to form mineralization and was non-toxic to the other cells. Owen *et al.* made GTR membranes from PLGA, plasticized with either MePEG or a diblock copolymer, which were loaded with tetracycline base (or hydrochloride salt) and cast by solvent evaporation [98]. At room/body temperature, this membrane is elastic and easy to handle, while it becomes swollen and stiffened in water so that it can suppress down migration of epithelium and gingival tissue.

3.3. The Cytokines/Growth Factors and Regulation Mechanism in Periodontal Ligament Regeneration

The combination of PDGF-BB and IGF-1 was the first reported use of growth factors for periodontal regeneration. A clinical trial also showed that the combination of growth factors in methylcellulose vehicle was safe and that it enhanced periodontal regeneration.

Recently, efforts to keep PDLSCs undifferentiated at early stages of cell culture have been made to ensure better multipotency of stem cells to differentiate into osteoblasts/cementoblasts and fibroblasts at later culture stages. According to Liu *et al.*'s research, nuclear expressions of Sox-2 and Oct-4 in PDLSCs were maintained until passage 3. Meanwhile, Sox-2 and Oct-4 mRNA in PDLSCs decreased and disappeared after passage 3. Surprisingly, BMP-4 not only could enhance proliferation, but also could reverse the decrease in Sox-2 and Oct-4 expression and promote their nuclear translocation even at passage 7. The beneficial effects of BMP-4 on PDLSCs may be due to the significant overlap in responsive genes between BMP-4 and Oct-4. BMP-4 may be an effective way to maintain the stemness of PDLSCs during a long-term culture.

As the periodontium is comprised of cementum, alveolar bone, and the functional periodontal ligaments between them, so use of various growth factors to induce PDLSCs differentiation into different directions is of interest. For example, BMP-2 and -7 and VEGF have been verified to enhance osteogenic differentiation of PDLSCs and promote the repair of bony defect in animal models. In contrast, TGF-β1 and its downstream protein connective tissue growth factor (CTGF) accelerated fibroblastic differentiation of PDLSCs through upregulation of type I collagen, α-smooth muscle actin, and periostin. Furthermore, FGF-2 promoted proliferation of PDLSCs but reversed the beneficial effects of BMP-2 and VEGF on osteogenic differentiation. Unexpectedly, sequential use of FGF-2 followed by BMP-2 resulted in more bone formation than use of BMP-2 or FGF-2 alone. Similarly, sequential use of FGF-2 followed by TGF-β1 also promoted fibroblastic differentiation of PDLSCs.

Brain-derived neurotrophic factor (BDNF) distributes in the central nervous system, peripheral nervous system, endocrine system, bone and cartilage tissue, even in PDL cells. New cementum, connective fibers and alveolar bone were formed by BDNF used with a telocollagen sponge. The fibers connected the newly formed cementum and bone in a canine class III furcation defect model. BDNF also could induce blood capillary formation. It is reported that the regenerative capacity of BDNF was improved by scaffolds made from high-molecular-weight hyaluronic acid in a canine periodontal defect model.

4. DENTAL PULP REGENERATION

4.1. The Stem Cells Used in Dental Pulp Regeneration

A number of stem cells exist in the human tooth and its environment, including DPSCs, SCAPs, DFSCs, and PDLSCs. Among them, DPSCs, SCAPs, and DFSCs are suitable cells for dental pulp regeneration, and have been successfully used in some animal models.

4.1.1. DPSCs

In 1996, Mooney *et al.* firstly described a technique to engineer new pulp-like tissues utilizing cultured human pulp-derived fibroblasts and synthetic PGA matrix [99]. These cell–matrix constructs were transplanted into immunocompromised mice, and fibroblasts were able to survive and to produce ECM [100]. In 2002, Gronthos *et al.* characterized the self-renewal capability, multi-lineage differentiation capacity, and clonogenic efficiency of DPSCs. Hydroxyapatite/tricalcium phosphate (HA/TCP) particles seeded with DPSCs were implanted in immunocompromised mice, and formed a fibrous pulp-like tissue including blood vessels and human odontoblast-like cells [101, 102].

Although many promising results demonstrated the regenerative potency of DPSCs in animal studies, more standard and representative experimental approaches, such as the tooth slice scaffold model and the ectopic root transplantation model were needed in clinical work. Besides the investigation of dentinogenesis and the evaluation of pulp vascularization, the scaffold of human tooth slice can stimulate regeneration of dental pulp [103, 104]. Some studies transplanted the complex of tooth slices with DPSCs on a biodegradable scaffold in immunodeficient mice and found the formation of vascularized pulp-like tissue [105, 106]. The ectopic root transplantation model is to transplant the emptied human root canals to avoid the restriction of the diameter of the apical foramen to blood supply. In 2010, Huang *et al.* [107] firstly demonstrated that "pulp-like tissue could be regenerated de novo in emptied root canal space by SCAPs and DPSCs". In an ectopic root transplantation model, SHEDs combined with a self-assembling peptide hydrogel could also form vascularized pulp/dentin-like tissue. However, the 2 ends of root canal were both open, which led to more potential influx of nutrients and cells through the root canal [108]. When sealing one apical end and leaving an apical opening of 1 mm of the human root canal, the transplantation of DPSCs or DPSC-HUVECs cocultures could form vascularized pulp-like tissue after 4 weeks [109]. Yang *et al.* transplanted the DPSCs in a silkfibroin scaffold, and found the similar result after 8 weeks [110].

For the regeneration potency of different subsets of DPSCs, a recent study sealed one end with MTA of human root canals, and found vascular pulp tissue in them after 21 days of subcutaneous transplantation. Comparing with CD105+DPSCs, ordinary DPSCs, and induced pluripotent stem cells (iPSCs), greater regeneration of vascularized pulp tissue was found in G-CSF-mobilized subpopulation of DPSCs group. It demonstrated that the regenerative potency of subpopulations of DPSCs is different, and determined by the intrinsic properties [111]. In 2006, Iohara and colleague have showed autogenous transplantation of the Bmp2-supplemented side population (SP) cells on the amputated pulp stimulated the reparative dentin formation [112]. Similar results have been found after autologous DPSCs transplantation in pulpectomized dog teeth [113, 114]. Iohara *et al.* [115] have assessed the efficacy and safety of pulp stem cells transplantation as a prelude for the impending clinical trials by the autologous transplantation of clinical-grade G-CSF-mobilized DPSCs in a dog pulpectomized tooth. This study demonstrated that "the combinatorial trophic effects of pulp stem cells and G-CSF are of immediate utility for pulp/dentin regeneration, suggesting the prerequisites of safety and efficacy critical for clinical applications" [115].

4.1.2. SCAPs

In 2006, Sonoyama *et al.* [116] transplanted the SCAPs seeded in HA/TCP particles into immunocompromised mice, and formed dentin/pulp-like complex. SCAPs combined with a PLGA scaffold were implanted in an ectopic root model and showed odontogenic potency. Comparing with DPSC-PLGA construct, more continuous and thicker dentin matrix had been observed in the regenerated dentin/pulp complex. "SCAPs based cell sheet-derived pellet with a mount of endogenous ECM have a strong capacity to form a heterotopic dental pulp/dentine complex in empty root canals" [117]. When the immature teeth suffer from apical periodontitis and/or apical abscesses, the root can continue its development after conservative treatment. This mainly owes to the SCAPs which differentiate into odontoblasts and promote their root formation. However, the studies about the dentin/pulp complex regeneration with autologous SCAPs are so few. Therefore, more elaborate researches are needed to clarify the differentiation potency of SCAPs.

4.2. The Scaffolds Used in Dental Pulp Regeneration

There are two kinds of materials usually used for dental pulp regeneration. One is synthetic polymers, such as PLA and PGA; the other is matrices from biological sources, such as collagen, alginate and chitosan. Synthetic polymersarenon-toxic and biocompatible, and degraded by hydrolysis. Collagen is biocompatible and degradable by enzymes, but natural polymers have difficulties in processing and are risky about spreading animal-associated pathogens or causing immunologic reaction. Alginate, a natural carbohydrate extracted from red algae, can be cross-linked and encapsulate cells in it when meets calcium ion. However, the degradation of alginate is not controllable *via* dissolution for its sensitivity to calcium chelating compounds [118]. Chitosan is obtained from deacetylation of chitin. For the biological functionality, compatibility, security, microbial degradation *via* naturally occurring enzymes and other excellent properties, it has been widely concerned and used in tissue engineering.

The DPSCs from deciduous teeth and third molars combined with hydroxyapatite/tricalcium phosphate (HA/TCP) formed bone, dentin and dentin-pulp like tissues. Nör *et al.* seeded SHED on a PLA scaffold and found the formation of a vascularized pulp-like tissue, odontoblast-like cells, and newly generated dentin. Huang *et al.* found soft tissue and regenerated dentin after implantation of SCAP onto PLGA in an empty root canal [119]. Researchers compared different materials, collagen I and III, alginate, and chitosan, and found that type I collagen was the best materials for cells proliferation and differentiation.

However, there is no polymer that has all the advantages of an ideal material, which should most closely imitate ECM of the cells. Polymers such as PLA have the advantages of being biocompatible, biodegradable, and inexpensive and easy to prepare. But they are lack of chemical information which is physiologically existed in the ECM. In contrast, collagen has the structural and chemical information of the ECM but has the difficulty in utilization for specific applications for the concerning about its biological origin, purity and immune reaction. An ideal scaffold should combine the best properties of each of these groups of biomaterials. For dental pulp regeneration, the ideal scaffold should contain at least these characteristics as follows: contamination control in the root canal, vascularization and innervation of a long and narrow space, and the possibility for creation of acellular matrix capable of recruiting resident stem cells in the respective tissues.

4.3. The Cytokines/Growth Factors and Regulation Mechanism in Dental Pulp Regeneration

Vascular Endothelial Growth Factor (VEGF) is a heparin-binding protein with specific affinity to endothelial cells and plays an important role in angiogenesis. The functions of VEGF involve the stimulation of endothelial cells proliferation and enhancement of their survival, promoting neovascularization in the area of injury. The VEGF family contains VEGF-A, VEGF-B, VEGF-C, VEGF-D and placenta growth factor. "Among these isoforms, VEGF-A is the most versatile in function. VEGF-A, also known as vascular permeability factor, and promotes cell migration, cell proliferation, vasodilatation and vascular permeability by binding to two tyrosine kinases receptors,VEGFR1 and VEGFR2.VEGF increases microvessel density of the dental pulp when VEGF treated tooth slices containing severed dental pulp were transplanted into subcutaneous tissues of severely combined immunodeficiency (SCID) mice. VEGF can stimulate the differentiation of human dental pulp cells into endothelial cells. Dental pulp cells treated by VEGF become positive for CD29, CD44, CD73, CD105, CD166, but negative for CD14, CD34, CD45. VEGF increases the expression of VEGFR1 (fms-like tyrosine kinase, Flt-1) and VEGFR2 (kinase-insert domain containing receptor, KDR) and microvessel formation in a three-dimensional fibrin mesh seeded with dental pulp cells. However, VEGF treatment does not appear to promote CD31, CD34,and CD144 positivity in dental pulp cells that are positive for CD29, CD90, CD105, CD166,CD146, and STRO-1" [120].

Stromal Cell-Derived Factor-1 (SDF-1), also known as chemokine (C-X-C motif) ligand 12 (CXCL12), is a chemoattractant involved in cell mobilization and homing by binding to the chemokine receptor CXCR4. "SDF-1 functions as a chemokine for hematopoietic stem cells, mesenchymal stromal cells, and immune cells.SDF-1 stimulates the migration and proliferation of CD31-/CD146-side population (SP) cells isolated from porcine tooth germ that are positive forCXCR4 and negative for hematopoietic markers. Furthermore, the CD31-/CD146-SP cells are good at migration and proliferation with localized SDF-1 expression in amputated canine dental pulp". Dental pulp-like tissue with capillaries and nerves regenerate in dog teeth following pulp extirpation and autologous implantation of the CD31-/CD146-SP cells or CD105+cells with SDF-1 into root canals [120].

CONFLICT OF INTEREST

The authors confirm that they have no conflict of interest to declare for this publication.

ACKNOWLEDGEMENTS

We thank Dr. Xiaoxiao Cai for the English editing. This work was funded by Funding of National Natural Science Foundation of China (81470721, 31170929), Sichuan Science and Technology Innovation Team (2014TD0001).

REFERENCES

[1] Du C, Moradian-Oldak J. Tooth regeneration: challenges and opportunities for biomedical material research. Biomed Mater 2006; 1(1): R10-7. [http://dx.doi.org/10.1088/1748-6041/1/1/R02] [PMID: 18458377]

[2] Chai Y, Slavkin HC. Prospects for tooth regeneration in the 21st century: a perspective. Microsc Res Tech 2003; 60(5): 469-79. [http://dx.doi.org/10.1002/jemt.10287] [PMID: 12619122]

[3] Cuozzo FP, Sauther ML. Tooth loss, survival, and resource use in wild ring-tailed lemurs (Lemur catta): implications for inferring conspecific care in fossil hominids. J Hum Evol 2004; 46(5): 623-31. [http://dx.doi.org/10.1016/j.jhevol.2004.02.004] [PMID: 15120269]

[4] Pihlstrom BL, Michalowicz BS, Johnson NW. Periodontal diseases. Lancet 2005; 366(9499): 1809-20. [http://dx.doi.org/10.1016/S0140-6736(05)67728-8] [PMID: 16298220]

[5] Sonoki K, Takata Y, Ansai T, *et al.* Number of teeth and serum lipid peroxide in 85-year-olds. Community Dent Health 2008; 25(4): 243-7. [PMID: 19149303]

[6] Ostberg AL, Nyholm M, Gullberg B, Råstam L, Lindblad U. Tooth loss and obesity in a defined Swedish population. Scand J Public Health 2009; 37(4): 427-33. [http://dx.doi.org/10.1177/1403494808099964] [PMID: 19141542]

[7] Jung SH, Ryu JI, Jung DB. Association of total tooth loss with socio-behavioural health indicators in Korean elderly. J Oral Rehabil 2011; 38(7): 517-24. [http://dx.doi.org/10.1111/j.1365-2842.2010.02178.x] [PMID: 21118289]

[8] Polzer I, Schimmel M, Müller F, Biffar R. Edentulism as part of the general health problems of elderly adults. Int Dent J 2010; 60(3): 143-55. [PMID: 20684439]

[9] Ferreira CF, Magini RS, Sharpe PT. Biological tooth replacement and repair. J Oral Rehabil 2007; 34(12): 933-9. [http://dx.doi.org/10.1111/j.1365-2842.2007.01785.x] [PMID: 18034675]

[10] Thesleff I, Tummers M. Stem cells and tissue engineering: prospects for regenerating tissues in dental practice. Med Princ Pract 2003; 12 (Suppl. 1): 43-50. [http://dx.doi.org/10.1159/000069840] [PMID: 12707500]

[11] Murray PE, Garcia-Godoy F. Stem cell responses in tooth regeneration. Stem Cells Dev 2004; 13(3): 255-62. [http://dx.doi.org/10.1089/1547328043230991811] [PMID: 15186721]

[12] Zhang YD, Chen Z, Song YQ, Liu C, Chen YP. Making a tooth: growth factors, transcription factors, and stem cells. Cell Res 2005; 15(5): 301-16.
[http://dx.doi.org/10.1038/sj.cr.7290299] [PMID: 15916718]

[13] Ulmer FL, Winkel A, Kohorst P, Stiesch M. Stem cells--prospects in dentistry. Schweiz Monatsschr Zahnmed 2010; 120(10): 860-83.
[PMID: 21207302]

[14] Woods EJ, Perry BC, Hockema JJ, Larson L, Zhou D, Goebel WS. Optimized cryopreservation method for human dental pulp-derived stem cells and their tissues of origin for banking and clinical use. Cryobiology 2009; 59(2): 150-7.
[http://dx.doi.org/10.1016/j.cryobiol.2009.06.005] [PMID: 19538953]

[15] Mauth C, Huwig A, Graf-Hausner U, *et al.* Restorative applications for dental pulp therapy topics in tissue engineering. In: Ashammakhi N, Reis R, Chiellini E, Eds. Topics in Tissue Engineering . 2007; 3: pp. 1-30.

[16] Laino G, d'Aquino R, Graziano A, *et al.* A new population of human adult dental pulp stem cells: a useful source of living autologous fibrous bone tissue (LAB). J Bone Miner Res 2005; 20(8): 1394-402.
[http://dx.doi.org/10.1359/JBMR.050325] [PMID: 16007337]

[17] Yang X, Walboomers XF, van den Beucken JJ, Bian Z, Fan M, Jansen JA. Hard tissue formation of STRO-1-selected rat dental pulp stem cells *in vivo.* Tissue Eng Part A 2009; 15(2): 367-75.
[http://dx.doi.org/10.1089/ten.tea.2008.0133] [PMID: 18652538]

[18] Yu V, Damek-Poprawa M, Nicoll SB, Akintoye SO. Dynamic hydrostatic pressure promotes differentiation of human dental pulp stem cells. Biochem Biophys Res Commun 2009; 386(4): 661-5.
[http://dx.doi.org/10.1016/j.bbrc.2009.06.106] [PMID: 19555657]

[19] Gronthos S, Mankani M, Brahim J, Robey PG, Shi S. Postnatal human dental pulp stem cells (DPSCs) *in vitro* and *in vivo.* Proc Natl Acad Sci USA 2000; 97(25): 13625-30.
[http://dx.doi.org/10.1073/pnas.240309797] [PMID: 11087820]

[20] El-Backly RM, Massoud AG, El-Badry AM, Sherif RA, Marei MK. Regeneration of dentine/pulp-like tissue using a dental pulp stem cell/poly(lactic-co-glycolic) acid scaffold construct in New Zealand white rabbits. Aust Endod J 2008; 34(2): 52-67.
[http://dx.doi.org/10.1111/j.1747-4477.2008.00139.x] [PMID: 18666990]

[21] Miura M, Gronthos S, Zhao M, *et al.* SHED: stem cells from human exfoliated deciduous teeth. Proc Natl Acad Sci USA 2003; 100(10): 5807-12.
[http://dx.doi.org/10.1073/pnas.0937635100] [PMID: 12716973]

[22] Ohazama A, Modino SA, Miletich I, Sharpe PT. Stem-cell-based tissue engineering of murine teeth. J Dent Res 2004; 83(7): 518-22.
[http://dx.doi.org/10.1177/154405910408300702] [PMID: 15218039]

[23] Ye L, Chen L, Feng F, *et al.* Bone marrow-derived stromal cells are more beneficial cell sources for tooth regeneration compared with adipose-derived stromal cells. Cell Biol Int 2015; 39(10): 1151-61.
[http://dx.doi.org/10.1002/cbin.10488] [PMID: 25965323]

[24] Chen YW, Yu YC, Chen L, *et al.* Human umbilical cord mesenchymal stem cells: a new therapeutic

option for tooth regeneration. Stem Cell Int 2015; 2015: 549432.
[http://dx.doi.org/10.1155/2015/549432]

[25] Wu G, Deng ZH, Fan XJ, *et al.* Odontogenic potential of mesenchymal cells from hair follicle dermal papilla. Stem Cells Dev 2009; 18(4): 583-9.
[http://dx.doi.org/10.1089/scd.2008.0066] [PMID: 18673019]

[26] Sharma B, Elisseeff JH. Engineering structurally organized cartilage and bone tissues. Ann Biomed Eng 2004; 32(1): 148-59.
[http://dx.doi.org/10.1023/B:ABME.0000007799.60142.78] [PMID: 14964730]

[27] Schröder U. Effects of calcium hydroxide-containing pulp-capping agents on pulp cell migration, proliferation, and differentiation. J Dent Res 1985; 64(Spec No): 541-8.
[PMID: 3857254]

[28] Anneroth G, Bang G. The effect of allogeneic demineralized dentin as a pulp capping agent in Java monkeys. Odontol Revy 1972; 23(3): 315-28.
[PMID: 4628280]

[29] Nakashima M. Dentin induction by implants of autolyzed antigen-extracted allogeneic dentin on amputated pulps of dogs. Endod Dent Traumatol 1989; 5(6): 279-86.
[http://dx.doi.org/10.1111/j.1600-9657.1989.tb00374.x] [PMID: 2640195]

[30] Nakashima M. An ultrastructural study of the differentiation of mesenchymal cells in implants of allogeneic dentine matrix on the amputated dental pulp of the dog. Arch Oral Biol 1990; 35(4): 277-81.
[http://dx.doi.org/10.1016/0003-9969(90)90043-A] [PMID: 2378581]

[31] Nakashima M. Induction of dentin formation on canine amputated pulp by recombinant human bone morphogenetic proteins (BMP)-2 and -4. J Dent Res 1994; 73(9): 1515-22.
[PMID: 7929986]

[32] Nakashima M. Induction of dentine in amputated pulp of dogs by recombinant human bone morphogenetic proteins-2 and -4 with collagen matrix. Arch Oral Biol 1994; 39(12): 1085-9.
[http://dx.doi.org/10.1016/0003-9969(94)90062-0] [PMID: 7717891]

[33] Tziafas D, Smith AJ, Lesot H. Designing new treatment strategies in vital pulp therapy. J Dent 2000; 28(2): 77-92.
[http://dx.doi.org/10.1016/S0300-5712(99)00047-0] [PMID: 10666965]

[34] Kitasako Y, Shibata S, Cox CF, Tagami J. Location, arrangement and possible function of interodontoblastic collagen fibres in association with calcium hydroxide-induced hard tissue bridges. Int Endod J 2002; 35(12): 996-1004.
[http://dx.doi.org/10.1046/j.1365-2591.2002.00606.x] [PMID: 12653318]

[35] Decup F, Six N, Palmier B, *et al.* Bone sialoprotein-induced reparative dentinogenesis in the pulp of rat's molar. Clin Oral Investig 2000; 4(2): 110-9.
[http://dx.doi.org/10.1007/s007840050126] [PMID: 11218498]

[36] Bohl KS, Shon J, Rutherford B, *et al.* Role of synthetic extracellular matrix in development of engineered dental pulp. J Biomater Sci Polymer Edn 1998; 9(7): 749-64.

[37] Goldberg M, Six N, Decup F, *et al.* Application of bioactive molecules in pulp-capping situations. Adv

Dent Res 2001; 15: 91-5.
[http://dx.doi.org/10.1177/08959374010150012401] [PMID: 12640750]

[38] Dobie K, Smith G, Sloan AJ, Smith AJ. Effects of alginate hydrogels and TGF-beta 1 on human dental pulp repair *in vitro*. Connect Tissue Res 2002; 43(2-3): 387-90.
[http://dx.doi.org/10.1080/03008200290000574] [PMID: 12489186]

[39] Nakashima M. Bone morphogenetic proteins in dentin regeneration for potential use in endodontic therapy. Cytokine Growth Factor Rev 2005; 16(3): 369-76.
[http://dx.doi.org/10.1016/j.cytogfr.2005.02.011] [PMID: 15878301]

[40] Watabe T, Miyazono K. Roles of TGF-beta family signaling in stem cell renewal and differentiation. Cell Res 2009; 19(1): 103-15.
[http://dx.doi.org/10.1038/cr.2008.323] [PMID: 19114993]

[41] Derynck R, Jarrett JA, Chen EY, *et al.* Human transforming growth factor-beta complementary DNA sequence and expression in normal and transformed cells. Nature 1985; 316(6030): 701-5.
[http://dx.doi.org/10.1038/316701a0] [PMID: 3861940]

[42] Cassidy N, Fahey M, Prime SS, Smith AJ. Comparative analysis of transforming growth factor-beta isoforms 1-3 in human and rabbit dentine matrices. Arch Oral Biol 1997; 42(3): 219-23.
[http://dx.doi.org/10.1016/S0003-9969(96)00115-X] [PMID: 9188992]

[43] Javelaud D, Mauviel A. Mammalian transforming growth factor-betas: Smad signaling and physio-pathological roles. Int J Biochem Cell Biol 2004; 36(7): 1161-5.
[http://dx.doi.org/10.1016/S1357-2725(03)00255-3] [PMID: 15109563]

[44] Lan HY, Chung AC. Transforming growth factor-β and Smads. Contrib Nephrol 2011; 170: 75-82.
[http://dx.doi.org/10.1159/000324949] [PMID: 21659760]

[45] Gold LI, Sung JJ, Siebert JW, Longaker MT. Type I (RI) and type II (RII) receptors for transforming growth factor-beta isoforms are expressed subsequent to transforming growth factor-beta ligands during excisional wound repair. Am J Pathol 1997; 150(1): 209-22.
[PMID: 9006337]

[46] Lambert KE, Huang H, Mythreye K, Blobe GC. The type III transforming growth factor-β receptor inhibits proliferation, migration, and adhesion in human myeloma cells. Mol Biol Cell 2011; 22(9): 1463-72.
[http://dx.doi.org/10.1091/mbc.E10-11-0877] [PMID: 21411633]

[47] Verrecchia F, Mauviel A. Transforming growth factor-beta signaling through the Smad pathway: role in extracellular matrix gene expression and regulation. J Invest Dermatol 2002; 118(2): 211-5.
[http://dx.doi.org/10.1046/j.1523-1747.2002.01641.x] [PMID: 11841535]

[48] Massagué J, Blain SW, Lo RS. TGFbeta signaling in growth control, cancer, and heritable disorders. Cell 2000; 103(2): 295-309.
[http://dx.doi.org/10.1016/S0092-8674(00)00121-5] [PMID: 11057902]

[49] Moses HL, Serra R. Regulation of differentiation by TGF-beta. Curr Opin Genet Dev 1996; 6(5): 581-6.
[http://dx.doi.org/10.1016/S0959-437X(96)80087-6] [PMID: 8939725]

[50] Melin M, Joffre-Romeas A, Farges JC, Couble ML, Magloire H, Bleicher F. Effects of TGFbeta1 on

dental pulp cells in cultured human tooth slices. J Dent Res 2000; 79(9): 1689-96. [http://dx.doi.org/10.1177/00220345000790090901] [PMID: 11023265]

[51] He H, Yu J, Liu Y, *et al.* Effects of FGF2 and TGFbeta1 on the differentiation of human dental pulp stem cells *in vitro*. Cell Biol Int 2008; 32(7): 827-34. [http://dx.doi.org/10.1016/j.cellbi.2008.03.013] [PMID: 18442933]

[52] Tziafas D, Papadimitriou S. Role of exogenous TGF-beta in induction of reparative dentinogenesis *in vivo*. Eur J Oral Sci 1998; 106 (Suppl. 1): 192-6. [http://dx.doi.org/10.1111/j.1600-0722.1998.tb02175.x] [PMID: 9541225]

[53] Howard C, Murray PE, Namerow KN. Dental pulp stem cell migration. J Endod 2010; 36(12): 1963-6. [http://dx.doi.org/10.1016/j.joen.2010.08.046] [PMID: 21092813]

[54] Takeda K, Oida S, Goseki M, *et al.* Expression of bone morphogenetic protein genes in the human dental pulp cells. Bone 1994; 15(5): 467-70. [http://dx.doi.org/10.1016/8756-3282(94)90268-2] [PMID: 7980956]

[55] Nakashima M, Toyono T, Murakami T, Akamine A. Transforming growth factor-beta superfamily members expressed in rat incisor pulp. Arch Oral Biol 1998; 43(9): 745-51. [http://dx.doi.org/10.1016/S0003-9969(98)00046-6] [PMID: 9783830]

[56] Nakashima M, Toyono T, Akamine A, Joyner A. Expression of growth/differentiation factor 11, a new member of the BMP/TGFbeta superfamily during mouse embryogenesis. Mech Dev 1999; 80(2): 185-9. [http://dx.doi.org/10.1016/S0925-4773(98)00205-6] [PMID: 10072786]

[57] Iohara K, Nakashima M, Ito M, Ishikawa M, Nakasima A, Akamine A. Dentin regeneration by dental pulp stem cell therapy with recombinant human bone morphogenetic protein 2. J Dent Res 2004; 83(8): 590-5. [http://dx.doi.org/10.1177/154405910408300802] [PMID: 15271965]

[58] Sloan AJ, Rutherford RB, Smith AJ. Stimulation of the rat dentinepulp complex by BMP7 *in vitro*. Arch Oral Biol 2000; 45(2): 173-7. [http://dx.doi.org/10.1016/S0003-9969(99)00131-4] [PMID: 10716622]

[59] Nakashima M, Mizunuma K, Murakami T, Akamine A. Induction of dental pulp stem cell differentiation into odontoblasts by electroporation-mediated gene delivery of growth/differentiation factor 11 (Gdf11). Gene Ther 2002; 9(12): 814-8. [http://dx.doi.org/10.1038/sj.gt.3301692] [PMID: 12040463]

[60] Six N, Lasfargues JJ, Goldberg M. Differential repair responses in the coronal and radicular areas of the exposed rat molar pulp induced by recombinant human bone morphogenetic protein 7 (osteogenic protein 1). Arch Oral Biol 2002; 47(3): 177-87. [http://dx.doi.org/10.1016/S0003-9969(01)00100-5] [PMID: 11839353]

[61] Hellberg C, Ostman A, Heldin CH. PDGF and vessel maturation. Recent Results Cancer Res 2010; 180: 103-14. [http://dx.doi.org/10.1007/978-3-540-78281-0_7] [PMID: 20033380]

[62] Bouletreau PJ, Warren SM, Spector JA, Steinbrech DS, Mehrara BJ, Longaker MT. Factors in the fracture microenvironment induce primary osteoblast angiogenic cytokine production. Plast Reconstr

Surg 2002; 110(1): 139-48.
[http://dx.doi.org/10.1097/00006534-200207000-00025] [PMID: 12087245]

[63] Alvarez RH, Kantarjian HM, Cortes JE. Biology of platelet-derived growth factor and its involvement in disease. Mayo Clin Proc 2006; 81(9): 1241-57.
[http://dx.doi.org/10.4065/81.9.1241] [PMID: 16970222]

[64] Yokose S, Kadokura H, Tajima N, *et al.* Platelet-derived growth factor exerts disparate effects on odontoblast differentiation depending on the dimers in rat dental pulp cells. Cell Tissue Res 2004; 315(3): 375-84.
[http://dx.doi.org/10.1007/s00441-003-0839-5] [PMID: 14740293]

[65] Denholm IA, Moule AJ, Bartold PM. The behaviour and proliferation of human dental pulp cell strains *in vitro*, and their response to the application of platelet-derived growth factor-BB and insulin-like growth factor-1. Int Endod J 1998; 31(4): 251-8.
[http://dx.doi.org/10.1046/j.1365-2591.1998.00161.x] [PMID: 9823114]

[66] Nakashima M. The effects of growth factors on DNA synthesis, proteoglycan synthesis and alkaline phosphatase activity in bovine dental pulp cells. Arch Oral Biol 1992; 37(3): 231-6.
[http://dx.doi.org/10.1016/0003-9969(92)90093-N] [PMID: 1375023]

[67] Kim JY, Xin X, Moioli EK, *et al.* Regeneration of dental-pulp-like tissue by chemotaxis-induced cell homing. Tissue Eng Part A 2010; 16(10): 3023-31.
[http://dx.doi.org/10.1089/ten.tea.2010.0181] [PMID: 20486799]

[68] Tsuboi T, Mizutani S, Nakano M, Hirukawa K, Togari A. Fgf-2 regulates enamel and dentine formation in mouse tooth germ. Calcif Tissue Int 2003; 73(5): 496-501.
[http://dx.doi.org/10.1007/s00223-002-4070-2] [PMID: 12958698]

[69] He H, Yu J, Liu Y, *et al.* Effects of FGF2 and TGFbeta1 on the differentiation of human dental pulp stem cells *in vitro.* Cell Biol Int 2008; 32(7): 827-34.
[http://dx.doi.org/10.1016/j.cellbi.2008.03.013] [PMID: 18442933]

[70] Kitamura C, Nishihara T, Terashita M, *et al.* Local regeneration of dentin-pulp complex using controlled release of fgf-2 and naturally derived sponge-like scaffolds. Int J Dent 2012; 2012: 190561.
[http://dx.doi.org/10.1155/2012/190561]

[71] Ishimatsu H, Kitamura C, Morotomi T, *et al.* Formation of dentinal bridge on surface of regenerated dental pulp in dentin defects by controlled release of fibroblast growth factor-2 from gelatin hydrogels. J Endod 2009; 35(6): 858-65.
[http://dx.doi.org/10.1016/j.joen.2009.03.049] [PMID: 19482186]

[72] Kikuchi N, Kitamura C, Morotomi T, *et al.* Formation of dentin-like particles in dentin defects above exposed pulp by controlled release of fibroblast growth factor 2 from gelatin hydrogels. J Endod 2007; 33(10): 1198-202.
[http://dx.doi.org/10.1016/j.joen.2007.07.025] [PMID: 17889689]

[73] Onishi T, Kinoshita S, Shintani S, Sobue S, Ooshima T. Stimulation of proliferation and differentiation of dog dental pulp cells in serum-free culture medium by insulin-like growth factor. Arch Oral Biol 1999; 44(4): 361-71.
[http://dx.doi.org/10.1016/S0003-9969(99)00007-2] [PMID: 10348363]

[74] Caviedes-Bucheli J, Canales-Sánchez P, Castrillón-Sarria N, *et al.* Expression of insulin-like growth factor-1 and proliferating cell nuclear antigen in human pulp cells of teeth with complete and incomplete root development. Int Endod J 2009; 42(8): 686-93. [http://dx.doi.org/10.1111/j.1365-2591.2009.01568.x] [PMID: 19467045]

[75] Caviedes-Bucheli J, Angel-Londoño P, Díaz-Perez A, *et al.* Variation in the expression of insulin-like growth factor-1 in human pulp tissue according to the root-development stage. J Endod 2007; 33(11): 1293-5. [http://dx.doi.org/10.1016/j.joen.2007.07.026] [PMID: 17963949]

[76] Caviedes-Bucheli J, Muñoz HR, Rodríguez CE, Lorenzana TC, Moreno GC, Lombana N. Expression of insulin-like growth factor-1 receptor in human pulp tissue. J Endod 2004; 30(11): 767-9. [http://dx.doi.org/10.1097/01.DON.0000134203.65706.8F] [PMID: 15505506]

[77] Davidson B, Reich R, Lazarovici P, *et al.* Expression of the nerve growth factor receptors TrkA and p75 in malignant mesothelioma. Lung Cancer 2004; 44(2): 159-65. [http://dx.doi.org/10.1016/j.lungcan.2003.11.014] [PMID: 15084380]

[78] Woodnutt DA, Wager-Miller J, O'Neill PC, Bothwell M, Byers MR. Neurotrophin receptors and nerve growth factor are differentially expressed in adjacent nonneuronal cells of normal and injured tooth pulp. Cell Tissue Res 2000; 299(2): 225-36. [http://dx.doi.org/10.1007/s004410050020] [PMID: 10741463]

[79] Byers MR, Wheeler EF, Bothwell M. Altered expression of NGF and P75 NGF-receptor by fibroblasts of injured teeth precedes sensory nerve sprouting. Growth Factors 1992; 6(1): 41-52. [http://dx.doi.org/10.3109/08977199209008870] [PMID: 1350451]

[80] Luukko K, Moshnyakov M, Sainio K, Saarma M, Sariola H, Thesleff I. Expression of neurotrophin receptors during rat tooth development is developmentally regulated, independent of innervation, and suggests functions in the regulation of morphogenesis and innervation. Dev Dyn 1996; 206(1): 87-99. [http://dx.doi.org/10.1002/(SICI)1097-0177(199605)206:1<87::AID-AJA8>3.0.CO;2-X] [PMID: 9019249]

[81] Arany S, Koyota S, Sugiyama T. Nerve growth factor promotes differentiation of odontoblast-like cells. J Cell Biochem 2009; 106(4): 539-45. [http://dx.doi.org/10.1002/jcb.22006] [PMID: 19170059]

[82] Balemans W, Van Hul W. Extracellular regulation of BMP signaling in vertebrates: a cocktail of modulators. Dev Biol 2002; 250(2): 231-50. [http://dx.doi.org/10.1006/dbio.2002.0779] [PMID: 12376100]

[83] Sloan AJ, Couble ML, Bleicher F, Magloire H, Smith AJ, Farges JC. Expression of TGF-beta receptors I and II in the human dental pulp by in situ hybridization. Adv Dent Res 2001; 15: 63-7. [http://dx.doi.org/10.1177/08959374010150011601] [PMID: 12640743]

[84] Zwijsen A, Verschueren K, Huylebroeck D. New intracellular components of bone morphogenetic protein/Smad signaling cascades. FEBS Lett 2003; 546(1): 133-9. [http://dx.doi.org/10.1016/S0014-5793(03)00566-0] [PMID: 12829249]

[85] Thesleff I. Epithelial-mesenchymal signalling regulating tooth morphogenesis. J Cell Sci 2003; 116(Pt 9): 1647-8.

[http://dx.doi.org/10.1242/jcs.00410] [PMID: 12665545]

[86] Zhou C, Yang G, Chen M, *et al.* Lhx8 mediated Wnt and TGFβ pathways in tooth development and regeneration. Biomaterials 2015; 63: 35-46.
[http://dx.doi.org/10.1016/j.biomaterials.2015.06.004] [PMID: 26081866]

[87] Li S, Hu J, Zhang G, *et al.* Extracellular Ca2+ Promotes Odontoblastic Differentiation of Dental Pulp Stem Cells via BMP2-Mediated Smad1/5/8 and Erk1/2 Pathways. J Cell Physiol 2015; 230(9): 2164-73.
[http://dx.doi.org/10.1002/jcp.24945] [PMID: 25656933]

[88] Sonoyama W, Coppe C, Gronthos S, Shi S. Skeletal stem cells in regenerative medicine. Curr Top Dev Biol 2005; 67(7378): 305-23.
[http://dx.doi.org/10.1016/S0070-2153(05)67010-X] [PMID: 15949539]

[89] Liu Y, Zheng Y, Ding G, *et al.* Periodontal ligament stem cell-mediated treatment for periodontitis in miniature swine. Stem Cells 2008; 26(4): 1065-73.
[http://dx.doi.org/10.1634/stemcells.2007-0734] [PMID: 18238856]

[90] Yalvac ME, Ramazanoglu M, Gumru OZ, Sahin F, Palotás A, Rizvanov AA. Comparison and optimisation of transfection of human dental follicle cells, a novel source of stem cells, with different chemical methods and electro-poration. Neurochem Res 2009; 34(7): 1272-7.
[http://dx.doi.org/10.1007/s11064-008-9905-4] [PMID: 19169817]

[91] Tobita M, Uysal AC, Ogawa R, Hyakusoku H, Mizuno H. Periodontal tissue regeneration with adipose-derived stem cells. Tissue Eng Part A 2008; 14(6): 945-53.
[http://dx.doi.org/10.1089/ten.tea.2007.0048] [PMID: 18558814]

[92] Hou LT, Yan JJ, Tsai AY, Lao CS, Lin SJ, Liu CM. Polymer-assisted regeneration therapy with Atrisorb barriers in human periodontal intrabony defects. J Clin Periodontol 2004; 31(1): 68-74.
[http://dx.doi.org/10.1111/j.0303-6979.2004.00436.x] [PMID: 15058377]

[93] Sakallioglu U, Yavuz U, Lütfioglu M, Keskiner I, Açikgöz G. Clinical outcomes of guided tissue regeneration with Atrisorb membrane in the treatment of intrabony defects: a 3-year follow-up study. Int J Periodontics Restorative Dent 2007; 27(1): 79-88.
[PMID: 17370665]

[94] Nygaard-Østby P, Bakke V, Nesdal O, Nilssen HK, Susin C, Wikesjö UM. Periodontal healing following reconstructive surgery: effect of guided tissue regeneration using a bioresorbable barrier device when combined with autogenous bone grafting. A randomized controlled clinical trial. J Clin Periodontol 2008; 35(1): 37-43.
[http://dx.doi.org/10.1111/j.1600-051X.2007.01160.x] [PMID: 18173400]

[95] Polson AM, Southard GL, Dunn RL, *et al.* Periodontal healing after guided tissue regeneration with Atrisorb barriers in beagle dogs. Int J Periodontics Restorative Dent 1995; 15(6): 574-89.
[PMID: 9601255]

[96] Coonts BA, Whitman SL, O'Donnell M, *et al.* Biodegradation and biocompatibility of a guided tissue regeneration barrier membrane formed from a liquid polymer material. J Biomed Mater Res 1998; 42(2): 303-11.
[http://dx.doi.org/10.1002/(SICI)1097-4636(199811)42:2<303::AID-JBM16>3.0.CO;2-J] [PMID: 9773827]

[97] Kim EJ, Yoon SJ, Yeo GD, Pai CM, Kang IK. Preparation of biodegradable PLA/PLGA membranes with PGA mesh and their application for periodontal guided tissue regeneration. Biomed Mater 2009; 4(5): 055001.
[http://dx.doi.org/10.1088/1748-6041/4/5/055001] [PMID: 19776491]

[98] Owen GR, Jackson JK, Chehroudi B, Brunette DM, Burt HM. An *in vitro* study of plasticized poly(lactic-co-glycolic acid) films as possible guided tissue regeneration membranes: material properties and drug release kinetics. J Biomed Mater Res A 2010; 95(3): 857-69.
[http://dx.doi.org/10.1002/jbm.a.32865] [PMID: 20824651]

[99] Mooney DJ, Powell C, Piana J, Rutherford B. Engineering dental pulp-like tissue *in vitro*. Biotechnol Prog 1996; 12(6): 865-8.
[http://dx.doi.org/10.1021/bp960073f] [PMID: 8983211]

[100] Buurma B, Gu K, Rutherford RB. Transplantation of human pulpal and gingival fibroblasts attached to synthetic scaffolds. Eur J Oral Sci 1999; 107(4): 282-9.
[http://dx.doi.org/10.1046/j.0909-8836.1999.eos107408.x] [PMID: 10467944]

[101] Gronthos S, Brahim J, Li W, *et al.* Stem cell properties of human dental pulp stem cells. J Dent Res 2002; 81(8): 531-5.
[http://dx.doi.org/10.1177/154405910208100806] [PMID: 12147742]

[102] Batouli S, Miura M, Brahim J, *et al.* Comparison of stem-cell-mediated osteogenesis and dentinogenesis. J Dent Res 2003; 82(12): 976-81.
[http://dx.doi.org/10.1177/154405910308201208] [PMID: 14630898]

[103] Sakai VT, Cordeiro MM, Dong Z, Zhang Z, Zeitlin BD, Nör JE. Tooth slice/scaffold model of dental pulp tissue engineering. Adv Dent Res 2011; 23(3): 325-32.
[http://dx.doi.org/10.1177/0022034511405325] [PMID: 21677087]

[104] Mullane EM, Dong Z, Sedgley CM, *et al.* Effects of VEGF and FGF2 on the revascularization of severed human dental pulps. J Dent Res 2008; 87(12): 1144-8.
[http://dx.doi.org/10.1177/154405910808701204] [PMID: 19029083]

[105] Cordeiro MM, Dong Z, Kaneko T, *et al.* Dental pulp tissue engineering with stem cells from exfoliated deciduous teeth. J Endod 2008; 34(8): 962-9.
[http://dx.doi.org/10.1016/j.joen.2008.04.009] [PMID: 18634928]

[106] Prescott RS, Alsanea R, Fayad MI, *et al. In vivo* generation of dental pulp-like tissue by using dental pulp stem cells, a collagen scaffold, and dentin matrix protein 1 after subcutaneous transplantation in mice. J Endod 2008; 34(4): 421-6.
[http://dx.doi.org/10.1016/j.joen.2008.02.005] [PMID: 18358888]

[107] Huang GT, Yamaza T, Shea LD, *et al.* Stem/progenitor cell-mediated de novo regeneration of dental pulp with newly deposited continuous layer of dentin in an *in vivo* model. Tissue Eng Part A 2010; 16(2): 605-15.
[http://dx.doi.org/10.1089/ten.tea.2009.0518] [PMID: 19737072]

[108] Rosa V, Zhang Z, Grande RH, Nör JE. Dental pulp tissue engineering in full-length human root canals. J Dent Res 2013; 92(11): 970-5.
[http://dx.doi.org/10.1177/0022034513505772] [PMID: 24056227]

[109] Dissanayaka WL, Hargreaves KM, Jin L, Samaranayake LP, Zhang C. The interplay of dental pulp stem cells and endothelial cells in an injectable peptide hydrogel on angiogenesis and pulp regeneration *in vivo*. Tissue Eng Part A 2015; 21(3-4): 550-63. [http://dx.doi.org/10.1089/ten.tea.2014.0154] [PMID: 25203774]

[110] Yang JW, Zhang YF, Wan CY, *et al.* Autophagy in SDF-1α-mediated DPSC migration and pulp regeneration. Biomaterials 2015; 44: 11-23. [http://dx.doi.org/10.1016/j.biomaterials.2014.12.006] [PMID: 25617122]

[111] Murakami M, Horibe H, Iohara K, *et al.* The use of granulocyte-colony stimulating factor induced mobilization for isolation of dental pulp stem cells with high regenerative potential. Biomaterials 2013; 34(36): 9036-47. [http://dx.doi.org/10.1016/j.biomaterials.2013.08.011] [PMID: 23988014]

[112] Iohara K, Zheng L, Ito M, Tomokiyo A, Matsushita K, Nakashima M. Side population cells isolated from porcine dental pulp tissue with self-renewal and multipotency for dentinogenesis, chondrogenesis, adipogenesis, and neurogenesis. Stem Cells 2006; 24(11): 2493-503. [http://dx.doi.org/10.1634/stemcells.2006-0161] [PMID: 16873765]

[113] Ishizaka R, Iohara K, Murakami M, Fukuta O, Nakashima M. Regeneration of dental pulp following pulpectomy by fractionated stem/progenitor cells from bone marrow and adipose tissue. Biomaterials 2012; 33(7): 2109-18. [http://dx.doi.org/10.1016/j.biomaterials.2011.11.056] [PMID: 22177838]

[114] Wang Y, Zhao Y, Jia W, Yang J, Ge L. Preliminary study on dental pulp stem cell-mediated pulp regeneration in canine immature permanent teeth. J Endod 2013; 39(2): 195-201. [http://dx.doi.org/10.1016/j.joen.2012.10.002] [PMID: 23321230]

[115] Iohara K, Murakami M, Takeuchi N, *et al.* A novel combinatorial therapy with pulp stem cells and granulocyte colony-stimulating factor for total pulp regeneration. Stem Cells Transl Med 2013; 2(7): 521-33. [http://dx.doi.org/10.5966/sctm.2012-0132] [PMID: 23761108]

[116] Sonoyama W, Liu Y, Fang D, *et al.* Mesenchymal stem cell-mediated functional tooth regeneration in swine. PLoS One 2006; 1(1): e79. [http://dx.doi.org/10.1371/journal.pone.0000079] [PMID: 17183711]

[117] Na S, Zhang H, Huang F, *et al.* Regeneration of dental pulp/dentine complex with a three-dimensional and scaffold-free stem-cell sheet-derived pellet. J Tissue Eng Regen Med 2013. [http://dx.doi.org/10.1002/term.1686] [PMID: 23365018]

[118] Boontheekul T, Kong HJ, Mooney DJ. Controlling alginate gel degradation utilizing partial oxidation and bimodal molecular weight distribution. Biomaterials 2005; 26(15): 2455-65. [http://dx.doi.org/10.1016/j.biomaterials.2004.06.044] [PMID: 15585248]

[119] Huang GT, Yamaza T, Shea LD, *et al.* Stem/progenitor cell-mediated de novo regeneration of dental pulp with newly deposited continuous layer of dentin in an *in vivo* model. Tissue Eng Part A 2010; 16(2): 605-15. [http://dx.doi.org/10.1089/ten.tea.2009.0518] [PMID: 19737072]

[120] Kim SG, Zhou J, Solomon C, *et al.* Effects of growth factors on dental stem/progenitor cells. Dent

Clin North Am 2012; 56(3): 563-75.
[http://dx.doi.org/10.1016/j.cden.2012.05.001] [PMID: 22835538]

[121] Potdar PD, Jethmalani YD. Human dental pulp stem cells: Aplication in future regenerative medicine. World J Stem Cells 2015; 7(5): 839-51.

SUBJECT INDEX

A

Adenoviruses 64, 75, 133
Adipocytes 5, 6, 93, 116, 118, 136, 139, 168, 235, 279, 353, 361
Adipogenic differentiation 153, 154, 317
Adiponectin 245, 280
Adipose-derived stem cells (ADSCs) 176, 319, 354, 361
Adipose tissue 3, 5, 7, 9, 132, 134, 138, 139, 140, 141, 142, 143, 167, 182, 237, 279, 352
  human 134, 140
Adult stem cells (ASCs) 13, 18, 49, 69, 130, 131, 134, 139, 140, 141, 142, 143, 144, 145, 146, 147, 148, 149, 153, 167, 170, 171, 172, 185, 237, 238
A-hydroxy acids 331, 332
Alginate 175, 313, 330, 331, 366, 367
Alveolar bone 44, 95, 97, 98, 109, 110, 131, 161, 163, 164, 168, 170, 326, 349, 350, 360, 361, 363, 364
Alveolar bone defect model 161, 164
Alveolar bone level (ABL) 362
Alveolar bone regeneration 163, 171, 271
Alveolar cleft 109, 110, 118
Alveolar periosteal cells (APCs) 144, 270
American society for testing and materials (ASTM) 39
Anatomical structures 42, 94, 97, 121, 122, 157
Angiogenesis 13, 14, 17, 93, 116, 139, 151, 171, 182, 183, 235, 236, 240, 279, 316, 357, 367
Application of induced ips cells 82
ASCs, porcine 148
Asphyxia 102, 103
Autogenous bone 114
Autologous bone transplantation 127, 163
Autologous tissue 92, 127
Autologous tissue transplantation 90, 92, 157, 167
Autologous tissue transplants 111, 113
Autologous transplantation 83, 93, 145, 268, 365

B

Bioactive molecules 315, 320, 322
Bioceramics 128, 148, 329
Biocompatibility 118, 126, 127, 129, 155, 156, 177, 178, 256, 313, 326, 327, 331, 333, 355, 361, 362
Biocomposites 125, 126
Biological activities 158, 312, 313
Biological safety 61, 156
Biometric identification 61
Biomimetic scaffolds 148, 149, 323
Biphasic calcium phosphate (BCP) 147, 148
Blood outgrowth endothelial cells (BOECs) 281
BMSCs and adipose-derived stem cells 354
BMSC-treated skin defects 152
Bone 9, 11, 45, 47, 48, 93, 98, 101, 104, 108, 113, 114, 115, 120, 122, 125,

Jing Wang & YunFeng Lin (Eds.)

129, 130, 132, 135, 138, 145, 146, 148, 162, 163, 172, 176, 185, 233, 273, 323, 324, 327, 330, 354, 356, 361, 364
lamellar 273, 323, 324
regenerated 114
tissue-engineered 135, 176
Bone/cartilage tissue 166
Bone compartment 45
Bone defects 90, 96, 97, 106, 110, 115, 118, 135, 145, 147, 158, 161, 162, 163, 164, 168, 169, 178, 180, 184, 234, 235, 272
alveolar 97, 161
jaw 96, 97, 110
Bone formation, tissue-engineered 184
Bone grafting, autogenous 163, 164, 362
Bone marrow cells 4, 135, 152
Bone marrow mesenchymal stromal cells (BMMSCs) 237, 238, 240, 270, 273
Bone marrow stromal cells 63, 135, 136, 146
Bone matrix 116, 156, 172, 272, 323, 329
demineralized 116, 156, 272
Bone morphogenetic proteins (BMPs) 13, 116, 146, 156, 356
Bone reconstruction 127
Bone regeneration 49, 90, 114, 115, 116, 131, 135, 139, 145, 146, 147, 149, 156, 160, 163, 170, 172, 173, 176, 179, 180, 181, 183, 184, 185, 243, 320, 332
Bone scaffolds 324, 334, 335
Bone tissue 114, 125, 166, 175, 263, 272, 324
Bone tissue regeneration 168, 183
Bone tissue repair 184
Bovine porous bone mineral (BPBM) 184
Brain-derived neurotrophic factor (BDNF) 364

C

Calcium ion 318, 329, 360, 366
Carbon nanotubes 126, 246
Cardiac function 240, 242, 279, 280
Cardiomyocytes 138, 245, 278, 279, 280
Cartilage, articular 266, 276, 277, 325
Cartilage cells 135, 183
Cartilage damage 276
Cartilage defects 90, 149, 177, 182, 183, 276
Cartilage formation 135, 181
Cartilage regeneration 13, 90, 135, 179, 181, 182, 277, 289
Cartilage tissue regeneration 149, 169
Cartilage tissues 175, 289, 324, 364
Cell adhesion and proliferation 178, 250, 251, 252
Cell-based tissue 93, 116, 171
Cell behavior 313, 318, 323, 355
Cell detachment 233, 252, 259
Cell differentiation 61, 74, 78, 80, 323, 356, 358, 360
Cell growth 19, 157, 350, 352
Cell-instructive 312, 313, 318, 320
Cell lineages, endothelial 79, 80

Cells 3, 4, 5, 6, 7, 8, 9, 10, 11, 14, 16, 19, 20, 33, 35, 37, 48, 49, 60, 61, 62, 63, 65, 66, 67, 68, 69, 70, 71, 72, 73, 74, 75, 76, 77, 78, 79, 80, 81, 82, 83, 116, 117, 118, 130, 131, 132, 133, 134, 135, 136, 137, 138, 140, 141, 142, 143, 144, 145, 147, 148, 149, 150, 151, 153, 154, 157, 158, 161, 167, 168, 169, 170, 171, 172, 174, 179, 180, 181, 185, 186, 232, 234, 235, 236, 238, 242, 243, 245, 246, 247, 248, 249, 251, 252, 253, 254, 255, 256, 257, 258, 259, 260, 261, 262, 263, 265, 266, 268. 269, 270, 272, 273, 274, 275, 276, 278, 279, 280, 282, 283, 284, 285, 286, 287, 313, 314, 317, 318, 319, 322, 323, 324, 327, 330, 334, 336, 337, 338, 350, 352, 353, 354, 356, 357, 358, 359, 360, 361, 363, 364, 365, 367, 368
adherent 4, 5, 246
adipose stem 133, 140, 278
adult 76, 131
adult stem 69, 130, 131, 134, 167
adventitial 137
attached 327, 334
chondrogenic 149, 150
co-cultured 236, 243, 273
confluent 248, 269, 275
cultured 232, 247, 261, 262
dental stem 134, 361
donor 138, 282
epithelial stem 285, 286
feeder 68, 235, 268
formed tissue 82
hematopoietic 74, 77, 79, 140, 143
hematopoietic stem 9, 147, 168, 368
hepatic 81, 83
hES 62, 79
inflammatory 16, 116, 245
isolated 9
isolated SVF 141, 144
living 33, 35, 48, 49, 314
low stem 170, 238
mature 60, 61, 138
mesenchymal stem cell/stromal 279
miPS 80
nerve growth factor receptor 10
neural 74, 138
non-hematopoietic stem 11
normal 83, 260
nucleated 7, 8
odontogenic 79
packaging 68
periodontal ligament stem 181, 360
reprogrammed 70, 71, 73
scaffold-free 49
seed 130, 131, 169, 172, 185, 234, 246, 272, 274, 284, 287
steady-state pluripotent 62, 67
suspended 143
target 68, 262, 357
transplanted 245, 276, 319
Cell sheet compose of single cells 264
Cell sheet manipulator technique 264
Cell sheet materials 232, 233
Cell sheet of periodontal ligament cells 237
Cell sheets 83, 231, 232, 233, 234, 235, 236, 241, 242, 243, 244, 245, 246, 247, 248, 249, 250, 251, 252, 254, 256, 257, 258, 259, 262, 263, 264, 265, 266, 268, 269, 270, 271, 273,

274, 275, 277, 279, 280, 281, 283, 284, 285, 286, 288, 289
adipocyte 245, 280
autologous myoblast 288, 289
automatic fabrication of 264, 266
contiguous urothelial 283, 284
epithelial 264, 275, 285, 286
implanted 285, 286
multilayered 233, 264
multi-layered 244, 262
nasal mucosal epithelial 270
oral mucosal epithelial 274, 286, 288
renal 283
tissue-derived stem 271
tissue-engineered epithelial 268
tissue-engineered RPE 269
urothelial 284
Cell sheet technology (CST) 231, 232, 233, 234, 235, 236, 240, 242, 243, 246, 265, 266, 267, 268, 269, 270, 271, 273, 274, 275, 276, 277, 278, 282, 283, 284, 285, 286, 288, 289
Cell sheet tissue engineering 247
Cell sheet transplantation 266, 275, 276, 277, 278, 279, 280
Cellular biological properties 61, 70
Cellular reprogramming 61, 74
Chemoattractants 15, 16, 17, 368
Chemotaxis, stimulated 12, 13, 14, 15, 16, 17
Chimeras 63, 64, 74, 76
Chitosan 175, 260, 261, 262, 313, 319, 330, 366, 367
surface-sulfonated 261, 262
Chitosan surfaces 260, 262
Chondrocytes 5, 81, 93, 116, 118, 135, 136, 146, 149, 150, 168, 169, 181, 182, 183, 235, 324, 325, 330, 334, 353
Chondrocyte sheet 245, 276, 277
layered 276, 277
Chondrogenic differentiation 81, 151, 324
Chronic granulomatous disorder (CGD) 83, 84
Classification systems 97, 99, 108, 101, 111
soft tissue defect 101
Class IV defects 98, 99
Cleft lip 105, 106, 107, 109, 115, 118, 166
bilateral 106, 107
Cleft lip and palate 105, 111, 115, 118, 166
Cleft palate 105, 106, 107, 153, 351
Clinical attachment level (CAL) 362
Clones, iPS cell 76
Collagen 125, 130, 148, 151, 160, 175, 243, 251, 252, 270, 287, 313, 319, 323, 330, 355, 363, 366, 367
Collagen complexes 70
Collagen scaffolds 167, 175
Complex tissue flap repair techniques 112
Complex tissues 135, 266, 353
Composite materials 111, 125, 127, 129, 158, 174, 313, 318
Composites, functional 126
Composite scaffolds 177, 178
Cone-bean computed tomography (CBCT) 46, 47
Congenital malformations 48, 91, 92, 130

Connective tissue growth factor (CTGF) 363
Copolymers 241, 251, 313, 332, 334, 362
Cranial defects 128, 159, 181
Craniofacial defects, complex 92, 121
Craniofacial region 91, 94, 112, 113, 115, 122, 123, 157, 167, 168, 177
Craniofacial tissues 33, 48, 111, 116, 131
Critical-size defect (CSD) 117, 158, 159, 160, 163, 273
CST transplantation 245, 286, 287

D

DBM/cell sheet 272
Deciduous teeth, human exfoliated 133, 353
Defect animal models 266, 267
Defect reconstruction 90, 112, 113, 115, 117, 123, 125, 127
  craniofacial 90, 112, 113, 115
Defect repair, craniofacial bone 117
Defects 91, 98, 125, 153
  correction of deformities and reconstruction of 91
  low maxillectomy 98
  mucocutaneous 153
  skeleton tissues 125
Deformities and reconstruction of defects 91
Demethylation 70, 72
Demineralized bone matrix (DBM) 116, 156, 272
Dental follicle cell (DFCs) 244, 271
Dental follicle stem cells (DFSCs) 361, 364
Dental implants 47, 48, 125, 351
Dental pulp 7, 95, 167, 237, 349, 351, 352, 353, 354, 355, 358, 359, 365, 368
  soft connective tissue of 349
Dental pulp cells 356, 357, 358, 359, 368
Dental pulp stem cells (DPSCs) 133, 173, 174, 237, 352, 353, 360, 364, 365, 367
Dental pulp tissue 356, 358
Dental stem cells (DSCs) 134, 354, 361
Dental tissues 133, 352
DFCs sheets 271, 272
Differentiated cells 61, 73, 74, 82, 151
Differentiation 136, 138, 354, 355, 354, 356, 357, 358
  multilineage 136, 138
  odontoblast 354, 355, 357, 358
  odontoblastic 356, 357, 360
  odontogenic 354
Differentiation abilities 172, 353
Differentiation capabilities 149, 237
Differentiation factors 356, 357
Digital model surgery (DMS) 43
Donor site complications 115, 124, 125, 127, 167

E

EBL system 248, 249
ECM proteins and proliferate stem cells in CST 240

Electron beam lithographic (EBL) 247, 248
Embryoid bodies (EB) 74, 78, 132, 239, 248, 249
Embryonic development 14, 15, 74, 105, 106, 107, 109, 131, 331, 357
Embryonic stem cells (ESCs) 9, 61, 62, 66, 71, 72, 75, 130, 131, 132, 133, 134, 150, 153, 167, 352
Enamel, calcified tissues of 349
Endoscopic submucosal dissection (ESD) 273, 274, 287, 288
Endothelial cells (ECs) 17, 74, 78, 79, 80, 136, 138, 141, 173, 239, 242, 243, 255, 279, 281, 367, 368
Endothelial colony-forming cells (ECFCs) 49
Engineered tissues 242, 246
Enzymatic treatment 261, 262, 263, 275
Epidermal growth factor (EGF) 13, 14, 79, 81, 184
Epithelial cells 136, 138, 235, 264, 271, 274, 275, 276, 282, 285, 286, 287, 361
  oral mucosal 268, 274
Epithelial stem cell deficiencies 286
ESCs and iPS cells 72
Extracellular matrix 49, 114, 174, 181, 232, 268, 284, 312, 314, 350, 354, 356

F

Facial processes, adjacent 105, 107
Factors 139, 156
  morphogenetic 156
  trophic 139
Factors of cell-instructive scaffolds for cell behavior 313, 318
Fiber-reinforced composites (FRCs) 128, 129
Fibroblast growth factor (FGF) 14, 79, 139, 357, 358
Fibroblast-like cells (FLCs) 70, 162
Fibroblasts 12, 17, 62, 84, 93, 135, 140, 243, 251, 287, 327, 334, 338, 363, 364
Follicle dermal papilla mesenchymal cells (FDPMCs) 354
Form cell sheets 251, 253, 255
Fracture healing 13, 128
Fused deposition modeling 34, 36
Fused deposition modeling (FDM) 34, 36, 37, 42

G

Gas exchange function 270
Gingival recession (GR) 110, 362

H

Healthy ocular surfaces 286, 289
Heart diseases 279, 280
Hepatocyte growth factor (HGF) 13, 14, 18, 81, 139, 235, 239, 240, 279
Hepatocytes 14, 74, 81, 83, 138, 255, 280, 281, 282
Hepatocyte transplantation 281
HiPS cells 61, 62, 78, 79, 81
  differentiated 81
Homing of mesenchymal stem cells 11
Homing-related molecules 137, 138
Human iPS cells 65, 71, 72, 73, 74

Human MSCs 6, 12, 150, 152, 169, 319
Human umbilical vein endothelial cells (HUVECs) 243, 255, 258

I

Indium tin oxide (ITO) 255, 256, 260
Induced IPS cells 82
Inkjet printing 37, 38, 50, 317
Inorganic materials 158, 177, 178
Insulin-like growth factors (IGFs) 12, 49, 358
Interactions, cell-material 312, 313
Investigation of drugs 61
IPS cell application 60, 62
IPS cell differentiation 77, 78, 79
IPS cell lines 62, 71, 79, 82, 84
IPS cells 60, 61, 62, 63, 64, 65, 66, 67, 69, 70, 71, 72, 73, 74, 75, 76, 77, 78, 79, 80, 81, 82, 83, 84
IPS CELLS 62, 70, 77, 282
IPS cells 64, 65,66, 67, 69, 70, 71, 72, 74, 77, 78, 81, 82, 83
  acquired 84
  applied human 83
  colonies of human 71
  differentiating 78
  first studied 81
  generated 70
  generating 65
  generation of 64, 65, 66, 69, 70
  induction of 66, 67, 77
  injecting 74
  mouse 71, 72, 74
  obtaining 82
  primitive 69
  protein-based 78
IPS CELLS, remaining 280
IPS cells, repaired 82
IPS cells acquisition 66, 67, 69
IPSCs, differentiation of 149, 151

L

Laminated object manufacturing (LOM) 34, 36, 38
Lateral maxillary dysplasia 109, 110
Layers 249, 250, 253
  conductive polymer 249, 250
  polymerization electrolysis 253
Lower critical solution temperature (LCST) 247, 249, 250, 252

M

Magnetite cationic liposome (MCLs) 262, 263
Mandibular 40, 44, 162, 163, 176
Materials, nonautologous tissue 115
Maxillary defects 97, 135
Maxillary process 106, 107
Maxillofacial 91, 96, 100, 101, 107, 110, 111
Maxillofacial defects 90, 91, 92, 112, 113, 114, 130
  large 112, 113
Maxillofacial deformities 107, 108, 112
Maxillofacial fractures 100, 103, 104, 108, 111
Maxillofacial injuries 100, 101
Maxillofacial region 90, 91, 92, 100, 103, 105, 106, 107, 111, 159, 169

Mechanical properties 47, 125, 126, 148, 179, 180, 313, 318, 324, 326, 329, 331, 336, 339, 355
Mechanical strength 128, 175, 176, 178, 285, 316, 318, 324, 327, 328, 332, 335
Mental foramen 102, 103
Mesenchymal cells 131, 134, 151, 350, 357
Mesenchymal Stem Cells 49, 93, 354
  derived 354
  encapsulated human 49
  marrow-derived 93
Mesenchymal stem/stromal cells 81, 231
Mesenchyme 350, 354, 359
Microvascular free tissue transfer 121, 124
Microvascular free tissue transfer (MFTT) 121, 124
Midface dysplasia 109, 111
Migration, stimulated 12, 17, 18
Migration response 17, 18, 19
Mitogen-activated protein kinase (MAPK) 19
Monocytes 16, 18, 139
Mouse embryonic fibroblasts (MEFs) 61, 67, 68
MSC differentiation 6, 152, 181, 183, 318
MSC frequency 7, 8
MSC isolation 4, 5, 6, 7, 10, 137
MSC markers 5, 136
MSC migration 11, 20, 137
MSC recruitment 13, 15, 20
MSCs, functions of 137
MSCs cell sheets 235, 236
MSC sheets 240, 244
Mucosa, esophageal 273, 287, 288
Murine 14, 17, 79, 80
Murine BMSCs 15, 17
Muscle cells 13, 116, 136, 235, 243, 353
  smooth 13, 136, 243, 353
Muscle tissue defects 116
Myocardium 234, 278, 288, 289

N

Nanofibers 314, 315, 318, 324, 337
Nasal defects 122
Nasal dysplasias 108, 109, 111
Natural materials 146, 155, 313
Neotissues 327, 328
Nerve cells 116, 136
Nerve growth factor (NGFs) 359
Nerve regeneration 167, 169, 171, 172, 173
Nerve tissue defect 166
Neural crest cells 107, 134, 350
Neural stem cells 63, 133, 150
NIPAAm, plasma polymerization of 248, 250

O

Odontogenic differentiation potency 354
Organogenesis 13, 14, 352
Orthotropic liver transplantation (OLT) 280
Osseointegration 47, 48, 161, 170, 184

Osteoblasts 5, 6, 45, 93, 116, 118, 126, 131, 135, 136, 146, 168, 173, 183, 184, 234, 273, 334, 353, 354, 361
Osteogenesis 137, 138, 145, 146, 147, 163, 164, 166, 272, 273, 329, 353
  distraction 145, 163, 164, 166
Osteogenic differentiation 11, 139, 147, 148, 177, 178, 180, 237, 240, 329, 363, 364

P

PCL scaffolds 181, 332
PDGF-AB 12, 357
PDGF-BB 12, 357, 358, 363
Peptides 64, 253, 255, 258, 321, 339
Periodontal ligament 41, 44, 45, 95, 110, 134, 161, 350, 360, 361
Periodontal ligament cells 237, 270, 314
Periodontal ligament stem cells (PDLSCs) 181, 238, 314, 360, 361, 363, 364
Periodontal regeneration 44, 45, 237, 326, 362, 363
  potency of 362
Periodontal tissue defects 171
Periodontal tissue growth factor 95
Periodontal tissues 95, 131, 161, 181, 237, 266, 270, 271, 326, 351, 361
Periodontitis 161, 162, 245, 271
Periodontium defects 90, 95, 110, 111
Periosteum 113, 160
Phase separation 316, 337
Phosphate-buffered saline (PBS) 67, 140, 141, 142, 144, 153, 154
Photolithography 251, 255, 317, 319
Photomasks 317, 318, 322
PLA 176, 313, 332, 334, 362, 366, 367
Plasmids 63, 64, 75, 149
Platelet-derived growth factor (PDGF) 9, 12, 136, 146, 184, 279, 357, 358
Platelet-derived growth factor receptor-α (PDGFRα) 9, 357
Platelet-rich plasma (PRP) 146, 169, 170, 184, 185
PLGA scaffolds 176, 317, 362, 366
Pluripotency 3, 60, 61, 62, 65, 66, 70, 71, 73, 74, 75
Pluripotent stem cells 61, 60, 65, 67, 70, 71, 72, 73, 134, 139, 150, 167, 168
  natural 60, 65, 71
PMB-PL Surface 259, 260
Polyanhydrides 328, 334, 335
Polyesters 155, 313, 314
Polyhydroxyalkanoates 331, 334, 335
Polyjet printing 38
Polymer particles 336, 337
Polymers 125, 126, 127, 146, 155, 176, 233, 313, 325, 328, 329, 330, 331, 332, 334, 336, 337, 355, 362, 366, 367
  natural 329, 330, 331, 355, 366
  synthetic 176, 313, 325, 329, 331, 355, 366
Polymer scaffolds 113, 336
Porcine models 278, 279
Porous scaffolds 322, 327, 336, 337
Preadipocytes 139, 141, 142, 147, 153
Preincubation 18, 19
Probing pocket depth (PPD) 362

Products, tissue-engineered 264
Progenitor cells, endothelial 143, 272, 279
Pulp cells 355, 356, 368
  human dental 356, 368
Pulp stem cells transplantation 365
Pulp tissue 173, 350

R

Reconstruction 90, 91, 92, 93, 94, 95, 97, 98, 99, 113, 114, 118, 119, 120, 121, 122, 123, 124, 125, 127, 128, 135, 153, 167, 173, 178, 180, 183, 285, 286, 288, 289
  corneal 285, 286, 289
Regenerate periodontal tissues 270, 361
Regeneration 163, 174, 271, 272, 284, 355
  guided bone 163
  pulp 174, 355
  root 271, 272
  urothelial 284
Regeneration material 155, 156
Regenerative medicine 4, 35, 48, 49, 50, 60, 61, 62, 70, 75, 82, 90, 93, 117, 130, 131, 132, 138, 149, 232, 236, 266, 273, 278, 285, 312, 329
Removable partial denture (RPD) 43
Renal cells 283
Renal tissues 282, 283
Repair bone defects 115, 177
Repair tissue defects 184
Reparation 90, 91, 92, 94, 95, 97, 98
Reparation and reconstruction 90, 91, 95
Reprogramming efficiency 66, 70
Reprogramming iPS cells 72
Reprogramming process 65, 70, 72, 73, 76
Resection, tumor 91, 92, 93, 94, 99, 115, 153, 166, 169, 177
Restore motor function 82, 83
Retinitis pigmentosa epithelium (RPE) 83, 268
RGD peptides 254, 258
Root canal 365, 367, 368
RPE cells 269
RPE cell sheets 269

S

Scaffold architecture 45, 337
Scaffold biomaterials for bone 146
Scaffold design 135, 355
Scaffold materials 113, 158, 171, 175, 177, 246, 247, 326
Scaffold mechanics 318
Scaffold porosity 179
Scaffolds 13, 16, 33, 35, 40, 44, 45, 48, 49, 90, 93, 94, 113, 116, 118, 125, 128, 129, 130, 135, 147, 148, 157, 158, 161, 167, 169, 174, 175, 176, 177, 178, 179, 180, 181, 182, 185, 186, 232, 246, 271, 272, 276, 285, 312, 313, 314, 315, 316, 317, 318, 319, 320, 322, 323, 324, 325, 326, 327, 328, 329, 330, 331, 333, 334, 335, 337, 338, 339, 349, 352, 353, 354, 355, 364, 365
  artificial 35
  biodegradable 125, 232, 365
  biological 130, 271

engineering 174, 177, 325, 330
fiber-guiding 44, 45
hybrid 165, 175, 330
optimized 327, 331
synthetic 116, 328, 355
Scaffolds for ASCs in bone repair 147
Scaffolds in tooth tissue engineering 326
Scaffold structure 315, 316
Scar formation 125, 151, 153
Schwann cells 169, 171
Selective laser sintering (SLS) 34, 36, 37, 38, 42, 338
Self-assembled monolayer (SAM) 253, 255
Side population (SP) 365, 368
Single cells 4, 253, 264
SiO_2-based scaffolds 324
Skeletal muscle cells 74, 80
Skin wound healing 151, 152
SLS process 42
Soft tissue 96, 97, 108, 110, 113, 156, 239, 351, 367
Soft tissue defect repair 101
Somatic cells 4, 10, 61, 69, 70, 72, 73, 133, 167, 235, 239
Stem cell differentiation 319
Stem cells proliferation 263
Stem cells used in dentin regeneration 352
Stem cells used in periodontal ligament regeneration 360
Stereolithography 34, 36, 338
Stromal cells 5, 13, 78, 139, 140, 141, 279, 368
adipose tissue-derived 279
mesenchymal 5, 368
Stromal medium 142, 143
Stromal vascular fraction (SVF) 140, 141, 142, 143, 144, 147
Subcutaneous space 281, 282
Subcutaneous tissue flaps 112, 121
Swallowing 94, 96, 120
Synthetic materials 115, 313, 319, 321

T

TCP scaffolds 330
Technology 61, 62, 63, 130
iPS cell 61, 62, 63
stem cell 130
Telomerase activity 61, 71, 72
Telomere length, iPS cells regain 72
Temperature-responsive system 231, 247, 248
Tetraploid complementation 73, 74
Thermally Induced Phase Separation (TIPS) 316, 337
Thyroid gland 275, 276
Tissue chambers 114
Tissue culture polystyrene 248, 319
Tissue defect reconstruction 116
Tissue defect restoration 93
Tissue defects 90, 96, 101, 106, 109, 110, 112, 113, 114, 153, 174, 175, 179, 264, 285
large 96, 285
multiple 113
osteochondral 175
partial 109
repair tooth 174
severe 96

simple 264
soft 90, 96, 101, 106, 112, 153
Tissue engineering 4, 11, 33, 40, 48, 50, 82, 84, 90, 93, 94, 95, 114, 116, 117, 130, 131, 146, 157, 158, 160, 163, 164, 166, 167, 168, 169, 171, 173, 174, 176, 177, 178, 180, 231, 232, 234, 239, 240, 241, 242, 246, 247, 249, 250, 252, 253, 256, 259, 260, 262, 263, 265, 266, 272, 273, 274, 276, 277, 284, 287, 312, 313, 318, 322, 324, 326, 329, 330, 331, 349, 351, 352, 353, 367
Tissue formation 313, 352
Tissue fragments 140, 141, 142
Tissue function 93, 117, 130
Tissue growth 114
Tissue homeostasis 4, 145
Tissue injuries 18, 19, 100, 101, 111
soft 100, 101, 111
Tissue reconstruction 92, 136, 266, 275, 327
complex 266, 275
Tissue regeneration 15, 18, 49, 90, 94, 95, 113, 130, 132, 133, 145, 153, 156, 158, 164, 170, 171, 174, 175, 179, 180, 182, 183, 185, 186, 231, 234, 270, 284, 285, 314, 362
adipose 90, 153
fat 153, 182
guided 95, 270
periodontal 90, 132, 170, 171
soft 90, 185
Tissue restoration 119, 132
Tissues 3, 4, 5, 7, 9, 11, 34, 35, 40, 41, 44, 45, 47, 48, 49, 50, 61, 68, 74, 82, 84, 90, 91, 93, 94, 95, 106, 112, 113, 116, 117, 130, 132, 136, 137, 138, 141, 146, 147, 155, 157, 158, 160, 161, 163, 164, 167, 168, 169, 171, 172, 173, 174, 175, 176, 177, 178, 180, 186, 231, 232, 233, 234, 235, 237, 239, 240, 241, 242, 245, 246, 247, 249, 250, 252, 253, 256, 259, 260, 262, 263, 265, 266, 267, 271, 272, 273, 274, 276, 277, 278, 281, 284, 285, 287, 288, 312, 313, 318, 322, 323, 325, 326, 327, 328, 329, 330, 331, 336, 338, 349, 350, 351, 352, 353, 354, 356, 357, 359, 361, 364, 365, 367, 368
adjacent 41, 325, 327
adult 7, 234
avascular 136, 241
bone marrow 4, 7
cell-dense 241, 246
connective 3, 7
damaged 11, 157, 232, 284, 285
diseased 93, 157
functional 49, 93, 338
gingival 44, 363
host 175, 236, 245, 278
injured 18, 267
ischemic cerebral 15
living 155, 336
mature 61, 68, 322
missing 93, 266, 322
neural 74, 173
patients'oral mucosa 288
regenerate 93
regenerated 44, 178, 271, 328, 329
skeletal muscle 241, 263
subcutaneous 141, 172, 175, 368

traditional 265, 284
urinary tract 284
Tissue types 74, 157
TMJ ankylosis 165, 185
Toll-like receptors (TLR) 18
Tongue defect 96, 97
Tooth loss 90, 109, 349, 351
Tooth model 40, 41
Tooth tissue engineering 326
Transcription factors 64, 65, 66, 70, 72, 79, 80, 149, 244
Transplanted cell sheet 245, 267, 274, 276, 282
Treated dentin matrix (TDM) 271, 272
Tumor necrosis factor 16, 138, 139, 280
Tumors, primary 19, 96

U

Umbilical cord blood 7, 9, 10, 167, 172
Umbilical cord blood 9
Urothelial cells 284
UV irradiation 257, 258

V

Vascular endothelial growth factor (VEGF) 13, 14, 15, 18, 116, 139, 147, 152, 176, 183, 184, 235, 239, 278, 279, 322, 358, 363, 364, 367, 368
Vascularization 49, 116, 184, 231, 234, 235, 239, 241, 243, 279, 281, 286, 367
Vascularized Flaps 120, 121, 123
Vascular tissue 262, 338

W

Wound healing 12, 14, 16, 17, 42, 171, 182, 185, 259, 274, 287, 289, 355, 357, 358

Z

Zygomatic dysplasia 110, 111

Made in the USA
Las Vegas, NV
03 November 2026

57424583R00245